Sustainable Composite Construction Materials

Sustainable Composite Construction Materials

Editors

Krishanu Roy
G. Beulah Gnana Ananthi

Basel • Beijing • Wuhan • Barcelona • Belgrade • Novi Sad • Cluj • Manchester

Editors
Krishanu Roy
The University of Waikato
Hamilton
New Zealand

G. Beulah Gnana Ananthi
Anna University
Chennai
India

Editorial Office
MDPI
St. Alban-Anlage 66
4052 Basel, Switzerland

This is a reprint of articles from the Special Issue published online in the open access journal *Journal of Composites Science* (ISSN 2504-477X) (available at: https://www.mdpi.com/journal/jcs/special_issues/sustainable_construction_composite_materials).

For citation purposes, cite each article independently as indicated on the article page online and as indicated below:

Lastname, A.A.; Lastname, B.B. Article Title. *Journal Name* **Year**, *Volume Number*, Page Range.

ISBN 978-3-0365-9754-6 (Hbk)
ISBN 978-3-0365-9755-3 (PDF)
doi.org/10.3390/books978-3-0365-9755-3

© 2023 by the authors. Articles in this book are Open Access and distributed under the Creative Commons Attribution (CC BY) license. The book as a whole is distributed by MDPI under the terms and conditions of the Creative Commons Attribution-NonCommercial-NoDerivs (CC BY-NC-ND) license.

Contents

About the Editors vii

Krishanu Roy and Beulah Gnana Ananthi Gurupatham
Editorial for the Special Issue on Sustainable Composite Construction Materials
Reprinted from: *J. Compos. Sci.* 2023, 7, 491, doi:10.3390/jcs7120491 1

**Nisala Prabhath, Buddhika Sampath Kumara, Vimukkthi Vithanage,
Amalka Indupama Samarathunga, Natasha Sewwandi, Kaveendra Maduwantha, et al.**
A Review on the Optimization of the Mechanical Properties of
Sugarcane-Bagasse-Ash-Integrated Concretes
Reprinted from: *J. Compos. Sci.* 2022, 6, 283, doi:10.3390/jcs6100283 5

**Remya Elizabeth Philip, A. Diana Andrushia, Anand Nammalvar,
Beulah Gnana Ananthi Gurupatham and Krishanu Roy**
A Comparative Study on Crack Detection in Concrete Walls Using Transfer
Learning Techniques
Reprinted from: *J. Compos. Sci.* 2023, 7, 169, doi:10.3390/jcs7040169 24

**Ramkumar Durairaj, Thirumurugan Varatharajan, Satyanarayanan Kachabeswara
Srinivasan, Beulah Gnana Ananthi Gurupatham and Krishanu Roy**
Experimental Investigation on Flexural Behaviour of Sustainable Reinforced Concrete Beam
with a Smart Mortar Layer
Reprinted from: *J. Compos. Sci.* 2023, 7, 132, doi:10.3390/jcs7040132 46

Naveen Kumar Koppula, Jens Schuster and Yousuf Pasha Shaik
Fabrication and Experimental Analysis of Bricks Using Recycled Plastics and Bitumen
Reprinted from: *J. Compos. Sci.* 2023, 7, 111, doi:10.3390/jcs7030111 65

**Pradeep Sivanantham, Deepak Pugazhlendi, Beulah Gnana Ananthi Gurupatham
and Krishanu Roy**
Influence of Steel Fiber and Carbon Fiber Mesh on Plastic Hinge Length of RCC Beams under
Monotonic Loading
Reprinted from: *J. Compos. Sci.* 2022, 6, 374, doi:10.3390/jcs6120374 82

**Yoganantham Chinnasamy, Philip Saratha Joanna, Karthikeyan Kothanda,
Beulah Gnana Ananthi Gurupatham and Krishanu Roy**
Behavior of Pultruded Glass-Fiber-Reinforced Polymer Beam-Columns Infilled with
Engineered Cementitious Composites under Cyclic Loading
Reprinted from: *J. Compos. Sci.* 2022, 6, 338, doi:10.3390/jcs6110338 102

**Chinnasamy Samy Madan, Krithika Panchapakesan, Potlapalli Venkata Anil Reddy,
Philip Saratha Joanna, Jessy Rooby, Beulah Gnana Ananthi Gurupatham and Krishanu Roy**
Influence on the Flexural Behaviour of High-Volume Fly-Ash-Based Concrete Slab Reinforced
with Sustainable Glass-Fibre-Reinforced Polymer Sheets
Reprinted from: *J. Compos. Sci.* 2022, 6, 169, doi:10.3390/jcs6060169 120

**Pradeep Sivanantham, Beulah Gnana Ananthi Gurupatham, Krishanu Roy,
Karthikeyan Rajendiran and Deepak Pugazhlendi**
Plastic Hinge Length Mechanism of Steel-Fiber-Reinforced Concrete Slab under
Repeated Loading
Reprinted from: *J. Compos. Sci.* 2022, 6, 164, doi:10.3390/jcs6060164 141

Chinnasamy Samy Madan, Swetha Munuswamy, Philip Saratha Joanna, Beulah Gnana Ananthi Gurupatham and Krishanu Roy
Comparison of the Flexural Behavior of High-Volume Fly AshBased Concrete Slab Reinforced with GFRP Bars and Steel Bars
Reprinted from: *J. Compos. Sci.* **2022**, *6*, 157, doi:10.3390/jcs6060157 . . . . **160**

Matteo Sambucci, Abbas Sibai, Luciano Fattore, Riccardo Martufi, Sabrina Lucibello and Marco Valente
Finite Element Multi-Physics Analysis and Experimental Testing for Hollow Brick Solutions with Lightweight and Eco-Sustainable Cement Mix
Reprinted from: *J. Compos. Sci.* **2022**, *6*, 107, doi:10.3390/jcs6040107 . . . . **177**

About the Editors

Krishanu Roy

Dr.Krishanu Roy is is currently a Senior Lecturer in Civil Engineering at the University of Waikato (UoW). He obtained his PhD in Civil Engineering from the University of Auckland and master's in Earthquake Engineering from the Indian Institute of Technology Roorkee (IIT Roorkee). Before joining the UoW, he was a Lecturer at the University of Auckland. After completing his master's degree, Kris spent one year working for Geodata Spa, one of the world's leading engineering firms for underground structures. During this time, he specialised in tunnel portal design, steel construction, seismic design of steel structures, and the designing of underground structures.

After obtaining his PhD degree, Kris spent a further two years at Kiwi Steel Holding LTD as a Research and Development Manager before joining the University of Auckland as a Lecturer. Over the last 8 years, his research has continued to revolve around thin-walled structures, covering a wide range of topics, such as modular construction, built-up columns and beams, modal decomposition, cold-formed steel connections, cross-section optimisation of single and built-up sections, numerical methods, stainless steel, aluminium structures, steel and aluminium claddings, corroded steel members, 3-D printed structures, the durability of cold-formed steel members, weather tightness of metal claddings, design methodologies, general stability, and composite materials for sustainable construction. He is currently leading the cold-formed steel research group of New Zealand.

To date, Kris has authored and co-authored 2 book chapters, 15 conference papers, and more than 150 journal papers in prestigious Scopus-indexed international journals. In google scholar, his work has attracted more than 3500 citations with an h-index of 35.

G. Beulah Gnana Ananthi

Dr. G. Beulah Gnana Ananthi currently serves as an Associate Professor in the Department of Civil Engineering at Anna University in Chennai, India. She earned her PhD in Civil Engineering from Anna University, and her passion for structural engineering led her to obtain a master's degree in Structural Engineering from the renowned National Institute of Technology Tiruchirapalli (NIT-Trichy). With a teaching career that spans nearly two decades, she began as a Senior Lecturer at Tagore Engineering College in Chennai in 2004. In 2007, Dr. Ananthi expanded her academic journey, taking on the role of Lecturer at Crescent Engineering College in Chennai. Later that same year, she joined the prestigious Anna University as a lecturer in the Department of Architecture and continued as an Assistant Professor until 2014. Since then, she has been dedicated to fostering the growth of students and advancing the field through her commitment to both teaching and research.

Over the last 15 years, Dr. Ananthi's research has focused on various aspects of civil engineering, including cold-formed steel, steel structures, cross-section optimization, and sustainable construction materials. Her contributions extend beyond the classroom, as evidenced by her impressive publication record. Dr. Ananthi has authored and co-authored 6 book chapters, 11 conference papers, and 50 articles in leading Scopus-indexed international journals, with her work being widely cited and recognized in the academic community. In addition to her academic pursuits, Dr. Ananthi has assumed leadership roles, as transcript signing officer from 2014-2016 and also as deputy controller of examinations in Anna University from 2016-2018. Her dedication to excellence in education, coupled with her impactful research, reflects her commitment to advancing the field of civil engineering.

Editorial

Editorial for the Special Issue on Sustainable Composite Construction Materials

Krishanu Roy [1,*] and Beulah Gnana Ananthi Gurupatham [2,*]

1. School of Engineering, The University of Waikato, Hamilton 3216, New Zealand
2. Division of Structural Engineering, College of Engineering Guindy Campus, Anna University, Chennai 600025, India
* Correspondence: krishanu.roy@waikato.ac.nz (K.R.); beulah28@annauniv.edu (B.G.A.G.)

Citation: Roy, K.; Gurupatham, B.G.A. Editorial for the Special Issue on Sustainable Composite Construction Materials. *J. Compos. Sci.* **2023**, *7*, 491. https://doi.org/10.3390/jcs7120491

Received: 19 November 2023
Accepted: 24 November 2023
Published: 28 November 2023

Copyright: © 2023 by the authors. Licensee MDPI, Basel, Switzerland. This article is an open access article distributed under the terms and conditions of the Creative Commons Attribution (CC BY) license (https://creativecommons.org/licenses/by/4.0/).

Sustainable composite construction materials play a crucial role in creating more environmental friendly and energy-efficient buildings. Sustainable construction is a growing imperative in the face of global environmental challenges. As the construction industry seeks to reduce its carbon footprint, a pivotal focus has emerged in the development and utilization of sustainable composite construction materials. Traditionally, we neglected the impact of embodied and operational carbon generated during the manufacture of building materials and the operational stage of building, leading to significant carbon emissions over the last several decades, causing global warming and other related problems. The articles featured in this Special Issue of the Journal of Composites Science aim to provide engineers and scientists with a comprehensive understanding of the current challenges in sustainable construction. This Special Issue contains ten articles, including a review article and nine original research papers contributed by renowned scholars specializing in civil and construction engineering, civil and architectural engineering, structural engineering, and physical sciences; these are all appropriately cited. The topics covered in this Special Issue include principles related to lightweight bricks, eco-sustainable cement mixes, high-volume fly ash-based concrete slabs reinforced with GFRP bars and steel bars, the mechanism of steel-fiber-reinforced concrete beams and slabs, sugarcane–bagasse ash, engineered cementitious composites, bricks created using recycled plastics and bitumen, smart mortar layers, and transfer learning techniques for crack detection. These papers strike a harmonious balance between academic and industrial research, showcasing a collaborative synergy between the two sectors. The following parts of this editorial provide a summary of these ten publications.

Sambucci et al. [1] utilized recycled rubber as an aggregate in the design, modeling, and experimental characterization of lightweight concrete hollow bricks. The near-compliance of rubber concrete blocks with standard requirements and their value-added properties demonstrated significant potential for incorporating waste rubber as an aggregate for non-structural applications. This study confirmed that fractal geometries and waste aggregates can be successfully integrated into bricks to achieve eco-friendly solutions with enhanced structural and acoustic performances.

Sivanantham et al. [2] conducted an experimental investigation to examine the influence of steel fiber reinforcement on the plastic hinge length of a concrete slab under repeated loading. The results obtained through repeated loading applied to the slab indicate that the steel fibers employed at critical sections of the plastic hinge length provided similar strength, displacement, and performance outcomes as those of conventional RCC slabs and fully steel-fiber-reinforced concrete slabs. The study concludes that, rather than using steel fibers throughout the concrete slab, incorporating them specifically at the plastic hinge length is not only effective but also more economical.

Madan et al. [3] conducted a comparative study on the flexural behavior of an ordinary Portland cement (OPC) concrete slab and a high-volume-fly-ash (HVFA) concrete slab reinforced with GFRP rods and steel rods. In the fly ash concrete slabs, 60% of the cement used for casting the elements was replaced with class F fly ash, emerging as an eco-friendly and cost-effective alternative to OPC. The observation revealed that the GFRP rods, when utilized as a replacement for steel rods in both conventional and fly ash concrete, resulted in improved strength compared to that of the one-way slab. The findings of this investigation highlight the potential use of GFRP rods with fly ash concrete in slabs.

Madan et al. [4] investigated the flexural behavior of ordinary Portland cement (OPC) concrete slabs and high-volume-fly-ash (HVFA) concrete slabs reinforced with bi-directional GFRP sheets. Slab specimens were cast with 60% fly ash as a replacement for cement and were equipped with 1 mm thick GFRP sheets in two, three, and four layers. The experimental investigation demonstrated that HVFA concrete slabs reinforced with GFRP sheets present a more sustainable alternative compared to steel reinforcement, thereby contributing to sustainable construction. This study further underscores the potential use of high-volume fly ash as a cement replacement in concrete slabs, offering an effective means of mitigating the impact of greenhouse gas emissions and promoting sustainable construction.

Prabhath et al. [5] conducted a review of existing research on concrete containing various percentages of sugarcane bagasse ash (SCBA) as a partial replacement for OPC. The potential to minimize the cost of concrete in large-scale construction via the incorporation of suitable amounts of SCBA while meeting required standards and specifications was investigated as well. Based on the literature, it was concluded that SCBA shows promise as a viable partial replacement material for OPC. This research could be extended to explore additional cement replacement materials, which, when combined with SCBA, may contribute to the development of low-cost and high-performance concrete.

Chinnasamy et al. [6] presented the results of an experimental investigation on the cyclic response of a GFRP beam column infilled with high-volume-fly-ash engineered cementitious composites (HVFA-ECC), incorporating 60%, 70%, and 80% fly ash as a replacement for cement. Consequently, GFRP sections infilled with HVFA-ECC could serve as lightweight structural components in buildings intended for construction in earthquake-prone areas. The utilization of high-volume fly ash, a byproduct of coal-burning power plants, in conjunction with manufactured sand in ECC due to the scarcity of river sand, not only enhances the structural properties of the engineered cementitious composite (ECC) but also contributes to a reduction in CO_2 emissions.

Sivanantham et al. [7] investigated the effects of carbon fiber mesh jacketing and steel fiber reinforcement at the plastic hinge length of a concrete beam subjected to a vertical monotonic load. Rather than distributing steel fibers throughout the entire span of the beam, they concentrated them solely at the plastic hinge length. This approach yielded comparable performance outcomes under monotonic loading while reducing the number of fibers, making it a more economical alternative. Meanwhile, they used carbon fiber jacketing for the whole beam span with fiber being placed at the plastic hinge length, which showed the best performance outcome when compared to that of other techniques.

Durairaj et al. [8] conducted an experimental study on the flexural behavior of sustainable reinforced cement concrete (RCC) beams featuring a smart mortar layer incorporated into the concrete mixture. The experimental results demonstrated that the smart mortar layer could detect damage in the RCC beams and interpret the damage through electrical measurements, enhancing the sustainability of the beam. Notably, compared to the hybrid brass-carbon-fiber-incorporating mortar layer, the brass-fiber-incorporating mortar layer exhibited a substantial increase in the fractional change in electrical resistivity (f_{cr}) values.

Koppula et al. [9] maximized the use of plastic waste to manufacture bricks that match the properties of conventional bricks without negatively impacting the environmental or ecological balance. These bricks were produced using a well-balanced mixture of high-density polyethylene (HDPE), quartz sand, and certain additive materials such as bitumen. The incorporation of HDPE and quartz sand ensured that the bricks were void-free and free of alkalis, making them a suitable and environmentally friendly choice for the construction industry.

Philip et al. [10] conducted a comprehensive analysis of well-known pre-trained networks for the classification of cracks in concrete buildings. The classification performance outcomes of convolutional neural network designs, including VGG16, VGG19, ResNet 50, MobileNet, and Xception, was compared using a concrete crack image dataset. This study revealed that the features acquired through training are highly accurate when applied to various materials. Pre-trained networks emerge as an excellent choice for the application of convolutional neural networks (CNNs) in crack detection tasks, as they require fewer training samples and exhibit a faster convergence rate.

The editors of this Special Issue express sincere appreciation to all the authors who generously shared their scientific knowledge and expertise. Their contributions do not only enrich this Special Issue but also significantly advance research in the field. The meticulous evaluation of numerous submissions conducted by peer reviewers deserves recognition, as their valuable insights and constructive feedback have markedly improved the overall quality of the papers published in this Special Issue. Lastly, the editors extend their gratitude to the Managing Editors of the Journal of Composites Science for their steadfast support throughout the entire process. Their dedication and assistance played a crucial role in ensuring the successful completion of this Special Issue. We hope that 'Sustainable Composite Construction Materials' will become a valuable resource for researchers, practitioners, and students in science and engineering. We believe that the presented findings and insights will contribute to a deeper understanding of sustainable construction, and we anticipate that the innovative solutions discussed in these papers will inspire further research and advancements in this field.

Challenges and Considerations: While the potential benefits of sustainable composite construction materials are evident, challenges persist. The upfront cost, durability, and end-of-life considerations require careful attention. Sustainable composite construction materials stand at the forefront of a paradigm shift in the construction industry, offering a path towards the development of environmentally friendly resilient buildings and infrastructures. As innovation continues to drive the development of these materials, their integration into mainstream construction practices holds the promise of a more sustainable and regenerative built environment. The pursuit of sustainable construction is not merely a choice but an imperative for a resilient and environmentally conscious future.

Conflicts of Interest: The authors declare no conflict of interest.

References

1. Sambucci, M.; Sibai, A.; Fattore, L.; Martufi, R.; Lucibello, S.; Valente, M. Finite Element Multi-Physics Analysis and Experimental Testing for Hollow Brick Solutions with Lightweight and Eco-Sustainable Cement Mix. *J. Compos. Sci.* **2022**, *6*, 107. [CrossRef]
2. Madan, C.S.; Munuswamy, S.; Joanna, P.S.; Gurupatham, B.G.; Roy, K. Comparison of the Flexural Behavior of High-Volume Fly Ash Based Concrete Slab Reinforced with GFRP Bars and Steel Bars. *J. Compos. Sci.* **2022**, *6*, 157. [CrossRef]
3. Sivanantham, P.; Gurupatham, B.G.A.; Roy, K.; Rajendiran, K.; Pugazhlendi, D. Plastic Hinge Length Mechanism of Steel-Fiber-Reinforced Concrete Slab under Repeated Loading. *J. Compos. Sci.* **2022**, *6*, 164. [CrossRef]
4. Madan, C.S.; Panchapakesan, K.; Reddy, P.V.A.; Joanna, P.S.; Rooby, J.; Gurupatham, B.G.A.; Roy, K. Influence on the Flexural Behaviour of High-Volume Fly-Ash-Based Concrete Slab Reinforced with Sustainable Glass-Fibre-Reinforced Polymer Sheets. *J. Compos. Sci.* **2022**, *6*, 169. [CrossRef]
5. Prabhath, N.; Kumar, B.S.; Vithanage, V.; Samarathunga, A.I.; Sewwandi, N.; Maduwantha, K.; Madusanka, M.; Koswattage, K. A Review on the Optimization of the Mechanical Properties of Sugarcane-Bagasse-Ash-Integrated Concretes. *J. Compos. Sci.* **2022**, *6*, 283. [CrossRef]
6. Chinnasamy, Y.; Joanna, P.S.; Kothanda, K.; Gurupatham, B.G.A.; Roy, K. Behavior of Pultruded Glass-Fiber-Reinforced Polymer Beam-Columns Infilled with Engineered Cementitious Composites under Cyclic Loading. *J. Compos. Sci.* **2022**, *6*, 169. [CrossRef]
7. Sivanantham, P.; Pugazhlendi, D.; Gurupatham, B.G.A.; Roy, K. Influence of Steel Fiber and Carbon Fiber Mesh on Plastic Hinge Length of RCC Beams under Monotonic Loading. *J. Compos. Sci.* **2022**, *6*, 374. [CrossRef]
8. Koppula, N.K.; Schuster, J.; Shaik, Y.P. Fabrication and Experimental Analysis of Bricks Using Recycled Plastics and Bitumen. *J. Compos. Sci.* **2023**, *7*, 111. [CrossRef]

9. Durairaj, R.; Varatharajan, T.; Srinivasan, S.K.; Gurupatham, B.G.; Roy, K. Experimental Investigation on Flexural Behaviour of Sustainable Reinforced Concrete Beam with a Smart Mortar Layer. *J. Compos. Sci.* **2023**, *7*, 132. [CrossRef]
10. Philip, R.E.; Andrushia, A.D.; Nammalvar, A.; Gurupatham, B.G.A.; Roy, K. A Comparative Study on Crack Detection in Concrete Walls Using Transfer Learning Techniques. *J. Compos. Sci.* **2023**, *7*, 169. [CrossRef]

Disclaimer/Publisher's Note: The statements, opinions and data contained in all publications are solely those of the individual author(s) and contributor(s) and not of MDPI and/or the editor(s). MDPI and/or the editor(s) disclaim responsibility for any injury to people or property resulting from any ideas, methods, instructions or products referred to in the content.

Review

A Review on the Optimization of the Mechanical Properties of Sugarcane-Bagasse-Ash-Integrated Concretes

Nisala Prabhath [1], Buddhika Sampath Kumara [1,*], Vimukkthi Vithanage [1], Amalka Indupama Samarathunga [1], Natasha Sewwandi [2], Kaveendra Maduwantha [1], Madawa Madusanka [3] and Kaveenga Koswattage [1]

[1] Department of Engineering Technology, Sabaragamuwa University of Sri Lanka, Belihuloya 70140, Sri Lanka
[2] Sugarcane Research Institute, Udawalawe 70190, Sri Lanka
[3] Department of Materials Science and Engineering, University of Moratuwa, Moratuwa 10400, Sri Lanka
* Correspondence: buddhika@tech.sab.ac.lk

Abstract: Leading sugar-producing nations have been generating high volumes of sugarcane bagasse ash (SCBA) as a by-product. SCBA has the potential to be used as a partial replacement for ordinary Portland cement (OPC) in concrete, from thereby, mitigating several adverse environmental effects of cement while keeping the cost of concrete low. The majority of the microstructure of SCBA is composed of SiO_2, Al_2O_3, and Fe_2O_3 compounds, which can provide pozzolanic properties to SCBA. In this paper, literature on the enhancement of the mechanical properties of SCBA-incorporating concrete is analyzed. Corresponding process parameters of the SCBA production process and properties of SCBA are compared in order to identify relationships between the entities. Furthermore, methods, including sieving, post-heating, and grinding, can be used to improve pozzolanic properties of SCBA, through which the ideal SCBA material parameters for concrete can be identified. Evidence in the literature on the carbon footprint of the cement industry is utilized to discuss the possibility of reducing CO_2 emissions by using SCBA, which could pave the way to a more sustainable approach in the construction industry. A review of the available research conducted on concrete with several partial replacement percentages of SCBA for OPC is discussed.

Keywords: carbon footprint of cement; bagasse ash composites; green concrete; pozzolanic; sustainable construction

1. Introduction

Construction is always proceeding and reciprocal waste is generated in high volumes, as the demand for further construction has been rising throughout the past several decades. A 2.5% increase in cement production highlights the high consumption of concrete in recent times [1]. In 2005, global cement production was 2300 million tons and in 2020, it rose to 3500 million tons. The cement demand is expected to be around 4400 million tons in 2050 [2]. The global construction industry has been excessively dependent on ordinary Portland cement material (OPC), raising environmental, economic, and health concerns. An ideal solution for cement is needed to ensure greater sustainability for the construction sector as well [3].

OPC production is considered as one of the major global CO_2 emitting processes, largely thanks to being the third-most energy intensive operation for the production of one ton of produce. Therefore, the current process of OPC production cannot be considered as sustainable [4]. Health-related issues that emerge around dumpsites due to SCBA are notable [5]. Furthermore, the contribution of concrete to the total cost of a typical construction project is high, which can be reduced with suitable low-cost cement replacement materials [3]. In order for a material to be categorized as a suitable replacement for OPC, an initial requirement is that it has to possess pozzolanic characteristics.

Several replacement options have been mentioned in the literature [6–9], including fly ash, blast furnace slag, silica fume, wood ash, and ceramic waste. Sugarcane bagasse ash (SCBA) from the sugar industry is also considered as a potential replacement material for cement.

Fly ash contains aluminum and silica within its microstructure, which help to improve key aspects in concrete, such as workability, cohesiveness, ultimate strength, and durability. Blast furnace slag, which is a byproduct of the iron extraction process, provides concretes with enhanced workability properties and better resistance to adverse effects from chemicals, while reducing the early temperature rise after mixing the concrete. Silica fumes are generated during silicon production, which can be introduced into concrete mixtures and help to improve compressive strength, bond strength, and abrasion resistance and to reduce permeability. The workability and compressive strength of concrete structures are enhanced after wood ash from combustion boilers is introduced. Ceramic waste dust is generated from dressing and polishing of ceramic products. The strength and durability of concrete benefit when ceramic waste dust is present in the mix [6–9].

In this review (see Figure 1), the effects of SCBA on important mechanical properties (compressive strength, workability, split tensile strength) in concrete and evidence of a reduction in the cost of concrete by utilizing SCBA are discussed. Moreover, carbon footprint analysis of the cement industry is reviewed so that the importance of utilizing SCBA and other potential cement replacement materials can be highlighted as well. Furthermore, potential new research areas are identified so that future research can be systematically planned.

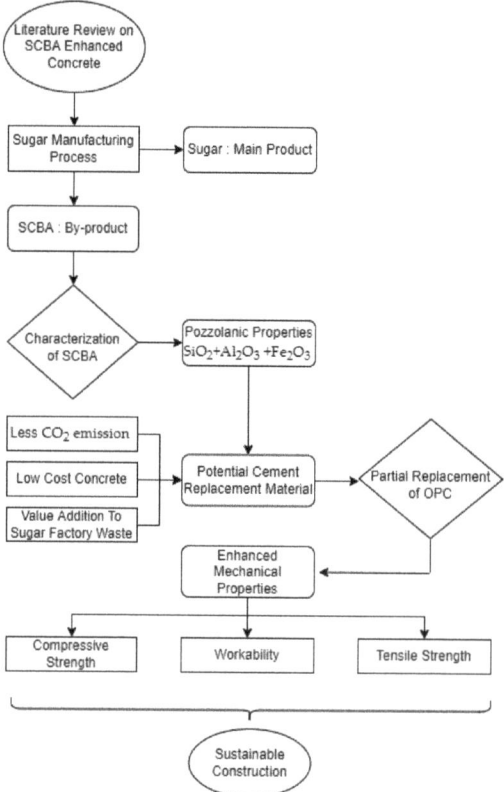

Figure 1. Scope of the review article (Sugarcane Bagasse Ash—SCBA).

2. Sugar Manufacturing Industries

Around 115 countries produce over 1850 million tons of sugarcane annually to supply sugar, alcohol, and paper to the global markets [7]. Sugarcane was also considered the world's second-largest crop-production industry during 2017–2018 [10]. The sugar industry in the global market is majorly dominated by nations, including Brazil, India, China, Thailand, etc. Brazil, the world's leading producer of sugarcane bagasse, produced more than 700 million tons in 2019, accounting for roughly 40% of global production. India and China, respectively, are the nations with the next-highest contributions to global sugarcane production [11–14]. In Thailand, it was recorded that 98 million tons of sugarcane were produced, placing Thailand in fourth place in global sugarcane production rankings in 2013 [14,15]. Nigerian sugarcane production recorded a high volume of over 15 million tons in 2013 [16].

Significant byproducts of the sugar industry are bagasse, molasses, SCBA, and filter press mud, which can be processed to the status of economically valuable byproducts in later processes of sugar production [17]. During 2017–2018, Indian sugarcane bagasse production reached a maximum of 30 million tons. China has an annual production of 1.2–2 million tons of SCBA from sugarcane bagasse [17]. It is commonly mentioned that high volumes of sugarcane bagasse and SCBA are produced in sugarcane manufacturing facilities around the world [11–14].

Over 25% of the initial sugarcane weight is converted to sugarcane bagasse during the sugar-making process [18,19]. SCBA can be generated up to an amount of 3–5% of the total weight of sugarcane bagasse used in the combustion chamber [7,8,13]. The burning process/technology and the content of materials that consist of crushed sugarcane bagasse contribute significantly to defining the final quality of SCBA. At the same time, quantum impurities are reduced from a complete and effective combustion while the level of crystallinity is altered simultaneously. In addition to that, the other most-critical factors that determine the properties of SCBA are the parameters of the post-processes, including grinding, post-burning, and sieving [17,20].

The initial stage of the SCBA production process (see Figure 2) is the harvesting of sugarcane crops. If paper mills are available for production, approximately 50% of the leftover bagasse is transferred to the paper mills to produce paper after the extraction of sugar juice from the sugarcanes [21]. Later, in the cogeneration area, sugarcane bagasse is used as a fuel to produce steam, which powers the generation of electricity driven by turbines. Burning of bagasse is performed in a controlled environment inside the boiler where the temperature is altered with care to achieve efficient complete combustion. The average temperature inside the boiler is set at or above 500 °C. A three-hour burning process at 600 °C calcination temperature produced the highest pozzolanic activity [11]. During this process, minimum silica, alumina, and iron oxide for natural pozzolans reach above 70% by weight, which is the requirement according to the American Society for Testing and Materials ASTM C618 [17]. Finally, the produced SCBA is collected from the bottom of the boiler and ash-contaminated air from the boiler can be filtered to collect another sample of SCBA [14,17]. SCBA is commonly used as a fertilizer in Brazil and India, where SCBA is commonly dumped in landfills [7,17]. In Sri Lanka, the majority of the bagasse is used as a biofuel to generate electricity needed to power the sugar production operation [21].

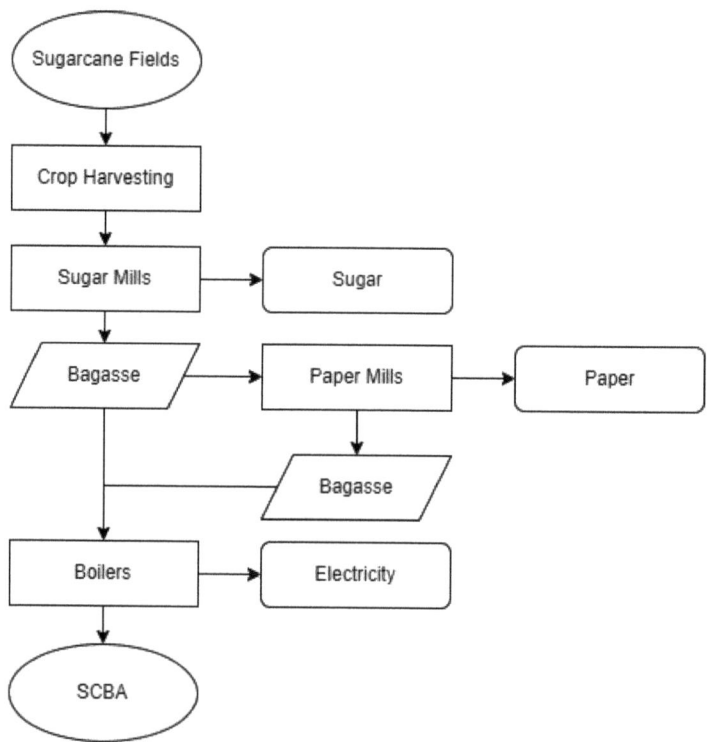

Figure 2. SCBA manufacturing process.

3. Sugarcane Bagasse Ash

Use of sugarcane as a biofuel for cogeneration of electricity is in practice at the moment [21,22] while it is mentioned that over 7% of Indian national electricity demand can be supplied using sugarcane bagasse as a fuel for steam turbines [23].

SCBA is generated as a byproduct during the sugarcane bagasse burning process. The main components that consist of sugarcane bagasse are cellulose (50%), hemicellulose (25%), and lignin (25%). Because sugarcane bagasse contains up to 50% moisture, it is dried before being introduced into boilers [24], although some sugar manufacturing plants do not contain a drying stage within their process [25].

Potential applications of SCBA byproducts of bagasse burning can be identified, including applications in glass-ceramic, Phillip site zeolite synthesis, geo polymers, Fe_2O_3-SiO_2 nanocomposites to remove chromium ions, sodium water glass, silica aerogels, and mesoporous silica as a catalyst silica and as an absorbent to clarify sugarcane juice [14,26,27]. The main requirement for SCBA to be used as a replacement material for OPC in concrete is its pozzolanic action. This depends on chemical properties and physical characteristics of SCBA produced from combustion (see Table 1).

Table 1. Physical properties of SCBA.

Reference	Calcination Temperature (°C)	Density (gcm^{-3})	Blaine Surface Area	Particle Size (μm)	Color
[7]	DNR	2.52	5140 cm^2/g	28.9	Reddish Grey
[14]	600–800	1.91	1450 cm^2/g	DNR	Varied with Temperature

Table 1. Cont.

Reference	Calcination Temperature (°C)	Density (gcm^{-3})	Blaine Surface Area	Particle Size (μm)	Color
[15]	DNR	2.35	274 cm^2/g	107.9	DNR
[28]	DNR	2.22	11,270 cm^2/g	12.97	DNR
[29]	DNR	2.23	4720 cm^2/g	DNR	Grey
[30]	900–1100	1.94	DNR	<45	DNR
[31]	DNR	2.16	296 m^2/kg	>300	Black
[32]	500	4.19	32.9708 m^2/g		
	600	3.17	32.3502 m^2/g	DNR	DNR
	700	3.24	31.6265 m^2/g		
[33]	DNR	2.86	DNR	40–90	Black
[34]	600	DNR	1960 (pre grind) 6400 (ground) cm^2/g	76	DNR
[35]	600 DNR	2.1	240 m^2/kg	5.0 DNR	Black
[36]	DNR	2.2	4710 cm^2/g	40.1	DNR

DNR—Data Not Recorded.

3.1. Physical Properties

The physical properties (see Table 1) of SCBA are defined (see Table 1 for available data on density, surface area, particle size, and color of SCBA), starting from the soil of the sugarcane plantation all the way to the SCBA collection method. The composition of the soil on which the crops are grown supplies nutrients to these sugar plants and the heavy metals present in the soil also rest within the plant bodies. Additional nutrients that are used as fertilizer by the farmers contribute to the composition of SCBA as well. Furthermore, the sugarcane variant and growth of plantation decide the internal composition of the sugarcane bagasse [14,37].

If the collected bagasse contains other impurities while it is inserted inside the boiler, the properties of such impurities will affect the properties of SCBA. It is crucial to be conscious about the location of plantations and the bagasse collection method. Combustion period and temperature inside the boiler affect the SCBA's physical properties significantly. SCBA is collected from leftover ash at the bottom of the boiler or from the air-filtration system in the plant. Finer SCBA particles are present in the filtration system with less carbon content as opposed to coarser SCBA particles collected from the bottom of the boiler. Samples from the boiler are likely to have more carbon from unburnt bagasse volumes. If the collected SCBA is milled, the physical properties can be determined based on the milling time period [28,37].

As per the findings of Qing Xu et al. [14], different morphologies were identified in three SCBA samples, which were processed at different calcination temperatures (600 °C, 700 °C, and 800 °C). Within each individual sample, the processing time period inside the boiler altered its morphology. In general, all three samples illustrated different textures after 1 h, 2 h, and 3 h calcination periods. Olubajo Olumide Olu et al. carried out similar research [37] for samples calcinated at 600 °C, 650 °C, and 700 °C for 60, 90, and 120 min and observed dissimilar compositions and morphologies in each sample.

Uncontrolled burning temperatures above 800 °C for longer periods convert the amorphous silica into crystalline silica phase. It was identified that temperatures below 800 °C are the most cost-effective process parameters in SCBA production, while performing grinding processes in the following step is performed to enhance the pozzolanic properties of SCBA even further [38].

3.2. Micromorphology

Several shapes can be found in fine SCBA particles, including prismatic, spherical, fibrous, and irregular. Spherical particles correspond to the melting of minor components, such as Mg, P, K, Si, Na, Fe, etc. Higher temperatures provide the thermal conditions required for spherical particle formation. Prismatic particles illustrate a crystallization effect within SCBA and it is disadvantageous to the pozzolanic properties present in SCBA. Large coarse fibrous particles indicate unburnt carbon bagasse components present in the SCBA profile [14,39,40]. The micromorphology of the final SCBA is influenced by the purity of the bagasse, the thermal conditions inside the boiler, and the biological profile of the sugarcane variant.

3.3. Chemical Properties

Individual samples' chemical compositions are different from one another and SCBA has a characteristic chemical composition to be generally categorized under class F pozzolan material based on ASTM C618-08a specification. That is, the sum weights of SiO_2, Al_2O_3, and Fe_2O_3 compounds are more than 70% of the total mass of the SCBA sample [7,14]. According to information listed in Table 2, the majority of SCBA found in the literature fulfills this requirement. Apart from the most-common chemical compounds that are listed in Table 2, trace amounts of Ag, As, Ba, Cd, Cr, Hg, Pb, and Mn heavy metals can also be found in SCBA [17,41].

Table 2. Chemical composition of SCBA.

(w/w) %	SiO_2	Al_2O_3	Fe_2O_3	CaO	MgO	SO_3	K_2O	Na_2O	LOI
[7]	60–65	4–5	6–8	10–12	2–3	1–2	2–4	DNR	4–6
[15]	65.26	6.91	3.65	4.01	1.10	0.21	1.99	0.33	15.34
[18]	62.43	4.28	6.98	11.8	2.51	1.48	3.53	DNR	4.73
[28]	36.58	8.3	4.0	2.71	0.51	DNR	0.45	DNR	DNR
[30]	54.4	9.1	5.5	12.4	2.9	4.1	1.3	0.9	9.4
[31]	75.9	1.55	2.32	6.25	1.77	DNR	8.4	0.12	4
[35]	77.08	1.46	2.42	6.22	1.6	DNR	5.36	0.3	4.2
[36]	72.85	1.07	6.96	9.96	6.49	DNR	6.76	1.96	4.23
[41]	78.34	78.34	3.61	2.15	0.12	DNR	3.46	DNR	0.42
[42]	87.97	1.84	2.65	2.65	0.72	0.15	0.32	0.28	10.45
[43]	78.34	8.55	3.61	2.15	DNR	DNR	3.46	0.12	DNR
[44]	63.62	18.82	7.48	2.30	1.74	0.20	2.29	1.42	DNR
[45]	55.97	12.44	6.5	0.84	0.48	1.00	0.9	0	17.98
[46]	35.17	0.281	5.22	2.07	0.91	0.03	3.75	0.01	DNR
[47]	35.168	0.281	5.217	2.071	0.908	0.027	3.745	0.012	DNR
[48]	72.3	5.52	10.8	1.57	1.13	DNR	DNR	DNR	1.52
[49]	71.4	3.39	3.50	6.73	DNR	2.24	8.18	DNR	4.38
[50]	78.34	8.55	3.61	2.15	DNR	DNR	DNR	0.12	0.42
[51]	73	6.7	6.3	2.8	3.2	DNR	2.4	1.1	0.9
[52]	64.15	9.05	5.52	8.14	2.85	DNR	1.35	0.92	4.90
[53]	63.3	8.1	3.6	4.6	3.8	2.6	3.8	DNR	3.2
[54]	72.40	1.83	2.29	12.50	1.95	3.10	3.05	0.56	1.89
[55]	35.17	0.281	5.22	2.07	0.91	0.03	3.75	0.01	DNR

LOI—Loss on Ignition.

3.4. Pozzolanic Activity

It is mentioned that application of pozzolans as supplementary cementitious material can improve the mechanical and durability properties in concrete [24,29,43,56]. Pozzolanic strength in concrete is a result of the pozzolanic reaction between calcium hydroxide compounds present in cement materials from cement hydration, silicates, and/or aluminates in the chemical composition of SCBA and water in the concrete. Calcium hydroxide formation is executed during the cement hydration process where chemicals in OPC (calcium silicates and calcium aluminates) interact with water in the mix to form calcium hydroxides as one of the products. The need for calcium hydroxide is that silicates and aluminates are only soluble in highly basic media [57,58].

Calcium hydroxide molecules are then transported through water to combine with aluminum/silicates. As a result of this chemical reaction, calcium silicates and aluminum silicates are synthesized, which are responsible for the enhanced physical properties in concrete. This phenomenon occurs over longer periods, from months to years [57].

It is mentioned that curing samples at elevated temperatures within the first five hours of post-mixing enhances the reaction rates in concrete, which results in even stronger concrete [57]. SCBA samples with high LOI values have to be post-treated prior to using them in concrete as they do not possess acceptable pozzolanic activity. Some unburnt compounds in SCBA could be amorphous in nature and they might have the potential to enhance the reactivity of SCBA, which requires further experimentation to arrive at a conclusion [39].

The fineness in SCBA is reportedly directly related to the pozzolanic activity. Muhammad Izhar Shah et al. [59] carried out an investigation to identify the effect of milling time period over an SCBA surface area by using samples that were passed through a 200-micron standard sieve. The samples were grinded in a ball mill using ceramic balls as the grinding media for grinding periods of 15, 30, 45, and 60 min. Later, investigations to find surface area were carried out under the guidelines of ASTM C204, which revealed that SCBA samples increased their specific surface areas to the sequence, with the 60 min processed sample having the highest surface area and the 15 min milled sample having the lowest surface area. T. Murugesan et al. [60] observed that the strength activity index (percentage) value of processed SCBA is higher than 75%, which is the margin separating pozzolans and non-pozzolans, while raw SCBA did not satisfy the requirement.

Bahurudeen et al. [38] investigated pozzolanic activity by varying the temperature in the boiler and discovered that 700 °C produces the highest pozzolanic activity in SCBA samples ground up for 120 minutes.

According to the experimentation carried out by Marcela M.N.S. de Soares et al. [48], on pozzolanic activity comparison between SCBA, amorphous silica and crystalline silica led to the conclusion that SCBA has pozzolanic properties at low levels that are more similar to the pozzolanic properties in crystalline silica than amorphous silica.

3.5. Mineral Composition

Sugarcane crops, which are grown in silicic-acid-rich water-based soil, absorb compounds into plantations and polymerization into amorphous silica occurs inside the plant cells. The combustion process converts silica to reactive amorphous silica, which is identified in SCBA. Crystalline silica in SCBA is precent due to an uncontrolled incineration process and the sand in the soil being taken inside the boiler together with sugarcane bagasse (silica from sand is 4–10% [61]). Therefore, high amounts of quartz are present in SCBA [6,14].

Other miner minerals that were identified via X-ray diffraction (XRD) analysis using SCBA samples are mentioned as Calcite, Corundum, Hematite, Fluorite, Halite, Bornite, etc. [42].

Air flow conditions during the calcination process also affect the morphology of SCBA. It was identified that calcination without controlled air flow does not break down long bagasse fibers and, as a result, the LOI value of such SCBA is relatively higher [62].

3.6. SCBA Characterization

A wide range of characterization methods have been utilized throughout the literature to examine the microstructure and to identify the chemical compounds within SCBA, including Scanning Electron Microscopy (SEM), Energy Dispersive spectroscopy (EDS), X-ray Diffraction analysis (XRD), Thermogravimetric/Differential Thermal Analysis (TG/DTA), Energy Dispersive X-ray (EDX), and Fourier-transmission infrared (FTIR). These studies provided a better understanding of SCBA's potential for improving concrete properties [4].

In Figure 3, Image (A)—[4] depicts pores in elongated oval-shaped particles, which absorb water and oxygen. Image (B)—[31] depicts unburnt, carbon-rich fibrous particles. Image (C)—[45] depicts filter bagasse ash prismatic particles from combustion fumes. Image (D)—[47] depicts well-defined burnt flakes of SCBA. Image (E)—[55] depicts well-burnt flakes of SCBA. Image (F)—[63] depicts SCBA particles with lamellar aspect of superimposed layers.

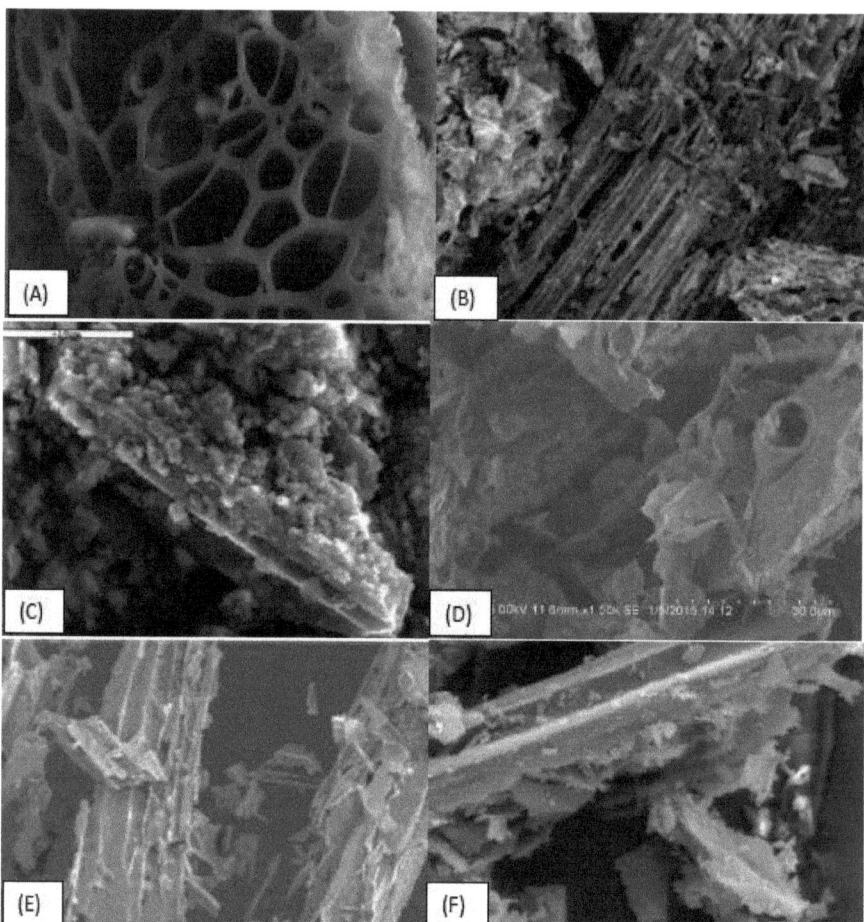

Figure 3. Microstructure of SCBA (source: (**A**)—[4], (**B**)—[31], (**C**)—[45], (**D**)—[47], (**E**)—[55], (**F**)—[63]).

From different characterizations, the presence of silica (SiO_2) has been highlighted in SCBA samples in both crystalline and amorphous phases. The roots of the sugarcane plants absorb soil, which then facilitates the formation of silica within the plant body. Pathogenic fungi that can potentially harm the plants are physically restricted from penetrating inside the plant by silica and water transportation within the plants is also facilitated by silica [63]. Depending on how the bagasse is collected, some sand from the fields may enter the boiler with the bagasse and the final SCBA material collected from the burner frequently contains this crystalline silica material.

Gritsada Sua-iam et al. [15] investigated microstructural properties of SCBA samples that were collected from an open dump site in Thailand. After drying, homogeneous samples were prepared and, later, XRD analysis on SCBA samples indicated the presence of quartz phases in SCBA. SEM images with ×1000 magnification confirmed the availability of crystalline silica from particles with distinguishable sharp edges. A Malvern Instruments Mastersizer 2000 particle size analyzer was used to analyze the particle size distribution of SCBA, OPC, and lime stone (LS) samples. SCBA particles (107.9 µm) were substantially larger compared to OPC particles (23.32 µm), while LS particles (15.73 µm) were slightly smaller than OPC.

Chidanand Patil et al. [20] carried out SEM and EDS investigations to analyze the microstructure and chemical composition of SCBA in comparison to OPC. SEM images of OPC revealed angular, irregular particles while EDS analysis indicated a high composition of calcium, oxygen, and silicon, whereas analysis of four SCBA samples that were collected from four different factories indicated unique characteristics compared to one another and OPC. Particle density, crystallinity, porosity, and particle shape varied depending on the source of the sample. Silicon, oxygen, and calcium were the predominant materials for all four samples of SCB.

Mao-Chieh Chi [30] conducted microscopic investigations on SCBA samples processed through a 900–1100 °C boiler and a no. 325 sieve. It was noted that SEM analysis presented particles of irregular shapes, rough surface texture, and high porous characteristics. XRD analysis provided evidence to confirm the presence of Silica (quartz), which was three-times higher than the Silica concentration in the reference OPC sample. Exothermal peaks of calcium silicate hydrate (C-S h) range between 115 and 225 °C, ettringite at 120–130 °C and calcium hydroxide (CH) 430–550 °C. After conducting TG tests in a range of 115 °C to 550 °C, it was observed that 10% SCBA substitution to cement provided mortar with accelerated hydration after 56 days of curing.

An investigation carried out by Daniel Véras Ribeiro et al. [32] compared the effect of calcination temperature on the pozzolanic activity of SCBA. Calcination temperatures (500 °C, 600 °C, and 700 °C) were chosen based on the SCBA TG curve. The XRD peaks of three samples indicated the presence of the amorphous silica phase and, from the samples collected at 600 °C and 700 °C, the presence of calcium silicate and calcium aluminate was observed, corresponding to the increased reactivity in SCBA. Density and grain size differed for each calcination temperature, while surface area remained unchanged.

Jijo James et al. [44] studied the microstructure of SCBA and found that SEM images of the samples consisted of well-defined burned flakes of bagasse. High temperatures inside boilers preserved the structure of bagasse. Crystalline bulky grains and pyrolyzed organic fractions were present in the microstructure. XRD analysis of SCBA indicated the presence of quartz, cristobalite, and calcite.

Moisés Frías et al. [45] conducted an analysis on three different types of SCBA samples (LBA—Laboratory Bagasse Ash, FBA—Filter Bagasse Ash, and BBA—Bottom Bagasse Ash) that were calcined at different temperatures and exhibited more than 75% SiO_2, Al_2O_3, and Fe_2O_3 combined percentages. Similar characteristics were identified in the XRD spectrum of three SCBA samples as well. Crystalline quartz is the major common component in the samples. TG/DTA indicated that 0 to 1000 °C heating resulted in a maximum weight loss in FBA while the minimum was in LBA. FTIR analysis aligned closely with the results from other techniques, indicating the presence of amorphous or not very crystalline substances. BBA and LBA analysis were very similar, while FBA had appreciable statistical differences in substance content. Different particle morphologies were identified in three samples from their SEM images as a result of the unique calcination process, from which each sample was collected. Particles were coarse in nature while particle sizes varied in the order of FBA < LBA < BBA. EDX analysis indicated that coarse particles were a result of quartz that were mentioned in XRD curves.

A solid waste study by Jijo James et al. [47] explained the presence of both crystalline and amorphous phases in the SCBA microstructure by using XRD patterns where peaks had high intensities but an increase in 2-theta angle corresponded to lowering of intensities. In SEM images of SCBA samples, charred remains of bagasse fibers were present, which could have affected the intensity peaks of the XRD curve.

Frequently mentioned information in the literature about the microscopic structure of SCBA is its pozzolanic property, unburnt particles, crystallinity, particle size, and shape, which contribute to the characteristic mechanical properties of SCBA-incorporated concrete. Relationships between the location of the SCBA samples and their microscopic properties can also be identified. SCBA samples collected from exhaust gas flirtation systems indicate better pozzolanic and microscopic properties to be used as binding materials.

3.7. Optimization of Mechanical Properties

Plenty of evidence for the productive use of SCBA in structural applications, such as concrete, bricks, soil, and steel, is found in the literature [12,42,46,50,51,64]. The chemical and microstructural properties of SCBA were utilized to explain the enhancement in mechanical properties of samples that were prepared with SCBA.

3.7.1. Workability of Concrete

The ability to transport, place, compact, fill, and resistance to segregation is generally defined as the workability of concrete. The common standard methods used to conduct testing on concrete to investigate workability are (ASTM C 143), American Association of State Highway and Transportation Officials (AASHTO T 119), or British Standards (BS EN 12350-2) [65]. It is important that concrete possesses low flow resistance as well, because this reduces safety issues, such as "white finger syndrome", and minimizes adverse environmental effects, including sound pollution, while concrete placement is being performed [15].

Priyesh Mulye et al. [7] carried out research work by replacing OPC of grade 53 with SCBA. Out of the concrete samples that were prepared with different SCBA percentages by weight (0%, 5%, 10%, 15%, and 20%), test results indicated the workability of concrete was increased significantly by 28% (slump increased from 70 mm to 90 mm) by replacing cement up to 15% and 20% with SCBA, compared to that of concrete with 0% SCBA in the mix.

As per the findings of R. Srinivasan et al. [43], it was observed that M20 grade concrete samples with SCBA percentages of 5%, 10%, 15%, 20%, and 25% by weight displayed higher slump values compared to the sample prepared with 100% OPC. In comparison to the slump of a 0% SCBA sample (60 mm), the 25% SCBA sample had the highest slump of 230 mm, while all the other samples also displayed slump values above 60 mm. This was noted as a clear indication of the positive effect SCBA has on the workability of concrete. It was suggested that the reason for this effect could be the high surface area of SCBA, which produced cement particles moistened with less water [66].

Sajjad Ali Mangi [67] conducted research work involving SCBA in M15 and M20 grade concrete. Six samples were prepared by replacing OPC with SCBA in amounts of 5%, 10%, and two control samples with 0% SCBA. It was observed that slump values of M15 grade 5%, 10% samples increased, respectively, by 15% and 28%, while the slump values of M20 grade concrete increased by 34% and 45%. It is recommended that use of super plasticizer is not essential because the recorded slum values can be categorized into low and medium degrees of workability.

3.7.2. Compressive Strength

One of the most important characteristics in concrete is its compressive strength properties, which contribute to the load-bearing capabilities without the occurrence of failure [67]. ASTM C109, BS EN 196-1, or AASHTO 106-02 are considered globally accepted standard methods of testing.

According to the work performed by Priyesh Mulye et al. [7], compressive strength enhancement can be achieved only up to a 15% partial replacement of SCBA. From the five-sample set prepared with 0%, 5%, 10%, 15%, and 20% SCBA, it was discovered that the sample with 15% SCBA has the highest average compressive strength compared to reference samples. Compressive strength values of samples were enhanced with longer curing periods, such that after 3 days of curing, compressive strength was increased by 6%, after 7 days of curing, a minor decrease was noticed and, again, after 28 days, compressive strength was enhanced by 2.1%. Samples with 20% SCBA in their composition indicated that the amount of SCBA negatively affects compressive strength properties.

Experimentation on SCBA by Pamela Camargo Macedo et al. [28] revealed that the compressive strength of concrete increases from the incorporation of SCBA and data from the research suggest that an enhancement in compressive stress occurs over longer periods. From the five sets of samples, including the reference, 3%, 5%, 8%, and 10% SCBA replacements, the highest values of compressive strength were recorded in the 10% specimen. A 23.23% increase in compressive strength was recorded after 56 days of curing for a 10% sample.

Prashant O Modani et al. [18] carried out research work with concrete samples prepared with 0%, 10%, 20%, 30%, and 40% SCBA and they observed that samples with 10% replacement had the highest compressive strength after 28 days of curing, Figure 4.

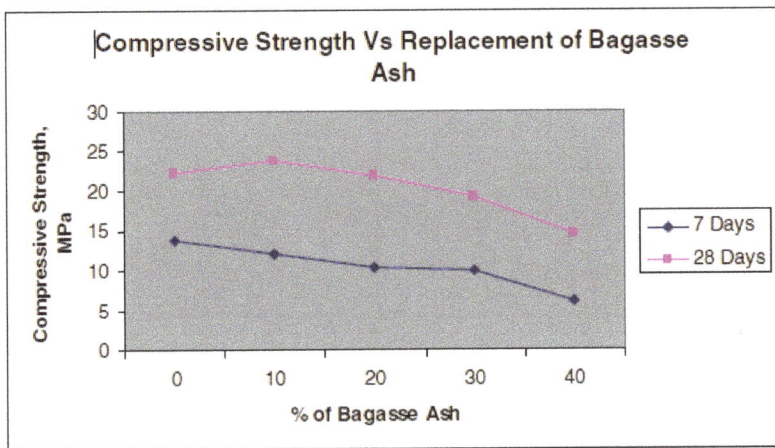

Figure 4. Compressive strength characteristic [18].

Mao-Chieh Chi [30] conducted research involving SCBA in mortars and his work indicated that the compressive strength of samples prepared from OPC replaced with 10% SCBA was increased after a 56-day curing period. Other samples with 20% and 30% replacements reduced their compressive strength compared to the control sample. It is also mentioned that the particle size of SCBA has the potential to fill the voids in concrete structures, which is associated with compressive strength enhancement.

T. Murugesan [31] et al. conducted research on the effects of SCBA and marble waste on concrete. From the five samples that were prepared with different percentages of SCBA, it was noticed that optimum compressive strength was achieved by the sample with 10% SCBA in its microstructure. The 20% replacement specimen also exhibited higher compressive strength than the control specimen, whereas the 30% SCBA sample had reduced compressive strength after 28-day and 56-day curing periods. A comparison of raw bagasse ash and sieved bagasse ash indicated that samples prepared with sieved bagasse ash had a higher strength activity index after 7 days and 28 days of curing.

According to the findings of S.Sanchana sri et al. [33], compressive strength of M20-grade concrete can be optimized by replacing OPC with 10% SCBA. Data indicated that compressive strength was gradually decreased below the compressive strength of the control specimen after 7 days and 28 days for the samples with 15% and 20% SCBA replacements.

Olubajo Olumide Olu et al. [37] investigated mortar with SCBA and observed that replacing 5% of OPC with SCBA resulted in a sample with the highest compressive strength, while samples with SCBA contents above 7.5% had lower compressive strength throughout the 60-day curing period.

In other research carried out by Jijo James et al. [46], soil block samples with SCBA and OPC were compared for their structural performance. It was noticed that 8% SCBA addition to the mix increased the compressive strength of soil blocks to 2.95 Mpa compared to 0% SCBA blocks, which had a compressive strength just above 2.5 Mpa.

Findings by K. Ganesan [52] et al. indicated that the compressive strength of concrete can be increased by replacing OPC with SCBA in a range of 5% to 20%. Samples with percentage replacements of 25% and 30% did not illustrate any positive influence on compressive strength. Maximum strength was attained when SCBA was added by replacing 10% OPC. Data were collected for a 90-day curing period and it was noticed that while compressive strength is increased with curing time, the percentage that increased relative to the control specimen was decreased for a 20% SCBA sample (see Figure 5).

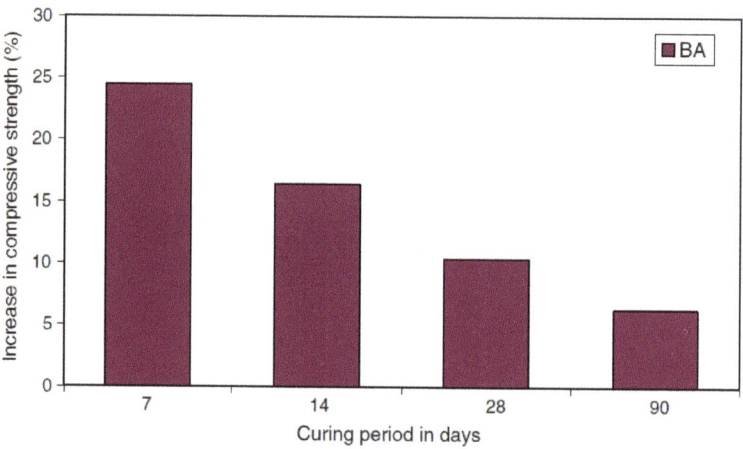

Figure 5. Relative increase in compressive strength of 20% SCBA blended concretes [52].

Research by Sajjad Ali Mangi et al. [67] provides evidence of an increase in compressive strength when SCBA is introduced by replacing 5% and 10% of OPC. The average compressive strength of M15-grade concrete was increased by 10.1% for a 5% SCBA sample and 4.8% for a 10% SCBA sample after 7 days of curing. The 5% SCBA sample of M20-grade concrete enhanced its compressive strength by 21.9%, while the 10% replacement sample had an increase of 12% after 7 days. A similar trend was observed with samples cured for 14 days and 28 days.

Noor-ul Amin [68] conducted research on concrete with SCBA and it was identified that the compressive strength of concrete can be increased by adding SCBA as a replacement to OPC in percentage amounts of 5%, 10%, 15%, and 20%. The optimum strength corresponded to the sample prepared by adding 10% SCBA. This work also revealed that the addition of 25% and 30% SCBA reduces the compressive strength below the values of the reference sample.

In an experiment by Chandan Kumar Gupta et al. [69], cement mortar samples were prepared with 0% to 25% SCBA replacements and the compressive strength of the 5% SCBA sample was noticeably higher than the sample prepared with 100% OPC. It is mentioned that samples with high percentages of SCBA indicated signs of very poor bonding between the materials due to insufficient water supplied to the mortar. This agrees with Noorwirdawati Ali et al. [70] who identified that introducing 20% of SCBA to be the optimum cement replacement value in earth bricks provided the maximum compressive strength and additional SCBA in 25% and 30% samples weakened the bond strength in the bricks.

Furthermore, it is reported that lightweight concrete has a more significant enhancement in its compressive strength when cement is partially replaced by SCBA in comparison to other concrete grades [1].

3.7.3. Split Tensile Strength

Tensile strength investigations of mortars from diametral compression were carried out by Pamela Camargo Macedo et al. [28]. It was identified that 3% replacement of OPC with SCBA was to be the optimal replacement content for enhanced tensile strength. Samples with SBCA content above 3% had lower tensile strengths than the control sample (see Figure 6).

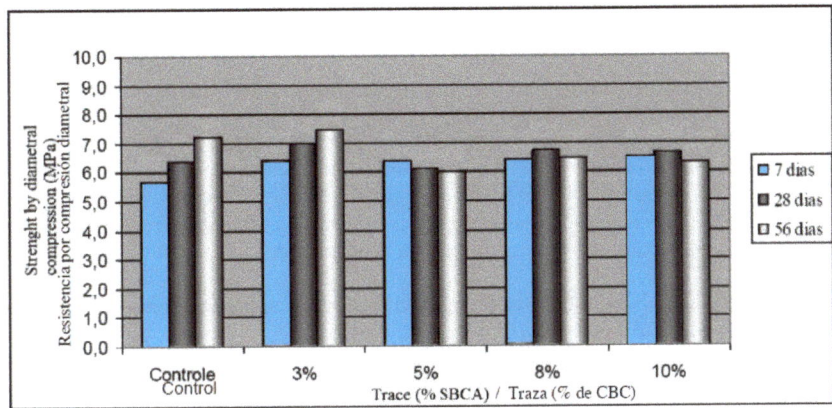

Figure 6. Tensile strength by diametral compression [28]. (Translation: 7 dias is 7 days, 28 dias is 28 days, and 56 dias is 56 days).

Work by S.Sanchana sri et al. [33] revealed that the tensile strength of concrete can be optimized by replacing 5% of OPC with SCBA. Compared to the control specimen, the compressive strength of a 5% SCBA sample was enhanced by 4.5%.

R. Srinivasan et al. [43] investigated concrete and SCBA, which indicated that SCBA can be effectively used in concrete up to 15% as a replacement for cement and the split tensile strength of such samples was been improved. A 39.9% increase in tensile strength was observed compared to that in the control sample for a sample with 5% SCBA in the mix.

Research and analysis by K. Ganesan [52] indicated that the slit tensile strength of concrete can be increased by incorporating SCBA into the mix design. It was noticed that replacing 15% of OPC with SCBA would optimize the tensile strength of concrete to a maximum, while it is possible to add SCBA up to 20% and have the sample's tensile strength enhanced. The addition of SCBA further reduced the tensile strength of the concrete samples (see Figure 7).

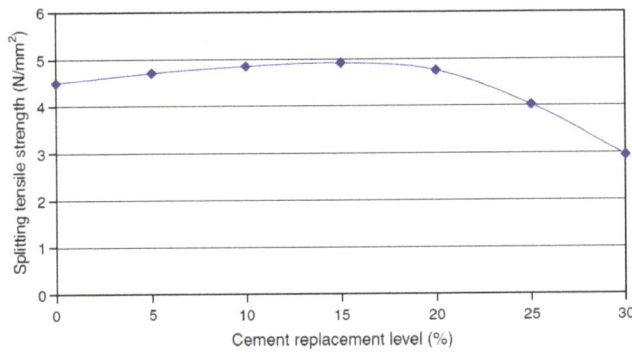

Figure 7. Splitting tensile strength of SCBA blended concretes at 28 days curing [52].

Findings by Noor-ul Amin [68] indicated that the tensile strength of concrete can be optimized by introducing 10% of SCBA to the mix. Compared with the control specimen, tensile strength was increased above 11%. It can be identified that with additional SCBA beyond 20%, tensile strength is considerably reduced.

The tensile strength of concrete with 20% SCBA in its composition increased to 4.81 MPa, whereas the value decreased to 3 MPa when the SCBA quantity was increased above 25% [71].

An experiment conducted by D. Patel [72] produced M20-grade concrete samples with enhanced flexural and split tensile strength properties after 28 days of curing with 10% cement replaced by SCBA. It is observed that tensile strength is increased by 4.1% after 28 days of curing. Tensile strength decreased when SCBA content reached 15% or higher.

Selvadurai Sebastin [73] conducted extensive research on the split tensile strength characteristics of mortars with SCBA. Eleven cylindrical and cubic sample sets with different SCBA contents (0–25%) were tested and it was noticed that both cylindrical and cubic samples have similar tensile strength characteristics. It is possible to identify data indicating enhancements in tensile strength of the samples, but a clear relationship with percentage SCBA and tensile strength cannot be identified.

Samples of M40-grade concrete prepared with reference to BIS: 10262-2009 (mix design ratio 1:1.56:2.42) showed enhanced flexural and split tensile properties after the introduction of SCBA to the mix (15% replacement) [13].

With the evidence available in the literature, it is emphasized that SCBA has physical and chemical properties to positively influence the enhancement in mechanical properties of concrete, mortar, soil, and brick materials. The pozzolanic characteristics of SCBA provided a meaningful explanation for this behavior. Manipulation of SCBA particle size and calcination temperature are mentioned as major parameters to be monitored in order to improve pozzolanic activity.

3.8. Cost Optimization

The economical aspect of concrete with SCBA was analyzed by Priyesh Mulye et al. [7] where they identified that normal concrete of grade M25 with mix design ratios 1:1.78:2.86 had a cost 12% more for 1 m^3 of concrete compared to the cost of concrete with 15% OPC replacement by SCBA, with the same mix design proportions.

Similar results were reported by Mangesh V. Madurwar et al. [74], where self-compacting concrete with SCBA cost 35.63% less for ingredients compared to the control concrete, while both had 34 Mpa in similar compressive strengths.

SCBA is often produced as a byproduct of the sugar industry, which has a very low economic value. Therefore, in the above literature, the possibility of producing low-cost concrete using SCBA as a partial replacement for OPC is mentioned while the final product is capable of satisfying the standards of defined quality management systems.

3.9. Carbon Footprint Analysis

The contribution of CO_2 to total global greenhouse gases stands out at 77%. The Earth System Research Laboratory from the US National Oceanic and Atmospheric Administration measurements indicated that in 1980, the mean CO_2 concentration was approximately 335 ppm, which later increased to 394 ppm in 2012. CO_2 concentrations have risen to 414.72 ppm in 2021 (see Figure 8) [75].

To avoid a +3 °C temperature increase, the International Panel on Climate Change announced that the global CO_2 concentration has to be maintained below the level of 450 ppm [76]. It is also mentioned that the average cost for CO_2 capture is in an estimated range of EUR 20 to 50 per ton of CO_2, without transportation and storage costs [77].

Several energy consumption rates and CO_2 emission values are mentioned in the literature for the production of a unit mass of concrete and cement. Such data are dependent upon several factors, including the weather, production site conditions, transportation distances, types of energy sources used, and the conditions of the plant equipment. Fossil fuel energy generation, approximately, gives rise to 80 g of CO_2 per 1 MJ, while natural gas based on 1 MJ only generates 55 g of CO_2 [78].

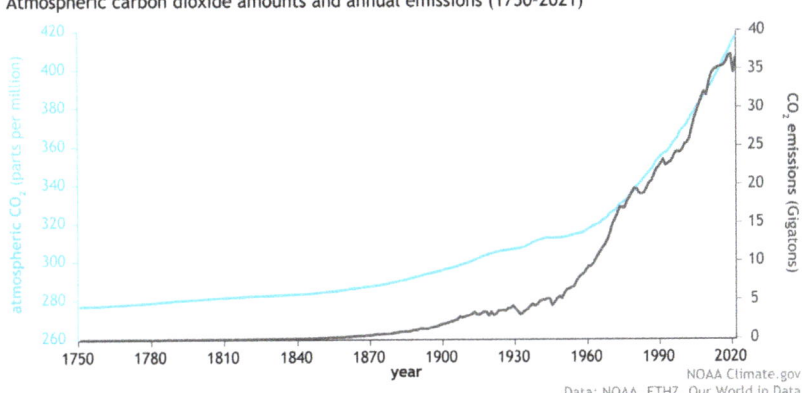

Figure 8. Atmospheric CO_2 amounts and annual emissions [75].

As far as the ordinary Portland Cement industry is concerned, cement is considered one of the most widely used essential materials in construction, which simultaneously contributes to 5–8% of annual global CO_2 emissions [7,13,73]. The reports indicate that approximately 1 kg of CO_2 is released into the atmosphere in the process of manufacturing 1 kg of Portland clinker [79]. For 1 kg of cement clinker, around 0.55 kg of CO_2 is generated inside the cement kiln while the calcination process of cement is occurring [78].

According to reports, OPC's 1 km long concrete pavement in China produced 8215.31 CO_2e (carbon dioxide equivalent), with the concrete processing stage accounting for 7.2% of the total emission [79]. The demand for OPC has been increasing and, as a result, the environmental concerns have intensified. The Kyoto Protocol commitments also urged the industry to move towards the implementation of clean development mechanisms [80].

Lightweight concrete blocks manufactured in Europe with 8–12% of cement additives are responsible for the emission of 239.7 kg of CO_2e for 1 kg of the product, while 1 kg of precast concrete emits 120.5 kg of CO_2e to the environment [81].

In an experiment carried out by Woubishet Zewdu Taffese et al. [82], it was identified that concrete, hollow concrete bars, and reinforcement bars were the major energy consumers and CO_2 emitters in a sample of five multi-storied buildings. They consumed 94% of the embodied energy while contributing to 98% of the CO_2 emissions.

The high CO_2 emissions in the cement and concrete industries are highlighted in the literature. The cost of recovering from the adverse effects of CO_2 is higher. Therefore, research towards low-cost, sustainable alternatives to cement is a better way to solve this problem.

3.10. Other Problems Affiliated with Sugar Industry

The world has identified the safe dumping of agricultural waste as another emerging issue in the field of agriculture. With respect to the sugar industry, both sugarcane bagasse and SCBA have the potential to cause adverse effects on the environment if they are discarded without a proper method. Unburned matter and oxides, such as silicon, aluminum, and calcium, have the potential to pollute soil, air, and water, posing environmental and social concerns. Harmful medical conditions can be observed in the lungs of factory workers and the public around dump sites (chronic lung condition pulmonary fibrosis) if processing and disposal of sugarcane bagasse ash are not conducted in a secure way [7].

Utilization of SCBA in concrete as a partial replacement to OPC will potentially reduce the amount of SCBA disposed to the environment. As a result, the aforementioned negative effects will be mitigated to some extent.

4. Conclusions

- There are a number of factors that define the microstructural properties of SCBA, including sugarcane variety, soil in the sugarcane fields, fertilizer, sugarcane collection method, bagasse burning process, and bagasse ash collection method. In order to obtain SCBA samples with sufficient pozzolanic activity, the burning process can be controlled within the boilers.
- Post-treatment methods, such as grinding, sieving, and post heating, positively affect the pozzolanic properties in SCBA and the parameters of such processes are directly related to the quality of the final SCBA.
- Greenhouse gas emissions during OPC production can be reduced by utilizing SCBA with suitable proportions in concrete. Since bagasse burning is generally conducted while electricity generation is performed using bagasse as a biofuel, neither any additional CO_2 emission nor extra energy consumption is required during SCBA synthesis. Controlled burning would reduce emissions and energy consumption even further.
- The cost of concrete in large-scale construction can be minimized by replacing OPC with suitable SCBA amounts while maintaining the required standards and specifications.
- From the information available in the literature, it can be concluded that SCBA has the potential to be used as a partial replacement for OPC. The performance of concrete can be enhanced while reducing the cost of cement as SCBA is available in high volumes.
- Future research can be conducted to identify other cement replacement materials, which can be used together with SCBA in concrete. Their properties and mix design parameters have to be major focus areas to develop low-cost, high-performance concrete.
- SCBA from an individual source possesses unique chemical and physical properties. Research can be carried out utilizing SCBA samples from various sugar manufacturing plants inside Sri Lanka to identify their potential to be used as a cement replacement material.

Author Contributions: The contribution of each author is mentioned here. Conceptualization, B.S.K., N.S., V.V. and K.K.; methodology, N.P., B.S.K., K.K. and V.V.; software, N.P.; validation, N.P., B.S.K., N.S., A.I.S. and M.M.; formal analysis, K.K., A.I.S. and N.S.; investigation, N.P., B.S.K., N.S., M.M. and A.I.S.; resources, B.S.K. and K.K.; data curation, V.V., N.S. and K.M.; writing—original draft preparation, N.P., B.S.K. and V.V.; writing—review and editing, B.S.K., V.V., N.S. and K.M.; visualization, N.P.; supervision, K.K.; project administration, B.S.K. and K.K.; funding acquisition, K.K. All authors have read and agreed to the published version of the manuscript.

Funding: This research was funded by SCIENCE AND TECHNOLOGY HUMAN RESOURCE DEVELOPMENT PROJECT (STHRDP)-ASIAN DEVELOPMENT BANK (ADB), grant number STHRD/CRG/R2/SB3 and the APC was funded by STHRD/CRG/R2/SB3.

Acknowledgments: The authors gratefully acknowledge the support provided by the Research Grant: Competitive Research Grant Round 2, Science and Technology Human Resource Development Project (STHRDP)-Project Output 03 (Grant No: STHRD/CRG/R2/SB3).

Conflicts of Interest: The authors declare no conflict of interest.

References

1. Zareei, S.A.; Ameri, F.; Bahrami, N. Microstructure, strength, and durability of eco-friendly concretes containing sugarcane bagasse ash. *Constr. Build. Mater.* **2018**, *184*, 258–268. [CrossRef]
2. Ahmad, W.; Ahmad, A.; Ostrowski, K.A.; Aslam, F.; Joyklad, P.; Zajdel, P. Sustainable approach of using sugarcane bagasse ash in cement-based composites: A systematic review. *Case Stud. Constr. Mater.* **2021**, *15*, e00698. [CrossRef]

3. Baig, M.; Das, A.; Ganvir, V.A.; Shelke, R.M. Review paper of potential of use of Sugar Cane Bagasse Ash in concrete. *Int. J. Innov. Res. Stud.* **2018**, *8*, 297–299. Available online: https://www.researchgate.net/publication/349367616 (accessed on 21 March 2022).
4. Jagadesh, P.; Ramachandramurthy, A.; Murugesan, R.; Sarayu, K. Micro-analytical studies on sugar cane bagasse ash. *Sadhana* **2015**, *40*, 1629–1638. [CrossRef]
5. Raje, K.S.; Rajgor, M. Bagasse Ash as an Effective Replacement in Fly Ash Bricks. *Int. J. Eng. Trends Technol.* **2013**, *4*, 4484–4489.
6. De Oliveira, D.C.G. Physical and Mechanical Performance of Mortars with Ashes from Straw and Bagasse Sugarcane, São Paulo. 2015. Available online: https://www.researchgate.net/publication/272176930 (accessed on 21 March 2022).
7. Mulye, P. Experimental Study on Use of Sugar Cane Bagasse Ash in Concrete by Partially Replacement with Cement. *Int. J. Res. Appl. Sci. Eng. Technol.* **2021**, *9*, 616–635. [CrossRef]
8. Zhang, P.; Liao, W.; Kumar, A.; Zhang, Q.; Ma, H. Characterization of sugarcane bagasse ash as a potential supplementary cementitious material: Comparison with coal combustion fly ash. *J. Clean. Prod.* **2020**, *277*, 123834. [CrossRef]
9. Chowdhury, S.; Mishra, M.; Suganya, O. The incorporation of wood waste ash as a partial cement replacement material for making structural grade concrete: An overview. *Ain Shams Eng. J.* **2015**, *6*, 429–437. [CrossRef]
10. Pandraju, S.; Kali, J.P.V.; Mondru, M. Energy Efficient Steam Boiling System for Production of Quality Jaggery. *Sugar Tech* **2020**, *23*, 915–922. [CrossRef]
11. Sousa, L.N.; Figueiredo, P.F.; França, S.; de Moura Solar Silva, M.V.; Borges, P.H.R.; Bezerra, A.C.D.S. Effect of Non-Calcined Sugarcane Bagasse Ash as an Alternative Precursor on the Properties of Alkali-Activated Pastes. *Molecules* **2022**, *27*, 1185. [CrossRef]
12. Solomon, S.; Li, Y.-R. Editorial-The Sugar Industry of Asian Region. *Sugar Tech* **2016**, *18*, 557–558. [CrossRef]
13. Trivedi, M.V.; Shrivastava, P.L.P. A study on geo polymer concrete using sugarcane bagasse ash: A Brief Review. *IJRTI* **2020**, *5*, 27–30. Available online: www.ijrti.org (accessed on 28 March 2022).
14. Xu, Q.; Ji, T.; Gao, S.-J.; Yang, Z.; Wu, N. Characteristics and Applications of Sugar Cane Bagasse Ash Waste in Cementitious Materials. *Materials* **2018**, *12*, 39. [CrossRef]
15. Sua-Iam, G.; Makul, N. Use of increasing amounts of bagasse ash waste to produce self-compacting concrete by adding limestone powder waste. *J. Clean. Prod.* **2013**, *57*, 308–319. [CrossRef]
16. Abdulkadir, T.S.; Oyejobi, D.O.; Lawal, A.A. Evaluation of sugarcane bagasse ash as a replacement for cement in concrete works. *Acta Tech. Corviniensis-Bull. Eng.* **2014**, *7*, 71.
17. James, J.; Pandian, P.K. A Short Review on the Valorisation of Sugarcane Bagasse Ash in the Manufacture of Stabilized/Sintered Earth Blocks and Tiles. *Adv. Mater. Sci. Eng.* **2017**, *2017*, 1706893. [CrossRef]
18. Modani, P.O.; Vyawahare, M. Utilization of Bagasse Ash as a Partial Replacement of Fine Aggregate in Concrete. *Procedia Eng.* **2013**, *51*, 25–29. [CrossRef]
19. Kishore, D.; Kotteswaran, S. Review on bagasse ash an effective replacement in fly ash bricks. *Int. Res. J. Eng. Technol.* **2018**, *5*, 176–179. Available online: www.irjet.net (accessed on 2 April 2022).
20. Patil, C.; Kalburgi, P.B.; Patil, M.B.; Prakash, K.B. SEM-EDS Analysis of Portland Cement and Sugarcane Bagasse Ash Collected from Different Boilers of Sugar Industry. *Int. J. Sci. Eng. Res.* **2018**, *9*, 632–637. Available online: http://www.ijser.org (accessed on 2 April 2022). [CrossRef]
21. Vijerathna, M.P.G.; Wijesekara, I.; Perera, R.; Maralanda, S.M.T.A.; Jayasinghe, M.; Wickramasinghe, I. Physico-chemical Characterization of Cookies Supplemented with Sugarcane Bagasse Fibres. *Vidyodaya J. Sci.* **2019**, *22*, 29. [CrossRef]
22. Kent, G.A. Issues Associated with Using Trash as a Cogeneration Fuel. *Sugar Tech* **2013**, *16*, 227–234. [CrossRef]
23. Solomon, S. Sugarcane By-Products Based Industries in India. *Sugar Tech* **2011**, *13*, 408–416. [CrossRef]
24. Santhanam, M.; Bahurudeen, A.; Vaisakh, K.S. Availability of Sugarcane Bagasse Ash and Potential for Use as a Supplementary Cementitious Material in Concrete. Available online: https://www.researchgate.net/publication/280100729 (accessed on 4 April 2022).
25. Arachchige, U.; Singhapurage, H.; Udakumbura, P.; Peiris, I.; Bandara, A.M.P.A.; Nishantha, P.G.U.; Anjalee, S.W.S.; Ruvishani, L.S. Jinasoma, N.; Pathirana, N.M.H.; et al. Sugar Production Process in Sri Lanka Boiler Operation and Maintainance View project Impacts of Air Pollution in Sri Lanka View project Sugar Production Process in Sri Lanka. *J. Res. Technol. Eng.* **2020**, *1*, 38–45. Available online: https://www.researchgate.net/publication/338791293 (accessed on 4 April 2022).
26. Tiwari, R.N.; Gandharv, C.A.; Dharamvir, K.; Kumar, S.; Verma, G. Scale Minimization in Sugar Industry Evaporators using Nanoporous Industrial Bio-solid Waste Bagasse Fly Ash. *Sugar Tech* **2018**, *21*, 301–311. [CrossRef]
27. Sewwandi, M.N.; Ariyawansha, S.; Kumara, B.S.; Maralanda, A. Optimizing pre liming ph for efficient juice clarification process in Sri Lankan sugar factories. *Int. J. Eng. Appl. Sci. Technol.* **2021**, *6*, 14–20. [CrossRef]
28. Macedo, P.C.; Pereira, A.M.; Akasaki, J.L.; Fioriti, C.F.; Payá, J.; Pinheiro, J.L. Performance of mortars produced with the incorporation of sugar cane bagasse ash. *Rev. Ing. Constr.* **2014**, *29*, 187–199. [CrossRef]
29. Vikram, V.; Soundararajan, A.S. Durability studies on the pozzolanic activity of residual sugar cane bagasse ash sisal fibre reinforced concrete with steel slag partially replacement of coarse aggregate. *Caribb. J. Sci.* **2021**, *53*, 326–344.
30. Chi, M.-C. Effects of sugar cane bagasse ash as a cement replacement on properties of mortars. *Sci. Eng. Compos. Mater.* **2012**, *19*, 279–285. [CrossRef]
31. Murugesan, T.; Vidjepriya, R.; Bahurudeen, A. Sugarcane Bagasse Ash-Blended Concrete for Effective Resource Utilization Between Sugar and Construction Industries. *Sugar Tech* **2020**, *22*, 858–869. [CrossRef]

32. Ribeiro, D.V.; Morelli, M.R. Effect of Calcination Temperature on the Pozzolanic Activity of Brazilian Sugar Cane Bagasse Ash (SCBA). *Mater. Res.* **2014**, *17*, 974–981. [CrossRef]
33. Sanchana, S.; Ramesh, M.T. Experimental Study on Strength and Durability of Concrete with Bagasse Ash And M-Sand. *Int. Res. J. Eng. Technol.* **2017**, *4*, 1885–1888. Available online: www.irjet.net (accessed on 14 April 2022).
34. Zaki, E.-S.I.; Rashad, A.M. Sugarcane Ash as Cement Pozzolana Binder in Cement Pastes. Available online: https://www.researchgate.net/publication/317004528 (accessed on 14 April 2022).
35. Kumar, D.S.S.; Chethan, K.; Kumar, B.C. Effect of Elevated Temperatures on Sugarcane Bagasse Ash-Based Alkali-Activated Slag Concrete. *Sugar Tech* **2020**, *23*, 369–381. [CrossRef]
36. Chandrasekar, S.; Asha, P. Use of sugar cane bagasse ash in fibre reinforced concrete—A Review. *Res. J. Adv. Eng. Technol.* **2018**, *4*, 3007–3012. Available online: http://www.irjaet.com (accessed on 16 April 2022).
37. Olu, O.O.; Aminu, N.; Sabo, L.N. The Effect of Sugarcane Bagasse Ash on the Properties of Portland Limestone Cement. *Am. J. Constr. Build. Mater.* **2020**, *4*, 77. [CrossRef]
38. Rosseira, A.; Sarbini, N.N.; Ibrahim, I.S.; Lim, N.H.A.S.; Sam, A.R.M.; Karim, N.F.N.A. Overview of the Influence of Burning Temperature and Grinding Time to the Properties of Cementitious Material based Agricultural Waste Products. *IOP Conf. Ser. Mater. Sci. Eng.* **2018**, *431*, 082004. [CrossRef]
39. Maldonado-García, M.A.; Montes-García, P.; Valdez-Tamez, P.L. A Review of the Use of Sugarcane Bagasse Ash with a High LOI Content to Produce Sustainable Cement Composites. *Acad. J. Civ. Eng.* **2017**, *35*, 597–603.
40. Cordeiro, G.; Filho, R.T.; Fairbairn, E. Effect of calcination temperature on the pozzolanic activity of sugar cane bagasse ash. *Constr. Build. Mater.* **2009**, *23*, 3301–3303. [CrossRef]
41. Tijore, N.A.; Pathak, V.B.; Shah, R.A. Utilization of Sugarcane Bagasse Ash in Concrete. *Int. J. Sci. Res. Dev.* **2013**, *1*, 1938–1941.
42. Saleem, M.A.; Kazmi, S.M.S.; Abbas, S. Clay bricks prepared with sugarcane bagasse and rice husk ash—A sustainable solution. *MATEC Web Conf.* **2017**, *120*, 03001. [CrossRef]
43. Srinivasan, R.; Sathiya, K. Experimental Study on Bagasse Ash in Concrete. *Int. J. Serv. Learn. Eng.* **2010**, *5*, 60–66. [CrossRef]
44. James, J.; Pandian, P.K. Bagasse Ash as an Auxiliary Additive to Lime Stabilization of an Expansive Soil: Strength and Microstructural Investigation. *Adv. Civ. Eng.* **2018**, *2018*, 1–16. [CrossRef]
45. Frías, M.; Villar, E.; Savastano, H. Brazilian sugar cane bagasse ashes from the cogeneration industry as active pozzolans for cement manufacture. *Cem. Concr. Compos.* **2011**, *33*, 490–496. [CrossRef]
46. James, J.; Pandian, P.K.; Deepika, K.; Venkatesh, J.M.; Manikandan, V.; Manikumaran, P. Cement Stabilized Soil Blocks Admixed with Sugarcane Bagasse Ash. *J. Eng.* **2016**, *2016*, 1–9. [CrossRef]
47. James, J.; Pandian, P.K. Chemical, Mineral and Microstructural Characterization of Solid Wastes for use as Auxiliary Additives in Soil Stabilization. *J. Solid Waste Technol. Manag.* **2018**, *44*, 270–280. [CrossRef]
48. de Soares, M.M.; Garcia, D.C.; Figueiredo, R.B.; Aguilar, M.T.P.; Cetlin, P.R. Comparing the pozzolanic behavior of sugar cane bagasse ash to amorphous and crystalline SiO2. *Cem. Concr. Compos.* **2016**, *71*, 20–25. [CrossRef]
49. Athira, G.; Bahurudeen, A.; Vishnu, V.S. Availability and Accessibility of Sugarcane Bagasse Ash for its Utilization in Indian Cement Plants: A GIS-Based Network Analysis. *Sugar Tech* **2020**, *22*, 1038–1056. [CrossRef]
50. Manikandan, T.; Moganraj, M. Consolidation and rebound characteristics of expansive soil by using lime and bagasse ash. *Int. J. Res. Eng. Technol.* **2014**, *3*, 2321–7308.
51. Narendran, B.; Mourigokul, P. Consumption of bagasse ash as an effective booming in brick material—Review. *Shanlax Int. J. Arts Sci. Humanit.* **2017**, *5*, 84–88.
52. Ganesan, K.; Rajagopal, K.; Thangavel, K. Evaluation of bagasse ash as supplementary cementitious material. *Cem. Concr. Compos.* **2007**, *29*, 515–524. [CrossRef]
53. Cordeiro, G.C.; Andreão, P.V.; Tavares, L. Pozzolanic properties of ultrafine sugar cane bagasse ash produced by controlled burning. *Heliyon* **2019**, *5*, e02566. [CrossRef]
54. Maldonado-García, M.A.; Hernández-Toledo, U.I.; Montes-Garcia, P.; Valdez-Tamez, P.L. The influence of untreated sugarcane bagasse ash on the microstructural and mechanical properties of mortars. *Mater. Construcción* **2018**, *68*, 148. [CrossRef]
55. James, J.; Pandian, P.K. Valorisation of Sugarcane Bagasse Ash in the Manufacture of Lime-Stabilized Blocks. *Slovak J. Civ. Eng.* **2016**, *24*, 7–15. [CrossRef]
56. Mehdizadeh, B.; Jahandari, S.; Vessalas, K.; Miraki, H.; Rasekh, H.; Samali, B. Fresh, Mechanical, and Durability Properties of Self-Compacting Mortar Incorporating Alumina Nanoparticles and Rice Husk Ash. *Materials* **2021**, *14*, 6778. [CrossRef]
57. Sargent, P. Pozzolanic Reaction The development of alkali-activated mixtures for soil stabilisation. In *Handbook of Alkali-Activated Cements, Mortars and Concretes*; Woodhead Publishing: Sawston, UK, 2015; pp. 1–15.
58. Ahmad Wani, T.; Kumar Sharma, P. Partial Replacement of Cement with Rice Husk Ash and its Pozzolanic Activity: A review. *Int. J. Innov. Res. Technol.* **2021**, *7*, 110–115. Available online: https://www.researchgate.net/publication/348431151 (accessed on 2 May 2022).
59. Shah, M.I.; Amin, M.N.; Khan, K.; Niazi, M.S.K.; Aslam, F.; Alyousef, R.; Javed, M.F.; Mosavi, A. Performance Evaluation of Soft Computing for Modeling the Strength Properties of Waste Substitute Green Concrete. *Sustainability* **2021**, *13*, 2867. [CrossRef]
60. Murugesan, T.; Athira, G.; Vidjeapriya, R.; Bahurudeen, A. Sustainable Opportunities for Sugar Industries Through Potential Reuse of Sugarcane Bagasse Ash in Blended Cement Production. *Sugar Tech* **2021**, *23*, 949–963. [CrossRef]

1. Velmurugan, S. Recovery of Chemicals from Pressmud—A Sugar Industry Waste Project. Available online: https://www.researchgate.net/publication/264547627 (accessed on 2 May 2022).
2. Soares, M.M.N.; Poggiali, F.S.; Bezerra, A.C.S.; Figueiredo, R.B.; Aguilar, M.T.P.; Cetlin, P.R. The effect of calcination conditions on the physical and chemical characteristics of sugar cane bagasse ash. *Rev. Esc. Minas* **2014**, *67*, 33–39. [CrossRef]
3. Sales, A.; Lima, S.A. Use of Brazilian sugarcane bagasse ash in concrete as sand replacement. *Waste Manag.* **2010**, *30*, 1114–1122. [CrossRef] [PubMed]
4. Maneela, M.; Ariff, S.; Manesh, P.; Rashimi, G.V.; Manohar, M.; Srinivasa, S. Compressive strength of fly ash bricks with addition of bagasse ash. *Int. J. Curr. Eng. Sci. Res.* **2019**, *6*, 5–9. [CrossRef]
5. ICTAD-Document-of-Roads-and-Bridges-Construction-and-Maintenance. Colombo, SCA/5. 2009. Available online: https://www.cida.gov.lk/pages_e.php?id=47 (accessed on 4 May 2022).
6. Ganesh Babu, M.; Priyanka, G.S. Investigational Study on Bagasse Ash in Concrete by Partially Substitute with Cement. *Int. J. Comput. Eng. Res.* **2015**, *2*, 1044–1048.
7. Mangi, S.A.; Jamaluddin, N.; Ibrahim, M.H.W.; Abdullah, A.H.; Awal, A.S.M.A.; Sohu, S.; Ali, N. Utilization of sugarcane bagasse ash in concrete as partial replacement of cement. *IOP Conf. Ser. Mater. Sci. Eng.* **2017**, *271*, 012001. [CrossRef]
8. Amin, N.-U. Use of Bagasse Ash in Concrete and Its Impact on the Strength and Chloride Resistivity. *J. Mater. Civ. Eng.* **2011**, *23*, 717–720. [CrossRef]
9. Gupta, C.K.; Sachan, A.K.; Kumar, R. Examination of Microstructure of Sugar Cane Bagasse Ash and Sugar Cane Bagasse Ash Blended Cement Mortar. *Sugar Tech* **2021**, *23*, 651–660. [CrossRef]
10. Ali, N.; Zainal, N.A.; Burhanudin, M.K.; Samad, A.A.A.; Mohamad, N.; Shahidan, S.; Abdullah, S.R. Physical and Mechanical Properties of Compressed Earth Brick (CEB) Containing Sugarcane Bagasse Ash. *MATEC Web Conf.* **2016**, *47*, 1018. [CrossRef]
11. Ali, N.; Sobri, M.H.A.M.; Hadipramana, J.; Samad, A.A.A.; Mohamad, N. Potential Mixture of POFA and SCBA as Cement Replacement in Concrete—A Review. *MATEC Web Conf.* **2017**, *103*, 01006. [CrossRef]
12. Patel, D. Study of partial replacement of bagasse ash in concrete. *Int. J. Res. Publ. Rev.* **2022**, 983–992. [CrossRef]
13. Madurwar, M.V.; Ralegaonkar, R.V.; Mandavgane, S.A. Application of agro-waste for sustainable construction materials: A review. *Constr. Build. Mater.* **2013**, *38*, 872–878. [CrossRef]
14. Barcelo, L.; Kline, J.; Walenta, G.; Gartner, E. Cement and Carbon Emissions—Final Draft Manuscript. Available online: https://www.researchgate.net/profile/Ellis-Gartner/publication/257895979_Cement_and_carbon_emissions/links/56cad3fb08aee3cee54075ff/Cement-and-carbon-emissions.pdf (accessed on 2 June 2022).
15. Schneider, M.; Romer, M.; Tschudin, M.; Bolio, H. Sustainable cement production—present and future. *Cem. Concr. Res.* **2011**, *41*, 642–650. [CrossRef]
16. Shi, C.; Jimenez, A.F.; Palomo, A. New cements for the 21st century: The pursuit of an alternative to Portland cement. *Cem. Concr. Res.* **2011**, *41*, 750–763. [CrossRef]
17. Nielsen, C.V. Carbon Footprint of Concrete Buildings Seen in the Life Cycle Perspective. Available online: https://www.researchgate.net/publication/268008020 (accessed on 2 June 2022).
18. Sebastin, S.; Priya, A.K.; Karthick, A.; Sathyamurthy, R.; Ghosh, A. Agro Waste Sugarcane Bagasse as a Cementitious Material for Reactive Powder Concrete. *Clean Technol.* **2020**, *2*, 476–491. [CrossRef]
19. Sizirici, B.; Fseha, Y.; Cho, C.-S.; Yildiz, I.; Byon, Y.-J. A Review of Carbon Footprint Reduction in Construction Industry, from Design to Operation. *Materials* **2021**, *14*, 6094. [CrossRef]
20. Ruuska. Carbon Footprint for Building Products ECO2 Data for Materials and Products with the Focus on Wooden Building Products. 2013. Available online: http://www.vtt.fi/publications/index.jsp (accessed on 2 July 2022).
21. Taffese, W.Z.; Abegaz, K.A. Embodied Energy and CO2 Emissions of Widely Used Building Materials: The Ethiopian Context. *Buildings* **2019**, *9*, 136. [CrossRef]
22. Climate. Gov. Climate Change: Atmospheric Carbon Dioxide. Available online: https://www.climate.gov/news-features/understanding-climate/climate-change-atmospheric-carbon-dioxide (accessed on 26 August 2022).

Article

A Comparative Study on Crack Detection in Concrete Walls Using Transfer Learning Techniques

Remya Elizabeth Philip [1], A. Diana Andrushia [1,*], Anand Nammalvar [2], Beulah Gnana Ananthi Gurupatham [3] and Krishanu Roy [4,*]

1. Department of Electronics and Communication Engineering, Karunya Institute of Technology and Sciences, Coimbatore 641114, India
2. Department of Civil Engineering, Karunya Institute of Technology and Sciences, Coimbatore 641114, India
3. Division of Structural Engineering, College of Engineering Guindy Campus, Anna University, Chennai 600025, India
4. School of Engineering, The University of Waikato, Hamilton 3216, New Zealand
* Correspondence: diana@karunya.edu (A.D.A.); krishanu.roy@waikato.ac.nz (K.R.)

Abstract: Structural cracks have serious repercussions on the safety, adaptability, and longevity of structures. Therefore, assessing cracks is an important parameter when evaluating the quality of concrete construction. As numerous cutting-edge automated inspection systems that exploit cracks have been developed, the necessity for individual/personal onsite inspection has reduced exponentially. However, these methods need to be improved in terms of cost efficiency and accuracy. The deep-learning-based assessment approaches for structural systems have seen a significant development noticed by the structural health monitoring (SHM) community. Convolutional neural networks (CNNs) are vital in these deep learning methods. Technologies such as convolutional neural networks hold promise for precise and accurate condition evaluation. Moreover, transfer learning enables users to use CNNs without needing a comprehensive grasp of algorithms or the capability to modify pre-trained networks for particular purposes. Within the context of this study, a thorough analysis of well-known pre-trained networks for classifying the cracks in buildings made of concrete is conducted. The classification performance of convolutional neural network designs such as VGG16, VGG19, ResNet 50, MobileNet, and Xception is compared to one another with the concrete crack image dataset. It is identified that the ResNet50-based classifier provided accuracy scores of 99.91% for training and 99.88% for testing. Xception architecture delivered the least performance, with training and test accuracy of 99.64% and 98.82%, respectively.

Keywords: transfer learning; crack detection; concrete wall; convolutional neural network; structural health monitoring

1. Introduction

Many buildings have reached their design life expectancy; therefore, it is critical to safeguard the facilities/amenities through routine maintenance. Extensive research has been conducted in order to improve the performance of concrete and thus the health of structures [1,2]. The concrete's structural integrity is severely affected by the development of cracks, increasing the risk of failure or collapse in buildings and structures. Crack inspection is an important but tedious maintenance task for buildings and other infrastructures. Cracks reduce the load-bearing capacity of the structural elements and accelerate the damage level. Cracks in concrete adversely affect durability by reducing the lifespan of the buildings. Cracks can create distress for the occupants and impair the building's appearance.

When the crack inspection is executed manually, the work becomes time-consuming, labor-intensive, and necessitates skills. Crack monitoring and digital image processing are viable alternatives for visual inspections [3]. Although digital image processing-based technologies have largely been successful, "false positive" results occur occasionally. Therefore, a comprehensive automated crack monitoring system must be enabled to identify cracks from surface images that contain natural cracks or non-cracks that appear to be natural cracks [4,5].

Studies have been conducted in masonry structures, along with crack modelling and crack pattern identification by using numerical models and finite element analysis [6–8]. Tan et al. [9] studies the possibility of using distributed sensing technology to detect, locate, trace, quantify, and visualize the crack using fiber optic sensors. Kim et al. [10] proposed a crack identification strategy combining RGB-D and sensors that measure cracks regardless of the angle of view. The authors have deployed high-resolution digital cameras as sensors.

Many technologies based on computer vision (CV) and artificial intelligence (AI) have evolved to help automate the process [11,12]. Machine learning (ML) is a subset of AI-based techniques that many researchers have used to detect cracks on concrete surfaces. For crack detection, various ML methods such as support vector machines (SVM), Bayesian decision trees (BDT), and random forests (RF) are used.

Traditional ML methods include SVM, artificial neural networks (ANN) [13–15], and RF [16]. Even though these algorithms reduce false positives, their accuracy is still hugely dependent on crack features obtained through specific image processing steps. A pre-defined feature extraction stage in all these methods necessitates an additional image processing stage to make the patterns clearer to the learning algorithms. It also has a negative impact on the model's performance. Another disadvantage of ML methods is that the learning algorithms cannot learn higher-order features with complex information in the dataset.

Deep learning (DL) is a promising technology for addressing the issues associated with handcrafted feature extraction. Deep learning is a branch of ML that uses neural networks as a framework for its algorithms. DL techniques include auto encoders (AEs), deep belief networks (DBNs), deep Boltzmann machines (DBMs), recurrent neural networks (RNNs), and convolutional neural networks (CNNs) [17]. CNNs are critical among DL methods, which are primarily used to analyze image-based data [18]. Many datasets are utilized to train CNNs for different types of damage and fault diagnoses. Non-contact sensors' capabilities are enhanced by using trained networks to build autonomous structural health monitoring (SHM) systems [19].

Concrete crack detection DL models can be administered in various health monitoring scenarios to identify and locate cracks in concrete structures. The practical and feasible applications of concrete crack detection DL models are building inspection, infrastructure maintenance, construction quality control, historical preservation, etc. Therefore, concrete crack detection DL models yield an effective and efficient way to identify cracks in concrete structures, helping to prevent accidents, improve safety, and ensure the prolonged endurance of critical infrastructure.

Laxman et al. [20] developed a binary-class convolutional neural network (CNN) model which automatically detects the crack on concrete surfaces. They also interface the CNN model, combining the convolutional feature extraction layers with that of the regression models such as Random Forest and XG-Boost, resulting in automatic predictions of the depth of cracks. This experimental study was validated on reinforced concrete slabs. Apart from concrete cracks, many studies have been carried out on road surface damage. Xu et al. [21] demonstrated the advantage of combining Faster R-CNN and Mask R-CNN in detecting road pavement cracks. The limitations of the architecture was degradation in the effectiveness of the bounding box detected by Mask R-CNN. Huyan et al. [22] proposed a new architecture called CrackU-net that achieved pixel-wise crack detection. It was found that the proposed model outperformed traditional U-Net and fully convolutional neural networks (FCN).

The primary design concept of CNNs' architecture is to deploy successive convolutional layers to the input, resample the spatial dimensions while increasing the number of feature maps, and then repeating it. These architectures serve as rich feature extractors for image classification, object recognition, image segmentation, and other more laborious tasks. AlexNet, VGG16, ResNet, MobileNet, Inception, and Xception are illustrations of CNN architectures that have been widely used as classifiers and segmenters. Many CNN-based deep learning models rely on these networks [23–25].

Contemporary studies make use of transfer learning mechanisms. The term "transfer learning" (TL) commonly refers to a procedure in which a model is developed primarily for one particular problem and then utilized in some capacity for secondary problems. Because it directly integrates pre-trained models into feature extraction preprocessing and comprehensive new models, this method is adaptable. It has the advantage of reducing neural network model training time, resulting in fewer generalization errors [26,27].

Figure 1 compares traditional DL models to that of models based on TL. The basic idea behind TL models is that the architecture can be reused as it is pre-trained on similar datasets. The TL can be used in two different ways: one is reusing the network structure, and the other reusing both the network structure and weights by either retraining only a few layers, retaining all layers, or adding a few layers on top. The idea behind image classification TL techniques is that if a model is trained and tested on a large and diverse dataset, the model will successfully obtain a critical visual overview of distinct features or attributes. This method has been widely used in the semantic segmentation and image classification stages of crack detection [28]. The significance and importance of TL is that it substantially mitigates the usage of huge datasets for training, as its pre-defines or pre-trained models show promising results due to its handling of huge datasets, for example the ImageNet dataset.

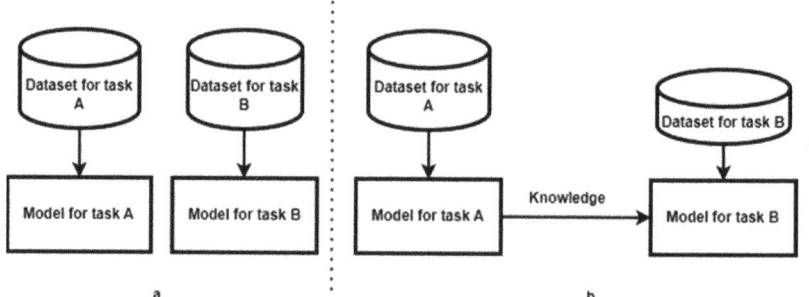

Figure 1. (a) Traditional deep learning model (b) Transfer learning-based deep learning model.

TL models have been used in various types of research. Su and Wang [29] compare the performances of the architectures MobileNetV2, DenseNet201, EfficientNetB0, and InceptionV3 for crack detection on concrete. It was discovered that EfficientNetB0 was efficient in terms of performance and generalization. Dung and Anh [30] used VGG-16 as the foundation of a fully convolutional neural network for crack classification. Zhong et al. [31] used an improvised variant of VGG16 for concrete pavement crack detection. TL techniques have been combined with fine-tuning procedures to achieve high accuracy in pre-trained models. Sun et al. [5] used the Xception architecture's pre-trained weights and biases to detect cracks and holes in concrete surfaces. Joshi et al. [32] used the ResNet50 architecture and a segmentation-based approach to detect cracks. Doğan and Ergan [33] used MobileNet architecture as a backbone for pixel-wise crack detection in lightweight mobile applications.

The backbone framework in most architectures is VGG16, ResNet, MobileNet, and Inception or Xception networks, with some fine tuning [24]. This study compares the most commonly used CNN architectures in order to determine the accuracy of these networks in classifying crack and non-crack concrete surface images. For TL, the pre-trained weights of the VGG16, ResNet50, MobileNet, and Xception architectures are used. For all architectures, the dataset used for classification is similar. Thus, TL takes less duration than building a network from the ground up. The availability of datasets is greatly reduced because all the networks are trained in classifying the 1000 different object categories that we encounter on day-to-day basis.

This study is executed in such a way that different types of DL models were compared for classifying and identifying cracks in concreate structures by implementing pre-trained architectures such as VGG16, VGG19, ResNet50, MobileNet, and Xception. Sizable concreate image datasets were used for validating and training these frame works.

Conducting research such as this contributes immensely to the progressive development of DL methods for structures in multitudinous ways. Primarily, the study proposes an effective approach for the automatic detection and classification of cracks in concrete surfaces using transfer learning methods, which saves time and reduces the need for manual inspection. Moreover, as all the architectures used in this study were already trained on thousands of image datasets, the need for new image datasets is also reduced. Secondly, it provides a critical comparison of different transfer learning frameworks that can be used as feature extraction or backbone architecture, depending on the availability of memory and time, which will be beneficial for future research in this area.

This analysis demonstrates the viability and potential of DL methods in scrutinizing the varied and complex structural data, resulting in progressive development, and extending the application of these models in other structural assessment areas. These models could help in determining the efficacy of categorizing and localizing different types of cracks, spalls, and other flaws in concrete buildings in natural settings by potentially automating the damage detection process.

The findings of this study will assist researchers in developing new technologies for efficiently maintaining the service life of infrastructures using unmanned aerial systems, automation systems for infrastructure monitoring by deploying various sensors such as high-resolution cameras, LIDAR systems, etc., and in the enhancement of SHM systems for constant monitoring of the serviceability and maintainability of infrastructures, thereby reducing costs [34–37].

This paper emphasizes the suitability of existing DL convolutional models for TL strategies. Previous research conducted by various researchers primarily focused on various topologies for classification and segmentation tasks based on these backbone models. This study compares and contrasts various backbone architectures to provide a comprehensive picture of the best backbone model, or TL model, to use when developing new systems for detecting and analyzing the formation of concrete cracks.

2. Methodology

The framework for comparing different CNN models is depicted in Figure 2. The datasets consist of raw images taken from residential buildings and datasets available online in data repositories. In this study, CNN models employ TL techniques to detect concrete surface cracks. The different models considered in this study are VGG16, VGG19, ResNet50, MobileNet, and Xception. The model's weight is learned on ImageNet, saved, and then applied to the models. As a result, the model has a higher starting point, substantially cutting training time and achieving improved performance. To be suitable for crack classification, the pre-trained CNN model must be retrained to find concrete surface cracks. On a sizeable dataset, a pre-trained model has already been trained and preserved; furthermore, the final fully connected layer of the original model is replaced by a new, fully connected layer. The steps of the experiment are outlined below.

TensorFlow is used to change the size of image datasets before training the model. The datasets are then loaded batch-wise and randomly for further operations. The structure of the crack detection model is then defined by loading and refining the pre-trained model. The final layer with complete connectivity is replaced with a bespoke layer. The number of classes in the custom layer has been set to 2, as per this investigation's pre-requisites. The weight values of other layers did not change. The model is then compiled and trained using the datasets. Before training the model, the network's structure-related hyperparameters are specified, and the optimal optimization technique is chosen. In this study, model training is guided by the adaptive-learning-rate optimization method Adam, and the cross-entropy loss function. After completing the training procedure, the model's performance is validated. The test dataset was then used to evaluate the model.

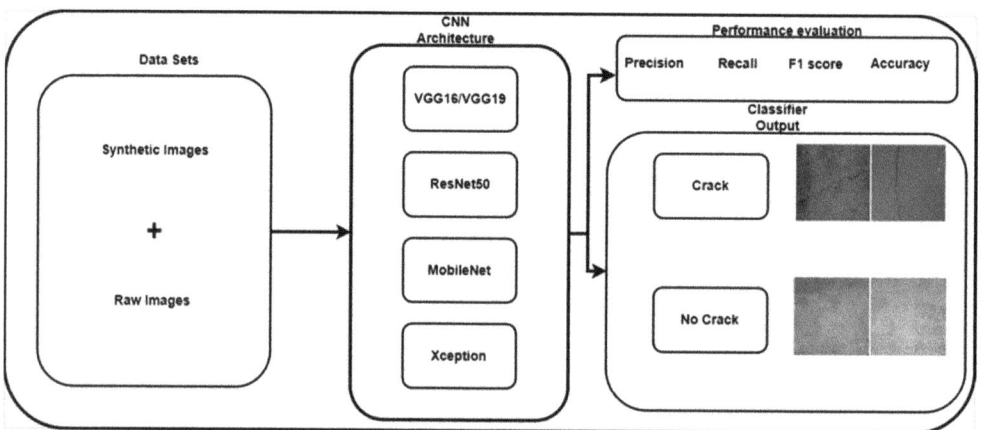

Figure 2. Framework for comparing different CNN models for classification of cracks.

2.1. VGG16

VGG16 is a convolutional neural network for classification and object detection purposes [31,38,39]. It is widely used for classifying images, and is uncomplicated to employ with transfer learning. It has 16 convolutional layers and 64 feature kernel filters, each measuring 3 × 3 pixels, making up the first and second convolutional layers. The input image's dimensions change to 224 × 224 × 64 as it is passed through the first and second convolutional layers (an RGB image with a depth of 3). The output is then sent to the maximum pooling layer with a stride of 2. The third and fourth convolutional layers are 124 feature kernel filters, and the filter size is 3 × 3. A maximum-pooling layer follows these two layers with stride 2, and the resulting output was reduced to 56 × 56 × 128. The fifth, sixth, and seventh layers are convolutional layers with a kernel size of 3 × 3. All three layers used 256 feature maps. A maximum pooling layer with stride 2 follows these layers. Eighth to thirteen are two sets of convolutional layers with a kernel size of 3 × 3. All these sets of convolutional layers had 512 kernel filters. A maximum-pooling layer follows these layers with a stride of 1. Layers fourteen and fifteen are fully connected to the hidden layers of 4096 units, followed by a softmax output layer (sixteenth layer) of 1000 units. Figure 3 shows the schematic diagram of the VGG16 architecture.

2.2. ResNet-50

The residual neural network (ResNet) proposed by He et al. [40] won the ImageNet Large Scale Visual Recognition Challenge (ILSVRC 2015). ResNet implemented residual connections between layers, which aids in mitigating the loss, preserving knowledge gain, and enhancing performance during the training phase. A residual link in a layer indicates that a layer's output is a convolution of its input and output. Figure 4 depicts a block schematic of the architecture of the ResNet model. A convolutional layer with a 7 × 7 kernel size and 64 different kernels makes up the first layer. The following layer is the maximum pooling layer. The following convolution layers consist of 1 × 1-sized kernels with 64 kernels, 3 × 3-sized kernels with 64 kernels, and 1 × 1-sized kernels with 256 kernels. There will be nine layers after repeating this layer thrice. The following are convolution layers with sizes of 1 × 1, 3 × 3, and 1 × 1 with 128, 512, and 128 kernels, respectively. This is repeated about four times to get a total of 12 layers. A convolution layer of 1 × 1 size follows this with 256 kernels and two additional kernels of 3 × 3, 256, and 1 × 1, 1024, repeated six times for a total of 18 layers. Finally, a layer of 1 × 1, 512 kernels, with two additional kernels of 3 × 3, 512, and 1 × 1, 2048, was repeated three times for a total of nine layers. Following that, average pooling concludes with a fully connected layer with 1000 nodes, and a softmax function adds another layer. In total, there were 50 layers.

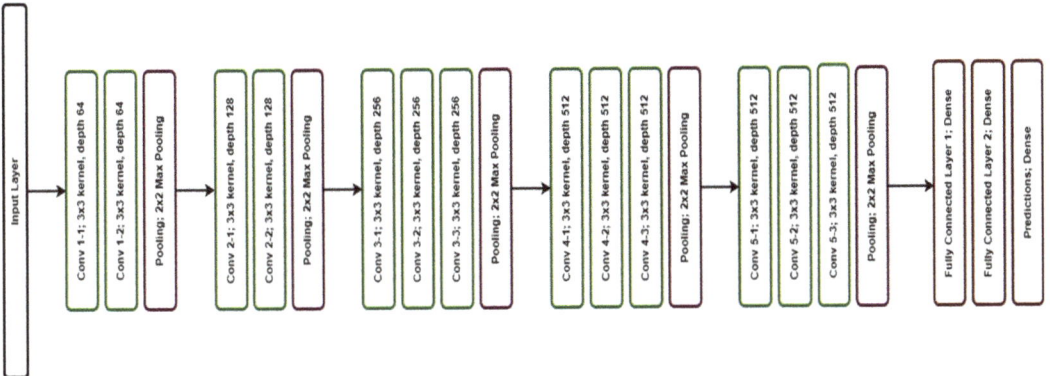

Figure 3. Block schematic of the VGG16 architecture.

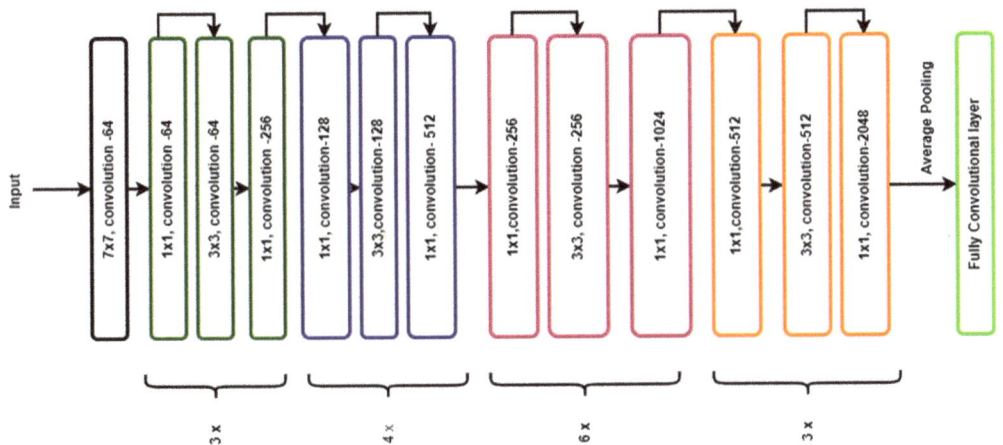

Figure 4. Block diagram of the ResNet-50 architecture.

2.3. MobileNet

MobileNet is a model primarily developed for use in mobile apps. MobileNet uses depth-wise separable convolutions. It dramatically decreases the number of parameters compared to that of networks with standard convolutions of the similar depth. Thus, lightweight deep neural networks are produced. The depth-wise separable convolution is accomplished by depth- and point-wise operations, making it suitable for embedded applications. The depth-wise convolution filter produces a single convolution on each input channel, whereas the point convolution filter linearly combines the depth-wise convolution output with 1 × 1 convolutions. MobileNet's architecture is illustrated in Figure 5 [41]. The depth-wise convolution filter produces a single convolution on each input channel, whereas the point convolution filter linearly combines the depth-wise convolution output with 1 × 1 convolutions, in the figure the depth wise separable convolution is highlighted in blue color which consist of point wise and depth wise layers. The average pooling layer is colored in red which is further directed to a fully convolution layer. The computational speed is advantageous in pointwise and depth-wise convolution [42].

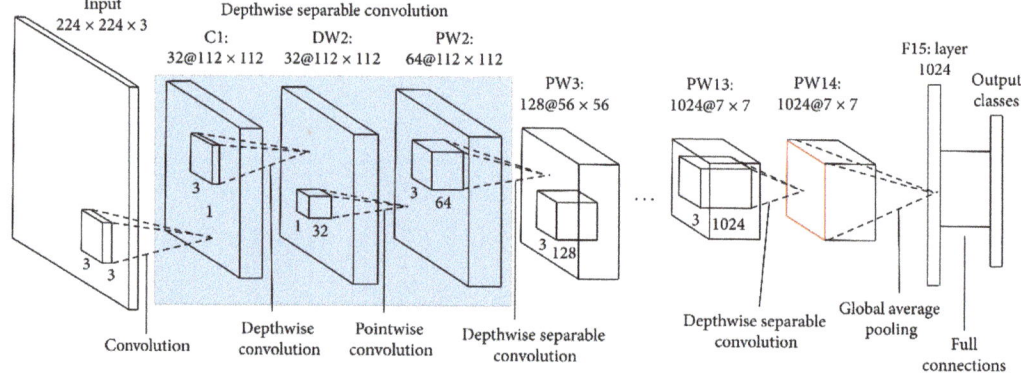

Figure 5. MobileNet architecture.

2.4. Xception

The Xception architecture is a variation of the Inception architecture that only employs depth-wise separable convolutional layers. Figure 6 displays Xception's architecture, which may be seen as a linear stack of depth-separable convolutional layers. The Xception architecture comprises of 36 convolutional layers organized across 14 modules with residual connections. The data flows through the input, the middle flow is then repeated about eight times, and the exit flow enters the Xception architecture. Batch normalization comes after all convolutional and separable convolution layers, which is not shown in the diagram. A depth multiplier of one is used for all separable convolution layers [43].

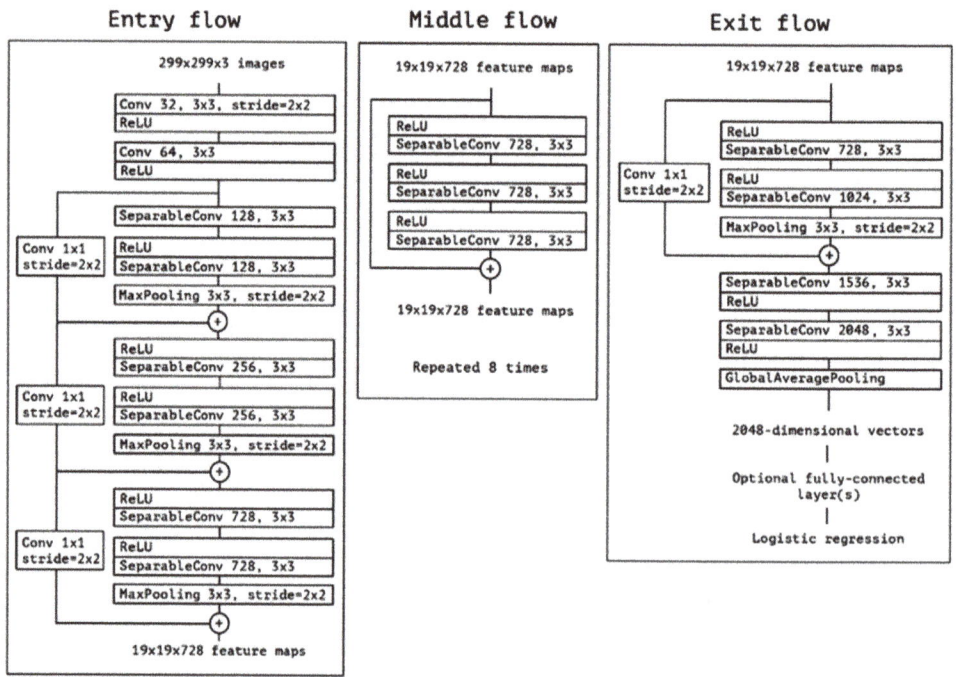

Figure 6. Architecture of Xception.

3. Details of the Experiment

The details of the experiment to classify cracked and non-cracked images are presented herewith. The experiments were conducted in two different stages: in the first stage, the models VGG16 and VGG19 were compared to analyze their performances. In the second stage, the VGG16 model was compared with the pre-trained models ResNet50, MobileNet, and Xception. All the pre-trained models were tested to determine which would generalize and perform better regarding the crack information contained in the image dataset.

3.1. Datasets

The database combines images in data repositories such as SDNET 2018 [23], Chun-data [44–46], and the data captured from residential buildings. The dataset was divided into two sets: positive images (cracked) and negative images (non-cracked). They were kept in separate folders named cracked and non-cracked, each containing 11,000 and 11,000 images, respectively, in RGB format. In total, 4400 images were used exclusively for testing and not for training. Sample images of cracked and non-cracked concrete walls are shown in Figure 7. The datasets were separated into training (80%) and testing (20%). Once again, the data was divided into training and validation data sets, as shown in Figure 8. The training dataset is the sample of data utilized to fit the model. The validation dataset is the sample used to offer an impartial evaluation of a model's fit to the training dataset while setting the hyperparameters of the model. As the validation dataset skill was added to the model design, the evaluation of the model started increasing. The test dataset is the data sample used to provide an unbiased evaluation of the final model's fit to the training dataset.

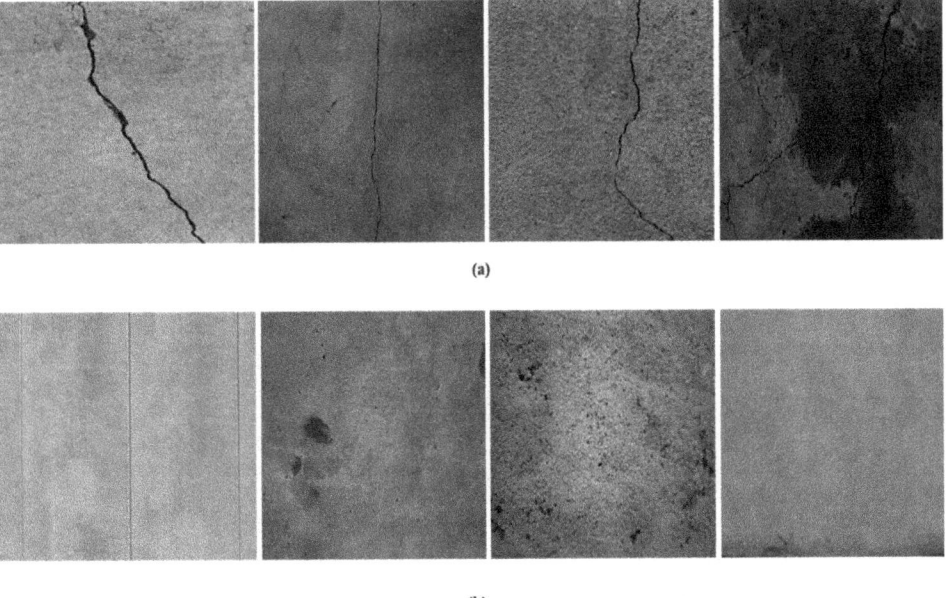

Figure 7. (a) Cracked concrete wall images (b) Non-cracked concrete wall images.

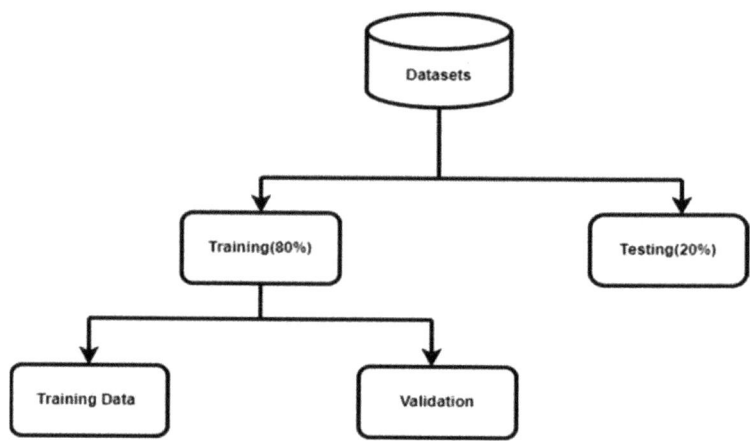

Figure 8. Dataset distribution.

3.2. Implementation Details

Python 3, the sklearn module, and the Keras neural network library, which contain the architecture and weights of VGG16, ResNet50, MobileNet, and Xception, were used to build the convolutional neural network. The experiments were performed in Google Colab. The target size was 100 × 100, the class mode was binary, the batch size was 64, and Adam, with an initial learning rate of 0.001, was used as the optimizer. The Adam optimization method employs stochastic gradient descent and adaptive estimation of first- and second-order moments. It is suitable for many data items and parameters since it is computationally efficient, memory-light, and invariant to the diagonal rescaling of gradients [47]. The maximum epoch was set to be 20. Table 1 shows the hyperparameter settings for each model.

Table 1. Hyperparameters used for training.

Parameter	Training
Initial learning rate	0.001
Batch size	64
Optimizer	Adam
Number of epochs	20
Steps per epoch	275

3.3. Performance Metrics

To examine the model's performance, evaluation metrics were needed. The model was accessed using precision, accuracy, recall, and F1 measures. Equations (1)–(4) show the accuracy, precision, recall, and F1 measures, respectively. Accuracy refers to the ratio of correct predictions to the total number of input images. Precision is the ratio of correct positive predictions to the total number of positive predictions. The recall is the proportion of accurate positive predictions compared to the total number of true positives. The F1 measure is the weighted harmonic mean of precision and recall [48].

$$Accuracy = \frac{true\ positive + true\ negative}{true\ positive + true\ negative + false\ positive + false\ negative} \quad (1)$$

$$Precision = \frac{true\ positive}{true\ positive + false\ positive} \quad (2)$$

$$Recall = \frac{true\ positive}{true\ positive + false\ negative} \quad (3)$$

$$F1 = \frac{2 \times precision \times recall}{precision + recall} \quad (4)$$

The classification report visualizer displays the model's precision, recall, F1, and support scores. A classification report is then used to evaluate the accuracy of a classification algorithm's predictions. The metrics of a classification report are assessed using true positives, false positives, true negatives, and false negatives [49].

4. Results and Discussion

This section summarizes the results of the trained networks used to categorize images using TL. The pre-trained models are assessed to determine which one would generalize and provide optimum results in the dataset images.

4.1. Comparison on VGG16 and VGG19 Architecture

In the first stage, VGG16 architecture is compared with that of its contemporaries, VGG19. All the hyper-parameters are set as given in Table 1, and it is found that the VGG16 architecture gave a test accuracy of 99.61% whereas VGG19 provided a test accuracy of 99.57%. Furthermore, the training duration was comparatively longer for VGG19 architecture, which was about 2.07 h. Sample datasets for the classification results of test images by the VGG16 and VGG19 architectures are shown in Figures 9 and 10, respectively. The model accuracy and loss of VGG19 is shown in Figure 11.

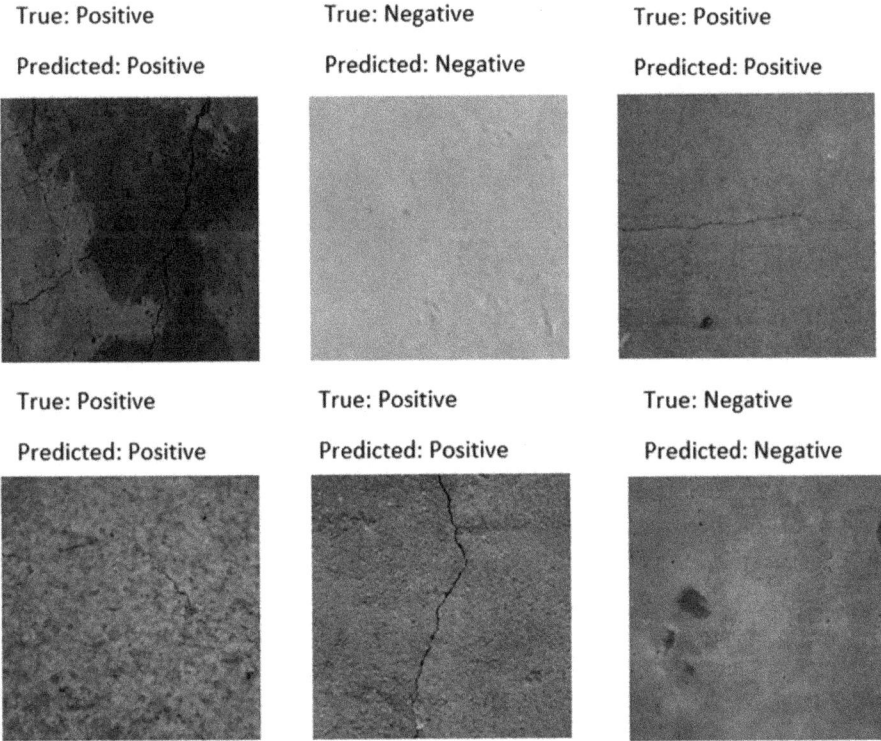

Figure 9. Sample classification result on test images by the VGG16 architecture.

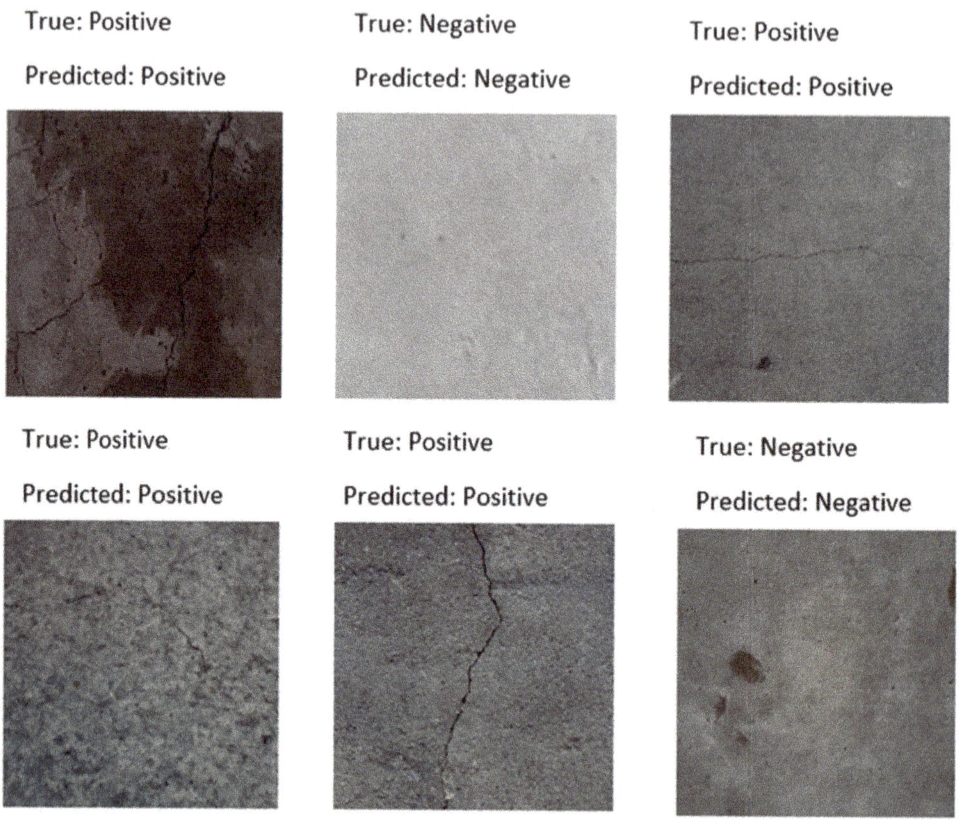

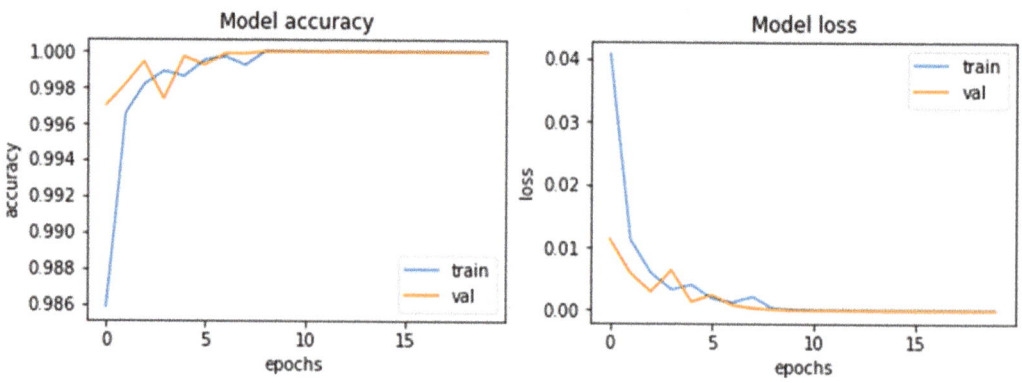

Figure 10. Sample classification result on test images by the VGG19 architecture.

Figure 11. Accuracy and loss curves of VGG19 architecture.

The comparison of the VGG16 and VGG19 models on training losses and training accuracies is provided in Figure 12a,b. Similarly, the comparison of the VGG16 and VGG19 models on validation loss and validation accuracy is provided in Figure 12c,d. It can be seen that the VGG19 architecture attained stability in validation accuracy a few epochs before the VGG16 architecture, which is an advantage of the network. The VGG16 model takes a few more epochs to settle down. However, considering the results of test accuracy and training duration, one can choose VGG16.

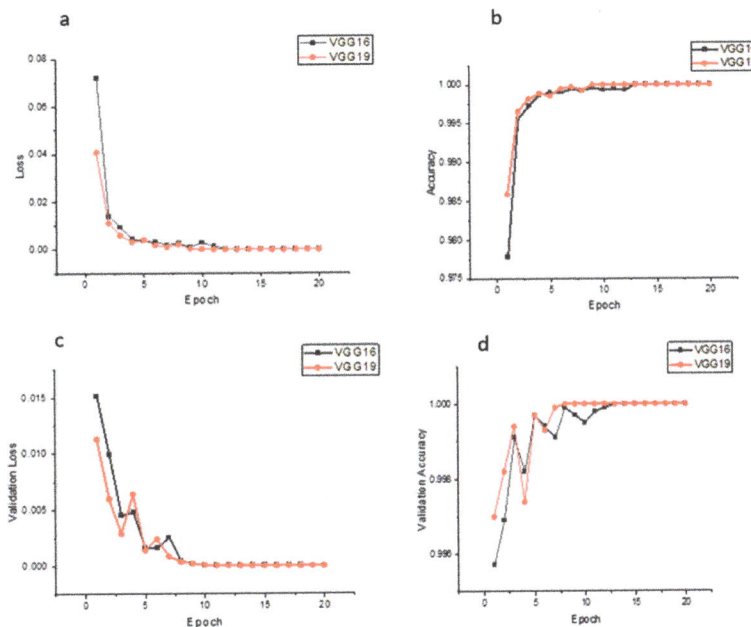

Figure 12. (a) Training loss of VGG16 and VGG19 (b) Training accuracy of VGG16 and VGG19 (c) Validation loss of VGG16 and VGG19 (d) Validation accuracy of VGG16 and VGG19.

4.2. Comparison on VGG16, ResNet50, MobileNet and Xception Architectures

The second comparison was made on the VGG16, ResNet50, MobileNet, and Xception architectures. The hyper-parameter details are the same as those mentioned in Table 2. A classification matrix is used to evaluate the models. For a balanced dataset, accuracy can be a good measure. Still, in the case of imbalanced datasets, precision, recall, and F1 measures need to be validated to measure the performance of the models. In applications where it is not critical to identify all positive samples, a high degree of precision over recall is acceptable; if precision exceeds recall, "false negatives" outnumber "false positives". The recall measures the maximum number of "true positives"; in this case, there are only negligible "false negatives" rather than "false positives". The F1 metric is the harmonic mean of precision and recall. In each of these metrics, one value indicates optimal performance [50]. Support is the number of images belonging to that particular class used to measure the metrics.

Table 2. Details and training accuracy of networks.

Network	Test Loss	Training Accuracy	No. of Epochs	Best Epoch
VGG16	0.00923	99.71%	20	11
VGG19	0.01592	99.67%	20	9
ResNet50	0.00073	99.91%	20	7
MobileNet	0.02300	99.72%	20	12
Xception	0.08611	99.64%	20	9

The training time taken for VGG16 was high compared to that of other architectures [38]. All the models gave more than 99% training accuracy results, as it was only a binary classification problem [51]. It was found that out of the four models, ResNet50, MobileNet, and VGG16 produced optimum results, 99.88%, 99.68%, and 99.61%, respectively, in terms of accuracy, on testing data. It was also found that the test accuracy was high for the ResNet50 architecture, followed by the MobileNet, VGG16, and Xception architectures.

The details and training accuracy of the models are mentioned in Table 2. Table 2 shows that the test loss was less for ResNet, followed by VGG16, MobileNet, and Xception architectures, which tells how well the architectures behave after each optimization iteration. Even though the MobileNet architecture training accuracy was 99.72%, the classification report shows that the precision, recall, and F1 metrics are lower than those of the VGG16 and ResNet50 architectures.

The statistical outcomes of the networks are displayed in Table 3. The measures ensure that the pre-trained architectures VGG16, VGG19, and ResNet50 accurately classified the crack images; only MobileNet and Xception gave some false negatives and false positives. From the statistical results mentioned in Table 3, it was clear that VGG16, VGG19, ResNet50, and MobileNet have a recall rate higher than the precision, showing the models have fewer "false negatives", which is necessary to identify the concrete cracks. In the Xception model, precision exceeded recall, indicating that the number of "false negatives" is greater than the number of "false positives".

Table 3. Statistical outcomes of the concrete cracks (crack or intact).

Architecture	Class	Precision	Recall	F1	Support
VGG16	Negative	1.00	0.99	1.00	2197
	Positive	0.99	1.00	1.00	2203
VGG19	Negative	1.00	0.99	1.00	2197
	Positive	0.99	1.00	1.00	2203
ResNet50	Negative	1.00	0.99	1.00	2197
	Positive	0.99	1.00	1.00	2203
MobileNet	Negative	0.99	0.99	0.99	2197
	Positive	0.99	0.99	0.99	2203
Xception	Negative	0.98	0.99	0.98	2197
	Positive	0.99	0.98	0.98	2203

The sample classification results from the test data are provided in Figure 13. The classification results for MobileNet and Xception in Figure 13c,d show that both architectures performed comparatively worse than others. It was noticed that the MobileNet and Xception architectures could not identify small cracks and images with crack-like features or background irregularities. The architectures misclassified crack-like features as a crack, which is a false positive, and those datasets with hairline cracks were not identified as cracks, which are the false negatives. Figure 14 is a breakdown of the training duration for each model. ResNet50, MobileNet, Xception, VGG16, and VGG19 architectures have relative training times of 34 min, 35 min to 38 s, 43 min to 44 s, 1.98 h, and 2.077 h, respectively. Due to the increase in the number of layers in the design, the training time for VGG19 was longer than that for VGG16, and it was much longer when compared to ResNet50 as illustrated in Figure 14 [52].

Model accuracy and loss are depicted in Figures 15–18 for the VGG16, ResNet50, Xception, and MobileNet architectures, respectively. These graphs show that all models, except for the Xception architecture, could appropriately learn the features. In Xception, it was discovered that validation accuracy and loss fluctuate, demonstrating the network's incapacity to easily fit the model during hyperparameter tuning. The response can be enhanced by fine-tuning the hyperparameters, which may increase the model's performance.

The training loss and accuracy—as well as the validation loss and accuracy—of all the architectures are depicted in Figure 19, which replicates Figures 15–18. The test data sets yielded better results. The accuracy of these classifications was higher than 98%. Therefore, deep learning classifiers could detect and categorize cracks, resulting in strong and dependable models using transfer learning. In addition to the accuracy metric, the evaluation metrics associated with classification tasks are shown in a classification report alongside the accuracy metric. Table 3 displays the report's key points. The models accurately categorize photos of cracked and uncracked concrete. The classification report demonstrates that the model evaluated the validation set effectively.

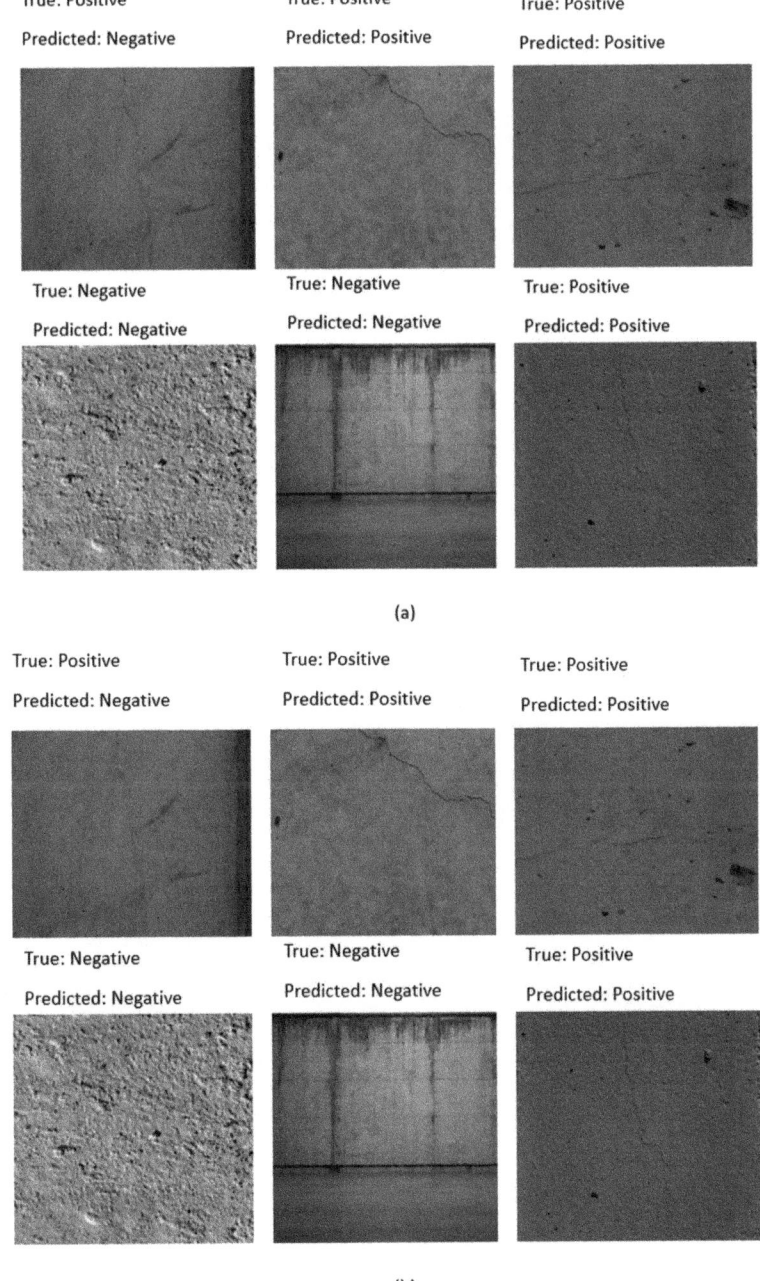

Figure 13. *Cont.*

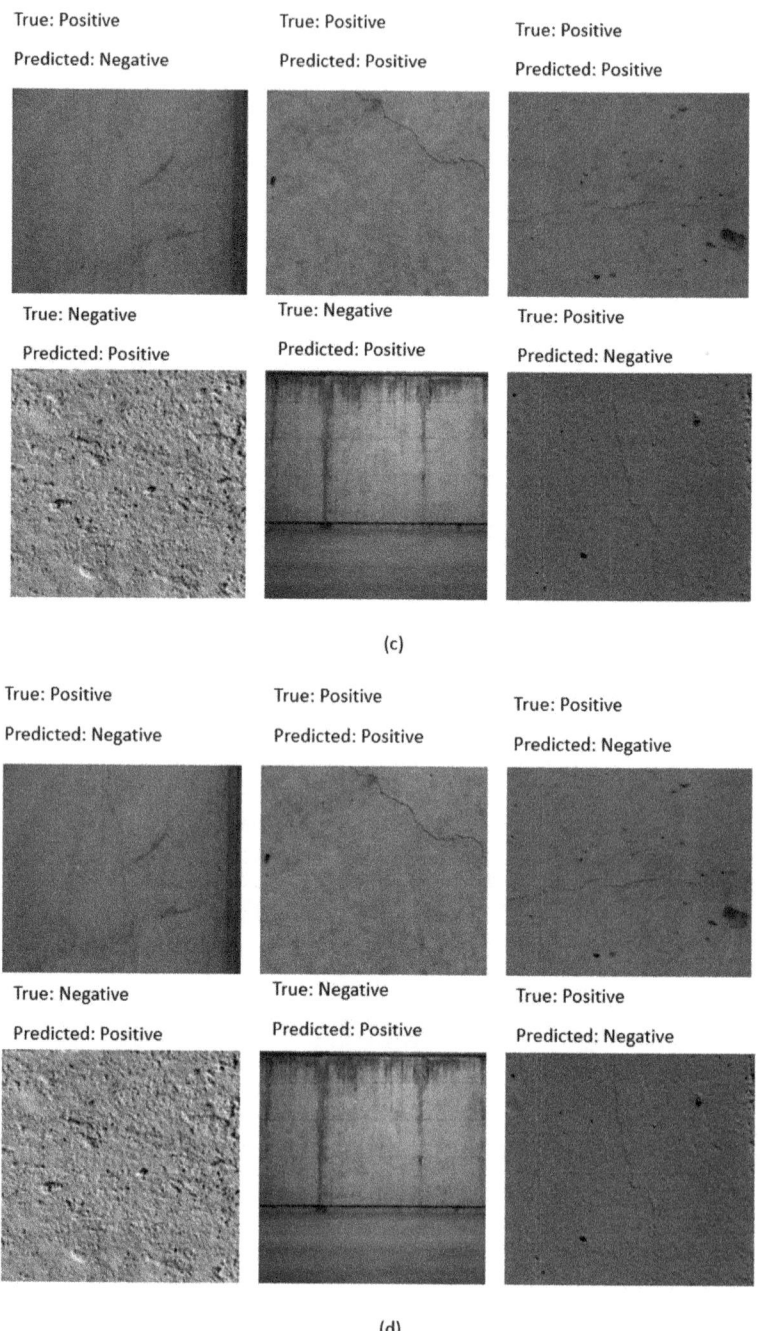

Figure 13. Classification result of test images (**a**) VGG16 (**b**) ResNet50 (**c**) MobileNet (**d**) Xception.

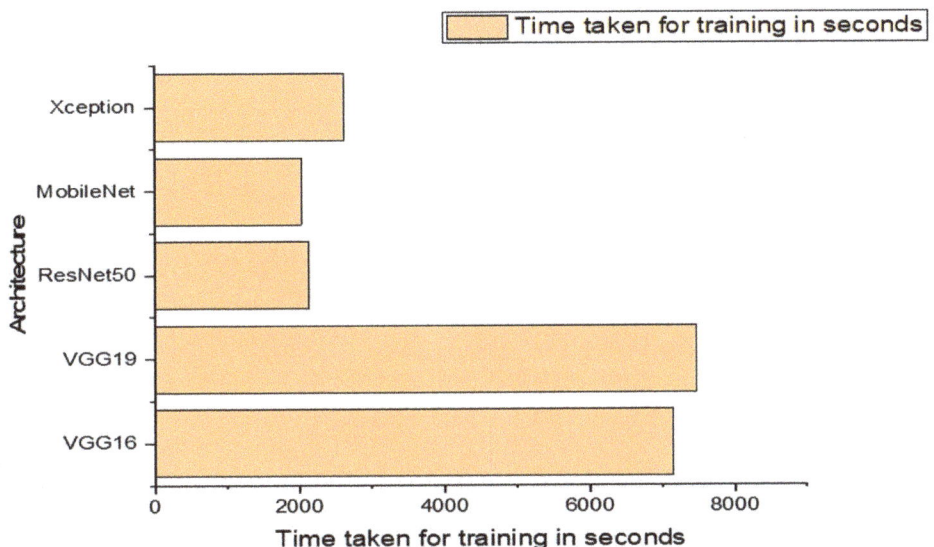

Figure 14. Time taken for training.

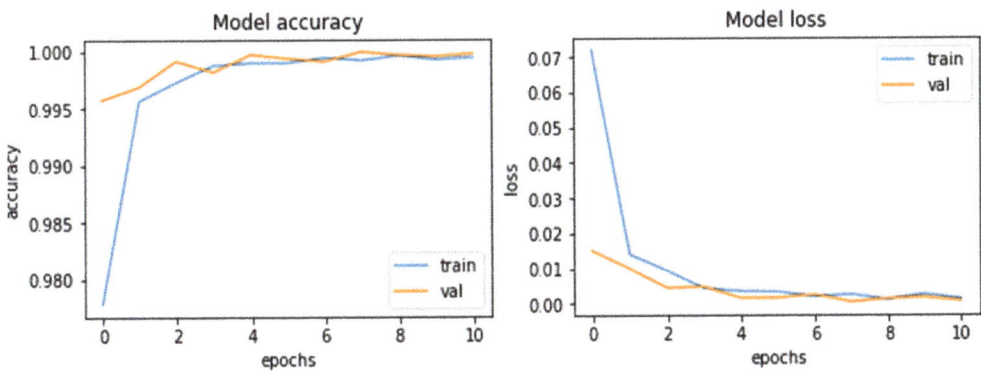

Figure 15. Accuracy and loss curves of VGG16 architecture.

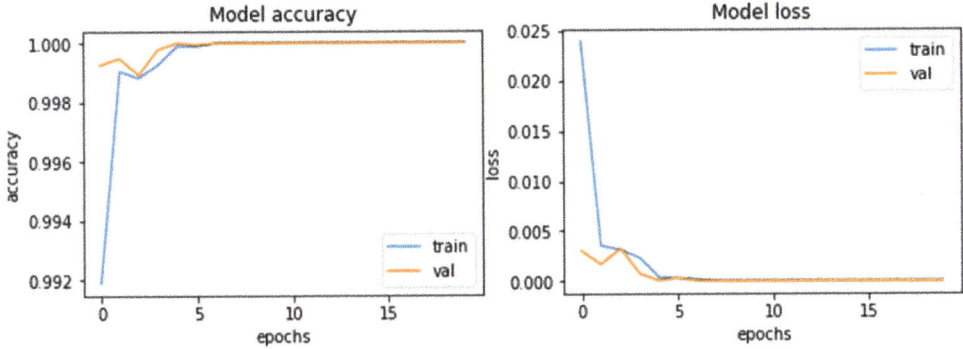

Figure 16. Accuracy and loss curves of ResNet50 architecture.

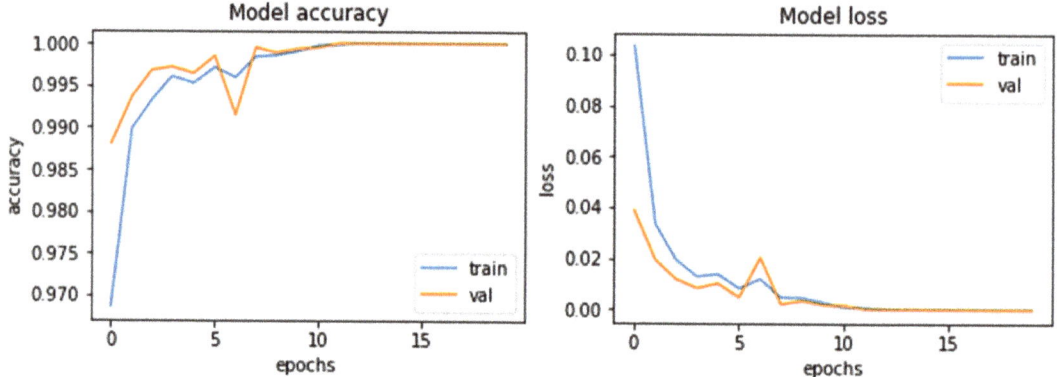

Figure 17. Accuracy and loss curves of Xception architecture.

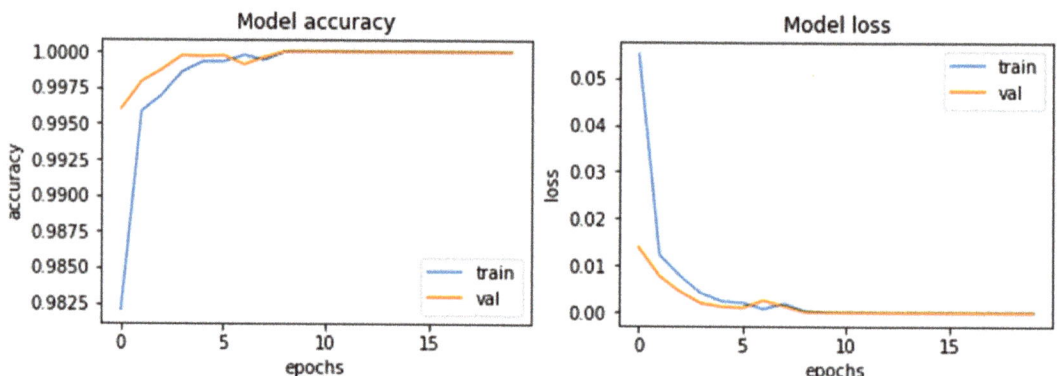

Figure 18. Accuracy and loss curves of MobileNet architecture.

Figure 20 compares the training and testing accuracy of all models considered in this study. The lack of variation in training and test accuracy suggests that the network may learn the features accurately. Because it has the highest training and testing accuracy, ResNet50 is ideal for identifying cracked images using deep learning models. Furthermore, the duration was much lower than for VGG16 and VGG19, despite having nearly the same precision, recall, and F1 scores. VGG16 and MobileNet have comparable training and testing accuracy. When time and complexity are not factors, the VGG16 is the cutting-edge network for classifying cracks on concrete wall images. However, when deploying mobile applications where time and complexity are constraints, the MobileNet architecture will provide nearly equal performance to VGG16 [48]. When compared to the other architectures, the performance of the Xception architecture was found to be inferior to others.

Deep learning models have several advantages over traditional machine learning models, but large data sets are still required. Transfer-learning approaches, in which the weights of pre-trained models trained on multiple datasets are used to transfer their knowledge when accessing new datasets, can reduce this to a greater extent. This research will help researchers identify the backbone network for classification networks and other hybrid architectures [30,39,53]. Five pre-trained models were tested and analyzed in this study. Model performance was assessed using the classification report metrics. The ResNet50 model outperformed the other four pre-trained models.

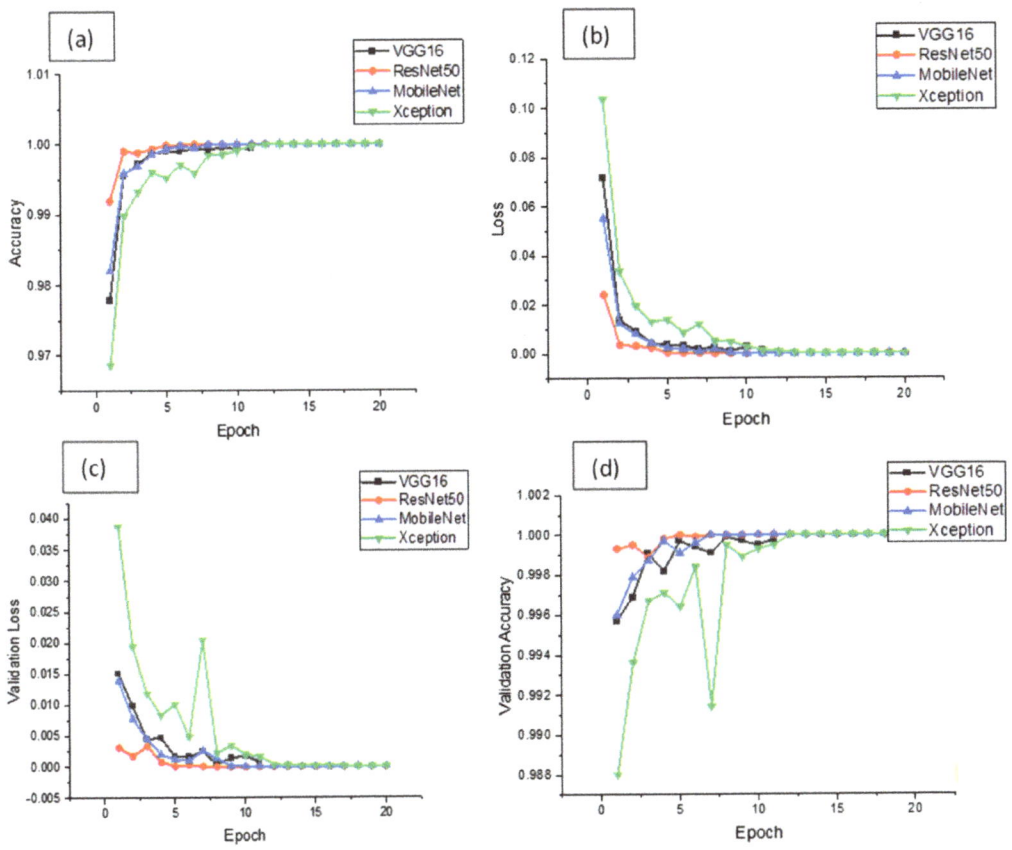

Figure 19. (**a**) Model accuracy (**b**) Model loss (**c**) Validation loss (**d**) Validation accuracy.

The distance at which the image is captured and its resolution become critical considerations in real-time approaches that use an autonomous vehicle to collect the image. In this investigation, raw photos were taken with a high-resolution camera, and the crack was visible to the naked eye.

4.3. Challenges in Deep Learning-Based Crack Classification of Concrete Walls

In this study, deep learning models are employed to classify cracks in concrete walls. However, the limited availability of datasets and the time-consuming process of labeling them can affect the accuracy and generalization of the models. To enhance the models' generalization capability, the models need to be trained on a wider range of datasets that reflect real-world environments with different lighting, moisture, and other conditions. Additionally, hyperparameter tuning is another challenge that requires time and effort, and there is no universal solution for it.

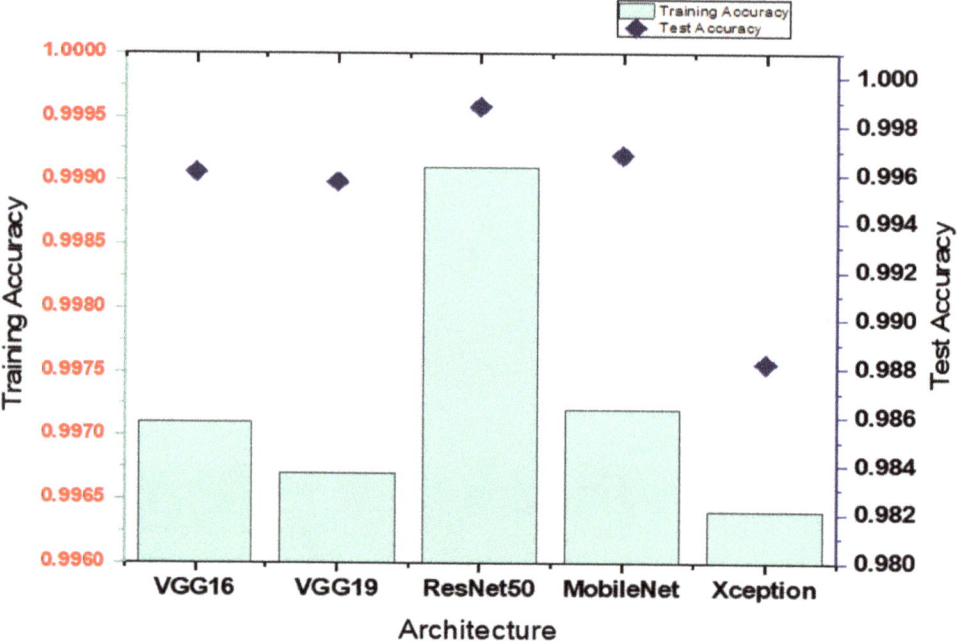

Figure 20. Comparison of training and test accuracies of all five architectures.

5. Conclusions

The traditional methods of detection and classification of cracks on different concrete structures take time, which is also labor-intensive and expensive when conducted manually. Therefore, automated study of damage in concrete structures is necessary for early diagnosis of the structures and for extending their service life. This study investigates the application of pre-trained deep learning networks for crack detection and classification.

In this paper, different backbone deep learning approaches were verified for automatically classifying concrete cracks. VGG16, VGG19, ResNet50, MobileNet, and Xception are the architectures considered.

- In the first stage, the performance of the VGG16 and VGG19 architectures was evaluated by comparing their results.
- It is ascertained that the VGG16 obtained a test and training accuracy of 99.61% and 99.71%, respectively, whereas the accuracy of the VGG19 was 99.57% and 99.67% for test and training accuracy, respectively.
- The speed and precision of VGG16 architecture were both better than those of VGG19 architecture.
- Secondly, the crack classification capabilities of the VGG16, ResNet50, MobileNet, and Xception architectures were evaluated. The ResNet50 architecture performed better than the other three architectures, with a test accuracy of 99.88%, and it required less training time than the other architectures, except for MobileNet.
- The training time for MobileNet was less than all other pre-trained models considered in this study.
- When compared to other architectures, the Xception architecture performed the worst. Furthermore, the generalization capability of the Xception architecture was less when compared to the other pre-trained models.
- Since the problem under examination was a binary classification problem, all models' accuracy was high, with an average test accuracy variation of only 0.22%.

Due to the low-level traits that cracks and other objects share with more abstract features, pre-trained networks had a high degree of applicability toward the identification of cracks, even when trained on wholly different datasets. It was found that the features acquired through the training are highly accurate when applied to other materials. Pre-trained networks are a good choice for deploying CNNs for the crack detection task, since they require fewer training samples and have a faster convergence rate.

Future attention will be placed on determining the efficacy of these structures in categorizing and localizing different types of cracks, spalls, and other flaws in concrete buildings in natural settings, which could automate the damage detection process.

Author Contributions: R.E.P. and A.D.A., Conceptualization, methodology, validation, and formal, analysis; A.D.A. and A.N., Data curation, visualization, supervising the research as well as the analysis of results; A.N., B.G.A.G. and K.R., Writing, review, submission, collaborating in and coordinating the research. All authors have read and agreed to the published version of the manuscript.

Funding: This research received no external funding.

Institutional Review Board Statement: Not applicable.

Informed Consent Statement: Not applicable.

Data Availability Statement: The data presented in this study are available on request from the corresponding author.

Conflicts of Interest: This manuscript has not been submitted to, nor is it under review by, another journal or other publishing venue. The authors have no affiliation with any organization with a direct or indirect financial interest in the subject matter discussed in the manuscript. The authors declare no conflict of interest.

References

1. Rajamony Laila, L.; Gurupatham, B.G.A.; Roy, K.; Lim, J.B.P. Effect of Super Absorbent Polymer on Microstructural and Mechanical Properties of Concrete Blends Using Granite Pulver. *Struct. Concr.* **2021**, *22*, E898–E915. [CrossRef]
2. Kanagaraj, B.; Kiran, T.; Gunasekaran, J.; Nammalvar, A.; Arulraj, P.; Gurupatham, B.G.A.; Roy, K. Performance of Sustainable Insulated Wall Panels with Geopolymer Concrete. *Materials* **2022**, *15*, 8801. [CrossRef] [PubMed]
3. Wei, W.; Ding, L.; Luo, H.; Li, C.; Li, G. Automated Bughole Detection and Quality Performance Assessment of Concrete Using Image Processing and Deep Convolutional Neural Networks. *Constr. Build. Mater.* **2021**, *281*, 122576. [CrossRef]
4. Kim, H.; Ahn, E.; Shin, M.; Sim, S.H. Crack and Noncrack Classification from Concrete Surface Images Using Machine Learning. *Struct. Health Monit.* **2019**, *18*, 725–738. [CrossRef]
5. Sun, Y.; Yang, Y.; Wei, F.; Wong, M. Autonomous Crack and Bughole Detection for Concrete Surface Image Based on Deep Learningused the Weights and Biases of the Xception Architecture. *IEEE Access* **2021**, *9*, 85709–85720. [CrossRef]
6. Lowe, D.; Roy, K.; Das, R.; Clifton, C.G.; Lim, J.B.P. Full Scale Experiments on Splitting Behaviour of Concrete Slabs in Steel Concrete Composite Beams with Shear Stud Connection. *Structures* **2020**, *23*, 126–138. [CrossRef]
7. Iannuzzo, A.; Angelillo, M.; De Chiara, E.; De Guglielmo, F.; De Serio, F.; Ribera, F.; Gesualdo, A. Modelling the Cracks Produced by Settlements in Masonry Structures. *Meccanica* **2018**, *53*, 1857–1873. [CrossRef]
8. Iannuzzo, A.; Serio, F.D.; Gesualdo, A.; Zuccaro, G.; Fortunato, A.; Angelillo, M. Crack Patterns Identification in Masonry Structures with a C° Displacement Energy Method. *Int. J. Mason. Res. Innov.* **2018**, *3*, 295–323. [CrossRef]
9. Tan, X.; Abu-Obeidah, A.; Bao, Y.; Nassif, H.; Nasreddine, W. Measurement and Visualization of Strains and Cracks in CFRP Post-Tensioned Fiber Reinforced Concrete Beams Using Distributed Fiber Optic Sensors. *Autom. Constr.* **2021**, *124*, 103604. [CrossRef]
10. Kim, H.; Lee, S.; Ahn, E.; Shin, M.; Sim, S.-H. Crack Identification Method for Concrete Structures Considering Angle of View Using RGB-D Camera-Based Sensor Fusion. *Struct. Health Monit.* **2021**, *20*, 500–512. [CrossRef]
11. Andrushia, A.D.; Thangarajan, R. RTS-ELM: An Approach for Saliency-Directed Image Segmentation with Ripplet Transform. *Pattern Anal. Appl.* **2020**, *23*, 385–397. [CrossRef]
12. Diana Andrushia, A.; Anand, N.; Prince Arulraj, G. A Novel Approach for Thermal Crack Detection and Quantification in Structural Concrete Using Ripplet Transform. *Struct. Control Health Monit.* **2020**, *27*, e2621. [CrossRef]
13. Cheng, H.D.; Wang, J.; Hu, Y.G.; Glazier, C.; Shi, X.J.; Chen, X.W. Novel Approach to Pavement Cracking Detection Based on Neural Network. *Transp. Res. Rec.* **2001**, *1764*, 119–127. [CrossRef]
14. Hoang, N.D. An Artificial Intelligence Method for Asphalt Pavement Pothole Detection Using Least Squares Support Vector Machine and Neural Network with Steerable Filter-Based Feature Extraction. *Adv. Civ. Eng.* **2018**, *2018*, 7419058. [CrossRef]

15. Wang, S.; Qiu, S.; Wang, W.; Xiao, D.; Wang, K.C.P. Cracking Classification Using Minimum Rectangular Cover–Based Support Vector Machine. *J. Comput. Civ. Eng.* **2017**, *31*(5), 04017027. [CrossRef]
16. Shi, Y.; Cui, L.; Qi, Z.; Meng, F.; Chen, Z. Automatic Road Crack Detection Using Random Structured Forests. *IEEE Trans. Intell. Transp. Syst.* **2016**, *17*, 3434–3445. [CrossRef]
17. Flah, M.; Suleiman, A.R.; Nehdi, M.L. Classification and Quantification of Cracks in Concrete Structures Using Deep Learning Image-Based Techniques. *Cem. Concr. Compos.* **2020**, *114*, 103781. [CrossRef]
18. Ali, R.; Chuah, J.H.; Talip, M.S.A.; Mokhtar, N.; Shoaib, M.A. Structural Crack Detection Using Deep Convolutional Neural Networks. *Autom. Constr.* **2022**, *133*, 103989. [CrossRef]
19. Kanagaraj, B.; Nammalvar, A.; Andrushia, A.D.; Gurupatham, B.G.A.; Roy, K. Influence of Nano Composites on the Impact Resistance of Concrete at Elevated Temperatures. *Fire* **2023**, *6*, 135. [CrossRef]
20. Laxman, K.C.; Tabassum, N.; Ai, L.; Cole, C.; Ziehl, P. Automated Crack Detection and Crack Depth Prediction for Reinforced Concrete Structures Using Deep Learning. *Constr. Build. Mater.* **2023**, *370*, 130709. [CrossRef]
21. Xu, X.; Zhao, M.; Shi, P.; Ren, R.; He, X.; Wei, X.; Yang, H. Crack Detection and Comparison Study Based on Faster R-CNN and Mask R-CNN. *Sensors* **2022**, *22*, 1215. [CrossRef] [PubMed]
22. Huyan, J.; Li, W.; Tighe, S.; Xu, Z.; Zhai, J. CrackU-Net: A Novel Deep Convolutional Neural Network for Pixelwise Pavement Crack Detection. *Struct. Control Health Monit.* **2020**, *27*, e2551. [CrossRef]
23. Dorafshan, S.; Thomas, R.J.; Maguire, M. SDNET2018: An Annotated Image Dataset for Non-Contact Concrete Crack Detection Using Deep Convolutional Neural Networks. *Data Br.* **2018**, *21*, 1664–1668. [CrossRef]
24. Loverdos, D.; Sarhosis, V. Automatic Image-Based Brick Segmentation and Crack Detection of Masonry Walls Using Machine Learning. *Autom. Constr.* **2022**, *140*, 104389. [CrossRef]
25. Dorafshan, S.; Thomas, R.J.; Maguire, M. Comparison of Deep Convolutional Neural Networks and Edge Detectors for Image-Based Crack Detection in Concrete. *Constr. Build. Mater.* **2018**, *186*, 1031–1045. [CrossRef]
26. Yosinski, J.; Clune, J.; Bengio, Y.; Lipson, H. How Transferable Are Features in Deep Neural Networks? In *Advances in Neural Information Processing Systems 27 (NIPS'14)*; MIT Press: Cambridge, MA, USA, 2014.
27. Ai, L.; Zhang, B.; Ziehl, P. A Transfer Learning Approach for Acoustic Emission Zonal Localization on Steel Plate-like Structure Using Numerical Simulation and Unsupervised Domain Adaptation. *Mech. Syst. Signal Process.* **2023**, *192*, 110216. [CrossRef]
28. Kang, D.; Benipal, S.S.; Gopal, D.L.; Cha, Y.J. Hybrid Pixel-Level Concrete Crack Segmentation and Quantification across Complex Backgrounds Using Deep Learning. *Autom. Constr.* **2020**, *118*, 103291. [CrossRef]
29. Su, C.; Wang, W. Concrete Cracks Detection Using Convolutional Neural Network Based on Transfer Learning. *Math. Probl. Eng.* **2020**, *2020*, 7240129. [CrossRef]
30. Dung, C.V.; Anh, L.D. Autonomous Concrete Crack Detection Using Deep Fully Convolutional Neural Network. *Autom. Constr.* **2019**, *99*, 52–58. [CrossRef]
31. Qu, Z.; Mei, J.; Liu, L.; Zhou, D.Y. Crack Detection of Concrete Pavement with Cross-Entropy Loss Function and Improved VGG16 Network Model. *IEEE Access* **2020**, *8*, 54564–54573. [CrossRef]
32. Joshi, D.; Singh, T.P.; Sharma, G. Automatic Surface Crack Detection Using Segmentation-Based Deep-Learning Approach. *Eng. Fract. Mech.* **2022**, *268*, 108467. [CrossRef]
33. Doğan, G.; Ergen, B. A New Mobile Convolutional Neural Network-Based Approach for Pixel-Wise Road Surface Crack Detection. *Measurement* **2022**, *195*, 111119. [CrossRef]
34. Rajamony Laila, L.; Gurupatham, B.G.A.; Roy, K.; Lim, J.B.P. Influence of Super Absorbent Polymer on Mechanical, Rheological, Durability, and Microstructural Properties of Self-Compacting Concrete Using Non-Biodegradable Granite Pulver. *Struct. Concr.* **2021**, *22*, E1093–E1116. [CrossRef]
35. Madan, C.S.; Munuswamy, S.; Joanna, P.S.; Gurupatham, B.G.; Roy, K. Comparison of the Flexural Behavior of High-Volume Fly AshBased Concrete Slab Reinforced with GFRP Bars and Steel Bars. *J. Compos. Sci.* **2022**, *6*, 157. [CrossRef]
36. Paul Thanaraj, D.; Kiran, T.; Kanagaraj, B.; Nammalvar, A.; Andrushia, A.D.; Gurupatham, B.G.A.; Roy, K. Influence of Heating–Cooling Regime on the Engineering Properties of Structural Concrete Subjected to Elevated Temperature. *Buildings* **2023**, *13*, 295. [CrossRef]
37. Madan, C.S.; Panchapakesan, K.; Reddy, P.V.A.; Joanna, P.S.; Rooby, J.; Gurupatham, B.G.A.; Roy, K. Influence on the Flexural Behaviour of High-Volume Fly-Ash-Based Concrete Slab Reinforced with Sustainable Glass-Fibre-Reinforced Polymer Sheets. *J. Compos. Sci.* **2022**, *6*, 169. [CrossRef]
38. Guzmán-Torres, J.A.; Naser, M.Z.; Domínguez-Mota, F.J. Effective Medium Crack Classification on Laboratory Concrete Specimens via Competitive Machine Learning. *Structures* **2022**, *37*, 858–870. [CrossRef]
39. Ye, W.; Deng, S.; Ren, J.; Xu, X.; Zhang, K.; Du, W. Deep Learning-Based Fast Detection of Apparent Concrete Crack in Slab Tracks with Dilated Convolution. *Constr. Build. Mater.* **2022**, *329*, 127157. [CrossRef]
40. He, K.; Zhang, X.; Ren, S.; Sun, J. Deep Residual Learning for Image Recognition. In Proceedings of the 2016 IEEE Conference on Computer Vision and Pattern Recognition (CVPR), Las Vegas, NV, USA, 27–30 June 2016; pp. 770–778. [CrossRef]
41. Wang, W.; Li, Y.; Zou, T.; Wang, X.; You, J.; Luo, Y. A Novel Image Classification Approach via Dense-Mobilenet Models. *Mob. Inf. Syst.* **2019**, *2020*, 7602384. [CrossRef]
42. Howard, A.G.; Zhu, M.; Chen, B.; Kalenichenko, D.; Wang, W.; Weyand, T.; Andreetto, M.; Adam, H. MobileNets: Efficient Convolutional Neural Networks for Mobile Vision Applications. *arXiv* **2017**, arXiv:1704.04861.

43. Chollet, F. Xception: Deep Learning with Depthwise Separable Convolutions. In Proceedings of the IEEE Conference on Computer Vision and Pattern Recognition (CVPR), Honolulu, HI, USA, 21–26 July 2017; pp. 1800–1807. [CrossRef]
44. Chun, P.J.; Izumi, S.; Yamane, T. Automatic Detection Method of Cracks from Concrete Surface Imagery Using Two-Step Light Gradient Boosting Machine. *Comput. Civ. Infrastruct. Eng.* **2021**, *36*, 61–72. [CrossRef]
45. Hoang, N.-D.; Huynh, T.-C.; Tran, V.-D. Concrete Spalling Severity Classification Using Image Texture Analysis and a Novel Jellyfish Search Optimized Machine Learning Approach. *Adv. Civ. Eng.* **2021**, *2021*, 5551555. [CrossRef]
46. Yamane, T.; Chun, P.J. Crack Detection from a Concrete Surface Image Based on Semantic Segmentation Using Deep Learning. *J. Adv. Concr. Technol.* **2020**, *18*, 493–504. [CrossRef]
47. Kingma, D.P.; Ba, J.L. Adam: A Method for Stochastic Optimization. In Proceedings of the 3rd International Conference on Learning Representations, ICLR 2015—Conference Track Proceedings, San Diego, CA, USA, 7–9 May 2015.
48. Fan, Z.; Li, C.; Chen, Y.; Di Mascio, P.; Chen, X.; Zhu, G.; Loprencipe, G. Ensemble of Deep Convolutional Neural Networks for Automatic Pavement Crack Detection and Measurement. *Coatings* **2020**, *10*, 152. [CrossRef]
49. Hossin, M.; Sulaiman, M.N. A Review on Evaluation Metrics for Data Classification Evaluations. *Int. J. Data Min. Knowl. Manag. Process* **2015**, *5*, 1–11. [CrossRef]
50. Bush, J.; Corradi, T.; Ninić, J.; Thermou, G.; Bennetts, J. Deep Neural Networks for Visual Bridge Inspections and Defect Visualisation in Civil Engineering. In Proceedings of the EG-ICE 2021 Workshop on Intelligent Computing in Engineering, Berlin, Germany, 30 June–2 July 2021; pp. 421–431.
51. Manjurul Islam, M.M.; Kim, J.M. Vision-Based Autonomous Crack Detection of Concrete Structures Using a Fully Convolutional Encoder–Decoder Network. *Sensors* **2019**, *19*, 4251. [CrossRef]
52. Paramanandham, N.; Koppad, D.; Anbalagan, S. Vision Based Crack Detection in Concrete Structures Using Cutting-Edge Deep Learning Techniques. *Trait. du Signal* **2022**, *39*, 485–492. [CrossRef]
53. Asadi, E.; Xu, C.; Rao, A.S.; Nguyen, T.; Ngo, T.; Dias-da-costa, D. Automation in Construction Vision Transformer-Based Autonomous Crack Detection on Asphalt and Concrete Surfaces. *Autom. Constr.* **2022**, *140*, 104316. [CrossRef]

Disclaimer/Publisher's Note: The statements, opinions and data contained in all publications are solely those of the individual author(s) and contributor(s) and not of MDPI and/or the editor(s). MDPI and/or the editor(s) disclaim responsibility for any injury to people or property resulting from any ideas, methods, instructions or products referred to in the content.

Article

Experimental Investigation on Flexural Behaviour of Sustainable Reinforced Concrete Beam with a Smart Mortar Layer

Ramkumar Durairaj [1], Thirumurugan Varatharajan [2,*], Satyanarayanan Kachabeswara Srinivasan [2], Beulah Gnana Ananthi Gurupatham [3] and Krishanu Roy [4,*]

[1] Research Scholar, Department of Civil Engineering, SRMIST, Kattankulathur 603203, India
[2] Department of Civil Engineering, SRMIST, Kattankulathur 603203, India
[3] Department of Civil Engineering, College of Engineering Guindy Campus, Anna University, Chennai 600025, India
[4] School of Engineering, The University of Waikato, Hamilton 3216, New Zealand
* Correspondence: assocdirector.cl@srmist.edu.in (T.V.); krishanu.roy@waikato.ac.nz (K.R.)

Abstract: This paper deals with an experimental study of the flexural behavior of sustainable reinforced cement concrete (RCC) beams with a smart mortar layer attached to the concrete mixture. In total, nine RCC beams were cast and tested. Two types of reinforced concrete beams were cast, and three different beams of sizes $1000 \times 150 \times 200$ mm and six different beams of sizes $1500 \times 100 \times 250$ mm were considered. The flexural behavior of these RCC beams was studied in detail. The electrical resistivity of these beams was also calculated, which was derived from the smart mortar layer. Research on the application of smart mortars within structural members is limited. The experimental results showed that the smart mortar layer could sense the damage in the RCC beams and infer the damage through the electrical measurement values, making the beam more sustainable. It was also observed that the relationship between the load and the fractional change in electrical resistance was linear. The fractional change in electrical resistivity was found to steadily increase with the increase in initial loading. A significant decrease in the fractional change in electrical resistivity was seen as the load approached failure. When a layer of mortar with brass fiber was added to the mortar paste, the ultimate load at failure was observed and compared with the reference beam specimen using Araldite paste. Compared to the hybrid brass-carbon fiber-added mortar layer, the brass fiber-added mortar layer increased the fractional change in the electrical resistivity values by 14–18%. Similarly, the ultimate load at failure was increased by 3–8% in the brass fiber-added mortar layer when compared to the hybrid brass-carbon fiber-added mortar layer. Failure of the beam was indicated by a sudden drop in the fractional change in electrical resistivity values.

Keywords: self-sensing; reinforced concrete beam; mortar; fibre; electrical resistance; carbon fibre; electrically conductive filler; brass fibre

1. Introduction

A reinforced cement concrete (RCC) beam is a structural member that carries all vertical loads from the slab and transfers them to the column. Structural beams can be made of different materials, such as steel, aluminum, wood, and concrete. RCC beams are the most common type of structural beam in the construction industry. Concrete is made up of cement, fine aggregate, and coarse aggregate. It is strong in compression, but weak in tension. To overcome the weakness in tension, reinforcement bars are introduced to the concrete, forming RCC members. Various studies have been conducted to improve the strength of RCC members by modifying their standard constituents with developing materials. These modifications aim to minimize structural failure and save lives. One innovation in the construction industry is smart concrete, which is a type of self-sensing, self-monitoring material that consists of conductive filler material along with the conventional constituents of concrete.

Carbon fibers are the most widely used type of conductive fillers, but their application in smart concrete or mortar is limited due to their high cost. Hence, they are uneconomical. To overcome this problem, brass fibers are introduced in the experiments. The primary focus of this paper is on the use of brass fiber as a conductive filler in cementitious mortar. Another main drawback of the use of carbon fibers is their high percentage of addition.

Research has been conducted extensively in this area to create a new technique that can replace steel and alloys while being reliable, affordable, and simple to handle. The use of FRP plates can be expanded to places where using steel would be impossible or impractical because they have several advantages over steel plates [1]. Glass fiber sheets performed slightly better than carbon fiber sheets due to their similar axial capacity [2]. CFRP increases the shear capacity to a greater extent for beams lacking sufficient shear reinforcement as compared to beams with adequate shear reinforcements [3]. FRP is increasingly in demand as a material to reinforce structural elements due to its high corrosion resistance and high strength-to-weight ratio [4]. One study reported that the CFRP beam specimen failed after the longitudinal steel reinforcement yielded concurrently with inclined cracks or splitting of the epoxy paste during the flexural test [5]. A strengthening system combining both the CFRP sheets and U-wrap anchors shows an increase in the initial stiffness of the RC beams [6]. The best results were obtained when the fibers reached their tensile failure, which occurred particularly when CFRP or steel plates were placed at the end of reinforcements in a direction perpendicular to the strengthening direction [7]. As a result, fiber-reinforced polymers that are externally bonded to concrete structures have gained widespread acceptance in the industry. However, this method reduces the beam's ductility [8]. Other researchers favor the use of U-wraps to avoid the debonding effect [9]. Adding U-wrap anchorage to the CFRP sheets can increase the strengthened beam's endurance without considerably increasing its capacity [10].

The FRP with the U-wraps could improve the beam's ability to support more weight in addition to reducing delamination [11]. Sectional dimensions, tension reinforcement ratio, shear reinforcement, load, and resistance all have an impact on the effectiveness of a reinforced concrete beam [12]. A concrete beam cast using high-strength concrete performed better in carrying the compressive strength, thus ensuring a good safety factor [13]. High-strength concrete is recommended in cases where weight reduction is crucial or where smaller load-bearing elements are required due to architectural concerns [14]. When the amount of reinforcement was unchanged, the beam ductility improved as concrete strength increased [15]. With various conductive additives, cement mortar produces a piezoresistive effect [16].

The electrical resistance showed a significant increase after seven days of air curing, whereas it decreased after 14 and 28 days of air curing [17]. An extremely sensitive strain sensor with a gauge factor of up to 700 is made of cement that contains short carbon fibers (0.24 volumetric percentage) [18]. The emergence and expansion of cracks significantly raise the electrical resistance of the cement-based composite. As a result, the cement-based composite has the potential to be utilized as a damage and strain sensor [19]. Electrical resistance variation and the perpendicular tensile strain showed a strong linear association [20]. Carbon fibers added in larger amounts resulted in fluctuations in a fractional change in resistivity (fcr), since the large amount being added reduced the even distribution of the fiber. This also reduced the workability of the material, thereby reducing its strength. Utilizing various conductive materials, researchers produced and tested piezoresistivity for strain and damage detection [21]. The steel fiber-reinforced cement-based composite can be used as a fire alarm detector due to the change in electrical resistance with temperature variation [22]. The self-sensing pavement was developed by embedding smart nickel particle-filled cement-based sensors into a concrete pavement. The smart cement-based sensor's high piezoresistive sensitivity enables the self-sensing pavement to precisely identify the passing of vehicles [23]. The wireless monitoring system, or nickel particle-reinforced cement composite, was developed for detecting vehicle movement [24]. Under a single compressive loading subjected to failure and within an elastic regime, the fractional change in electrical resistivity of a Portland cement-based composite containing nickel powder reached maximum values of 69.00% and 62.61% [25]. Sun et al. investigated [26] the application of cement-based composites as sensors in various structural components. Due to the significant strains that have led to the fracture of the concrete, carbon nanofiber–cement composites were unable to generate any type of damage-sensing mechanism when coupled with an RCC element [27]. When placed under uniaxial tension, embedded carbon-black-filled cement-based composite sensors exhibited good tensile strain sensing characteristics before being subjected to crushing failure [28]. Brass fibers are good electrically conductive filler materials used in mortar mixes to improve piezoelectric resistivity [29]. The piezoresistance was increased by adding 0.25% brass fibers to the standard mortar and by adding 95% brass fibers and 5% carbon fibers to the standard mortar [30].

Graphene nanoplatelets (GNPs) with relatively smaller surface areas and higher particle sizes form effective conductive paths and exhibit better piezoresistive characteristics [31]. As the sample size increases, the electrical resistivity also increases. However, the strain sensitivity decreases due to the obstruction of electrons by the aggregates. Additionally, large-scale bending test results verified the piezoresistivity of smart concrete, while crack formation and propagation dramatically increased the electrical resistance [32]. Kuralon fibers can further improve the strength and self-sensing properties of concrete. The mortar mixture with 8% graphite provided the best self-sensing properties to warn against the effects of cracking, and it also exhibited better mechanical properties [33]. The self-sensing behavior of mortar pavement was evaluated by the self-sensing of compression force, human motion detection, and vehicle speed monitoring. Moreover, the smart mortar slab could detect vehicle speed with high accuracy of traffic detection [34]. Experimental work on the mechanical properties of geopolymer concrete, mortar, and paste prepared using fly ash and blended slag resulted in an incremental improvement that was followed by a gradual reduction, and it finally reached a relatively consistent value with an increase in exposure temperature [35]. Recycled carbon fibers increase the flexural and tensile splitting strengths by up to 100%, whereas brass-coated steel fibers improve the compressive strength by 38%. Electrical conductivity tests show that recycled carbon fibers decrease the electrical resistivity of mortars [36].

Fiber-reinforced polymers (FRPs), which have a high strength-to-weight ratio and can withstand corrosive environments, can be used as structural components [37]. Based on the CHILE technique, the created numerical model accurately predicted the magnitude of spring-in deformation of L-shaped pultruded profiles [38]. In comparison to thermoplastic pultruded composites made from pre-consolidated tapes, pre-consolidated sheets enable pultrusion flat laminates with larger cross-sections, showing higher mechanical performance and surface roughness [39]. The mechanical properties degrade due to the development of micro-voids, fractures, and interface debonding [40]. In contrast to CFRP-strengthened specimens, which showed block splitting failure, fiber breakage, and buckling damage, the GFRP-strengthened specimens showed fiber bundle breakage, splitting, and buckling damage [41]. It was demonstrated that composite beams with wraps could attain the ductility and strengths of their counterparts with substantially higher beam depths [42].

Additional research can be performed by adding various other conductive fillers to the structural components in the future. Previous studies were carried out incorporating carbon fibers, carbon black, carbon nanofibers, etc. Since carbon fibers are uneconomical to be used as conductive fillers, an alternative brass fiber is introduced in the present study. The studies on the use of brass fiber, along with carbon fiber, as hybrid mortars with respect to structural applications on existing structural elements are limited. Hence, the newly developed smart mortar with the addition of brass fibers and hybrid brass-carbon fibers is introduced to the RCC beam specimen for damage detection.

The addition of fibres to the cementitious composites improves their flexural strength, flexural toughness, impact resistance, and tensile strength. Due to their high cost, carbon fibres are not widely used for practical purposes. Brass fibres are incorporated into the cementitious composites in this work as conductive fillers to improve the mortar's capacity for self-sensing. Due to the addition, fibres were also used to determine the self-sensing ability of brass fibres, and the bonding between the particles was observed within the cement matrix. Brass fibres were used to form thin mortar strips of 6 mm thickness. These strips were attached to the RCC beam to self-sense the damages occurring in the beam element, making it more sustainable. Further studies can also be carried out, implementing other conductive fillers to the structural elements.

2. Materials and Methods

To ascertain the flexural behaviour of the sustainable RCC beam, an experimental investigation was carried out. The self-sensing mortar layer was applied to the bottom of the beam to determine the failure of the beam through the self-sensing effect of the smart mortar.

2.1. Materials

The beam specimen was created using M30 concrete. The mix ratio for the M30 grade of concrete used for the beam specimen is shown in Table 1; the RCC beam's design and its reinforcing details were determined from the guidelines of IS 456-2000 [43]. The primary bars at the bottom were two numbers of 12 mm-diameter bars. The top reinforcement was made up of two numbers of 10 mm-diameter bars. Stirrups with a 100-mm centre-to-centre spacing were made of two-legged, 6-mm-diameter rods. Locally accessible river sand, OPC 53-grade cement, coarse aggregate measuring 12.5 mm, and portable water were used to create the RCC beams.

Table 1. Mix design for M30 grade concrete.

Cement (kg/m^3)	Fine Aggregate (kg/m^3)	Coarse Aggregate (kg/m^3)	Water (kg/m^3)
428.48	888.94	924.64	171.39

The smart mortar layer was made up of OPC 53-grade cement, locally available river sand of particle size less than 600 μm, potable water, and brass fibres of randomly varying length from 1 mm to 4 mm, with diameters varying from 0.1 mm to 1 mm (manufacturer: Sarda Industries Pvt. Ltd, Jaipur, Rajasthan, India). Carbon fibres of size 5 mm in length with a diameter of 10 μm were also used (manufacturer: Fibre Region Pvt. Ltd., Chennai, India). The pictorial representation of brass and carbon fibres is shown in Figure 1. A superplasticizer was used to improve the workability of the mortar mixture. Methyl cellulose (manufacturer: Southern India scientific Corporation Pvt. Ltd., Chennai, India) was additionally used for improving the dispersion of carbon fibres when carbon fibre was added to the cement mortar mixture. Silica fume (manufacturer: Astrra chemicals Pvt, Ltd., Chennai, India) was added to the smart mortar mixture to improve its strength. The fibres were randomly mixed to the mortar matrix. Modified MM 7.5 masonry mortar was adopted for smart mortar strips conforming to IS 2250-1981 [44]. The mix proportions along with the constituents are shown in Table 2.

In addition to the materials listed in Table 1, methylcellulose was added to the cement mortar mix at a rate of 1.872 kg/m^3 for the hybrid brass addition. The brass fibres were added at a dosage of 0.25% by the volume of the mortar. In the hybrid brass-carbon fibre addition, the fibres were added in a combination of 95% brass fibre and 5% carbon fibre. Carbon fibres were added at a rate of 0.24% by volume of mortar.

Table 2. Mix proportion and constituents for mortar strips.

Cement (kg/m^3)	Fine Aggregate (kg/m^3)	Water (kg/m^3)	Superplasticizer (kg/m^3)	Silica Fume (kg/m^3)
468	1427.5	234	7.02	70.20

(a) (b)

Figure 1. The pictorial representation of brass and carbon fibres. (**a**) Carbon fibre. (**b**) Brass fibre.

2.2. Casting of Specimen

Six beam specimens, measuring 1500 mm in length, 100 mm in width, and 250 mm in depth, were cast. The beams were designed by IS 456-2000 [43]. Two numbers of 12 mm steel bars were placed longitudinally at the bottom of the test beams, and two numbers of 10 mm steel bars were placed longitudinally at the top of the test beams as the tensile and compressive steel reinforcement, respectively. Two-legged stirrups of 6 mm diameter were used at 100 mm centre-to-centre spacing.

Two mortar layers were cast separately with dimensions of 1200 mm in length, 100 mm in width, and 6 mm in thickness. One mortar strip was made of brass fibre addition, while the other was made of hybrid brass–carbon fibre addition. These two mortar strips were pasted to the RCC beam with Araldite paste at the bottom. In the other two beam specimens, the freshly prepared mortar paste was applied to the existing beam, which was manufactured 28 days earlier. In the other two beam specimens, the mortar layer was freshly cast on the beam during the time of its manufacture. Detailed information on the beam specimens is given in Table 3.

Table 3. Details of the beam specimens.

S. No.	Beam	Variations Provided
1	Beam-1-FB-FM-BF	By incorporating brass fibres into the mortar mixture, a fresh mortar layer was cast on a fresh beam.
2	Beam-2-EB-FM-BF	By including brass fibres in the mortar mixture, a fresh mortar layer was cast on the top of the existing beam.
3	Beam-3-EB-EM-BF	Brass fibres were included in the mortar mixture, and a separate layer of cast mortar was placed over the existing beam.
4	Beam-4-FB-FM-HBC	By incorporating hybrid brass carbon fibres into the mortar mixture, a fresh mortar layer was cast on a fresh beam.
5	Beam-5-EB-FM-HBC	Hybrid brass-carbon fibres are added to the mortar mixture before casting a new mortar layer over an existing beam.
6	Beam-6-EB-EM-HBC	Hybrid brass carbon fibres were added to the mortar mixture, and a separate layer of cast mortar was applied to the existing beam with Araldite paste.

2.3. Test Method

The flexural behaviour of the beam was determined with the two-point method (loading frame manufacturer–SRM Institute of Science and Technology) of loading. Six beams with a smart mortar layer were tested with two-point loading. To measure the deflections, one deflectometer was positioned at the beam's midspan at the bottom. The dimensions of the beam with the smart mortar layer are illustrated in Figure 2. To measure the strains, three strain gauges were attached at the top, middle, and bottom of the beam.

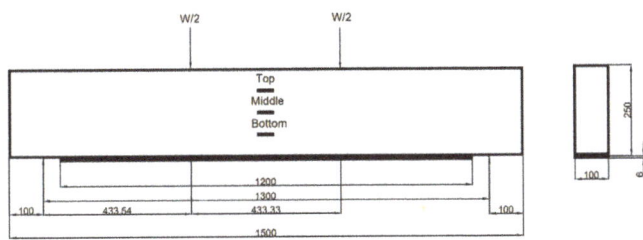

Figure 2. Pictorial representation of beam with smart mortar layer.

As depicted in Figure 3, the specimens were tested in a 400 kN loading frame. Using a load cell, the beam specimens were subjected to incremental loading until failure. The electrical measurements of the RCC beam were determined using the four-probe method. Silver paint was applied to the specimen at four points with intervals of 200 mm, as shown in Figure 4. Steel wires were wound along the applied silver paint to determine the flow of electricity across the specimen. Two multimeters were connected to the four probes, along with a DC power supply. Through the multimeter, the voltage and current readings were recorded. Using the voltage and current data that were obtained, the resistance and fractional change in electrical resistance were calculated. Under two-point loading, the load was applied in an increment of 5 kN. Readings were recorded using digital multimeters, a deflectometer, and strain gauges. The fractional change in electrical resistance (fcr) was determined using the resistivity value for each load increment of 5 kN.

Figure 3. Test setup of the beam with smart mortar layer.

Figure 4. Four–probe method with wire winding on smart mortar.

3. Discussion and Findings

3.1. A New Brass Fibre Added Mortar Layer Is Cast on a New 1.5-m-Long Beam

The ultimate load of fresh brass fibre added to the mortar layer cast on a fresh beam was observed at 145 kN. The stress observed at the bottom of the beam was higher when compared to the middle and top of the beam. At the initial stages of loading, the strain observed in the reinforced beam was minimal. As the loading was increased and the first crack was observed, the strain at the bottom increased gradually until failure. Figure 5 illustrates that, when the loading increased, flexural cracks developed in every reinforced concrete beam. From the beginning point of loading until failure, a progressive increase in the mid-span displacement of the beam was also noticed. The graphical representations of stress versus strain, fcr versus strain, load versus displacement, and load versus fcr are shown in Figure 6. The maximum bottom strain at failure was observed as 0.00131 for the brass fibre-added fresh mortar layer cast on a fresh concrete beam of length 1.5 m.

The fcr values were calculated with the current and voltage readings recorded through the multimeter connected to the beam through the four-probe winding method. The fractional change in electrical resistivity values was found to steadily increase with the initial increase in loading through the four-probe winding method. The fractional change in electrical resistivity values was found to steadily increase with the initial increase in loading. Whereas, when the loading was about to reach the failure load, a drastic change in the fractional change in electrical resistivity values was observed. This was caused by the formation and development of more cracks in the mortar layer.

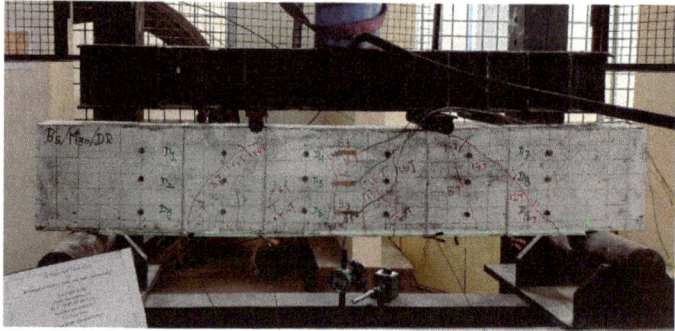

Figure 5. Crack patterns of the reinforced concrete beam with smart mortar layer.

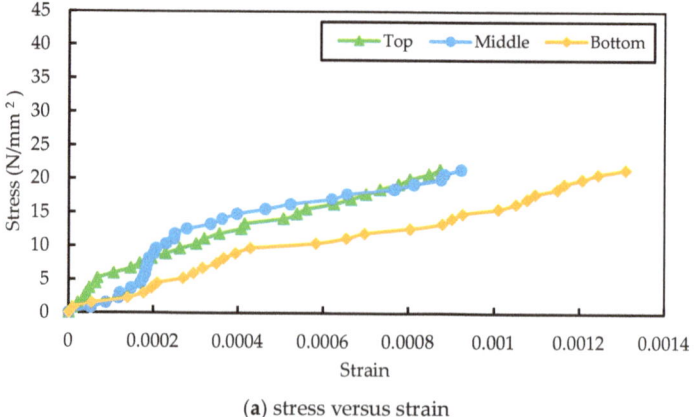

(**a**) stress versus strain

Figure 6. *Cont.*

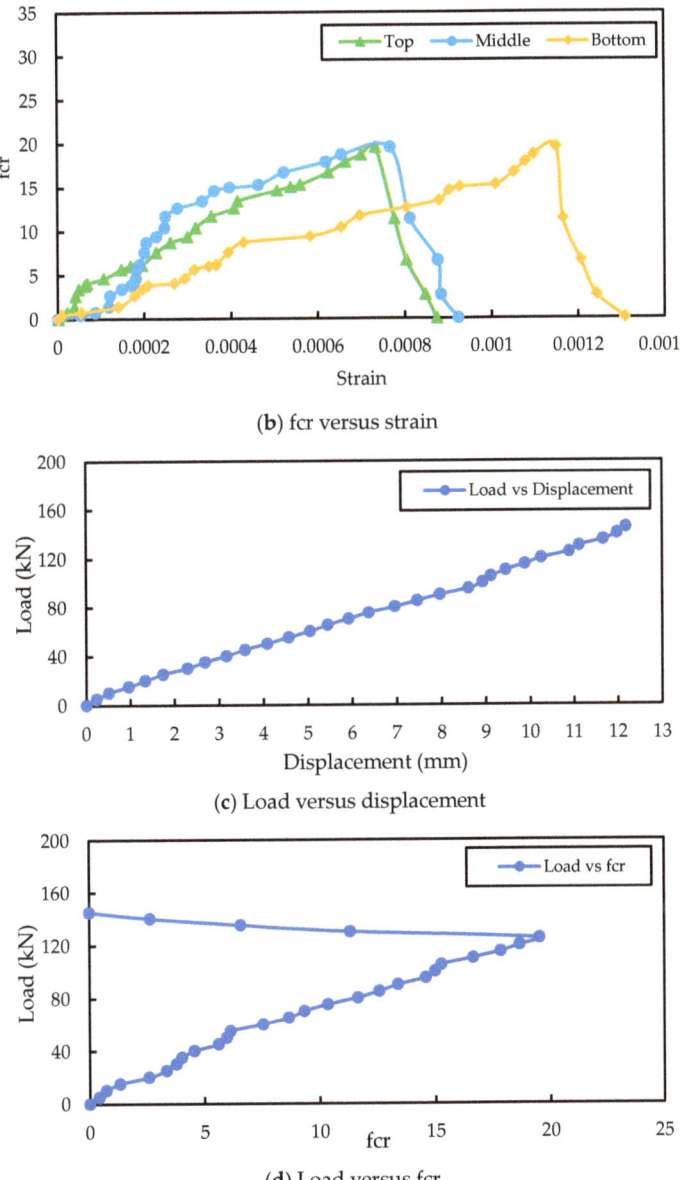

(b) fcr versus strain

(c) Load versus displacement

(d) Load versus fcr

Figure 6. Graphical representation of (**a**) stress vs. strain, (**b**) fcr vs. strain, (**c**) load vs. displacement and (**d**) load versus fcr behaviour of the brass fibre added fresh mortar layer cast on fresh concrete beam of length 1.5 m.

Similarly, the fractional change in electrical resistivity values decreased when the applied load approached the failure load and was at its minimum at the ultimate load. This occurred due to the splitting of the mortar layer at the bottom of the beam through the development of cracks in it. The electrical circuit formed through the four probes was disconnected by the breakage of the smart mortar layer, leading to the reduction in fcr values. The pictorial representation of the crack formed on the smart mortar layer at the bottom of the beam is shown in Figure 7.

Figure 7. Failure modes of the smart mortar layer.

3.2. A Fresh Mortar Layer of Brass Fibres Was Cast on an Existing 1.5-m-Long Beam

The ultimate load of the freshly added brass fibre mortar layer cast on an existing beam was measured at 140 kN. The behaviour of the beam with respect to displacement, strain, and stress was similar to that of a fresh mortar layer with added brass fibres cast on a fresh beam. When compared to a fresh brass fibre added mortar layer cast on a new beam, a fresh brass fibre added mortar layer cast on an existing beam reduced displacement by 20.37%. Similarly, fcr increased by 4.26% in a fresh brass fibre-added mortar layer cast on an existing beam when compared to a fresh brass fibre-added mortar layer cast on a fresh beam.

The failure of the smart mortar layer was caused by the formation and propagation of flexural cracks on the beam, which led to the disconnection of the self-sensing effect [27]. The maximum bottom strain at failure was observed as 0.00525 for the brass fibre-added fresh mortar layer cast on an existing concrete beam of length 1.5 m. The graphical representation of stress versus strain, fcr versus strain, load versus displacement, and load versus fcr is shown in Figure 8.

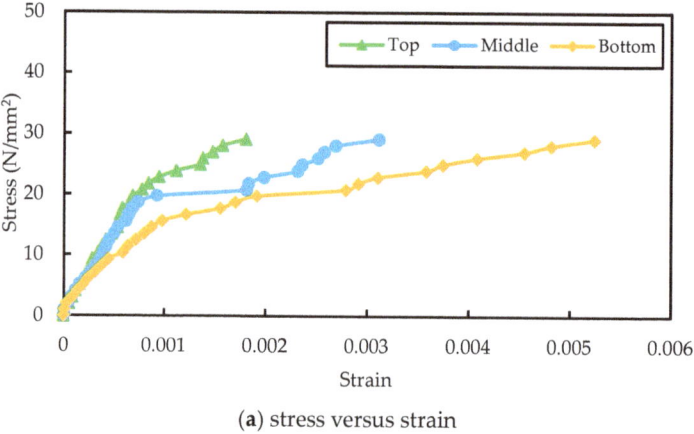

(**a**) stress versus strain

Figure 8. *Cont.*

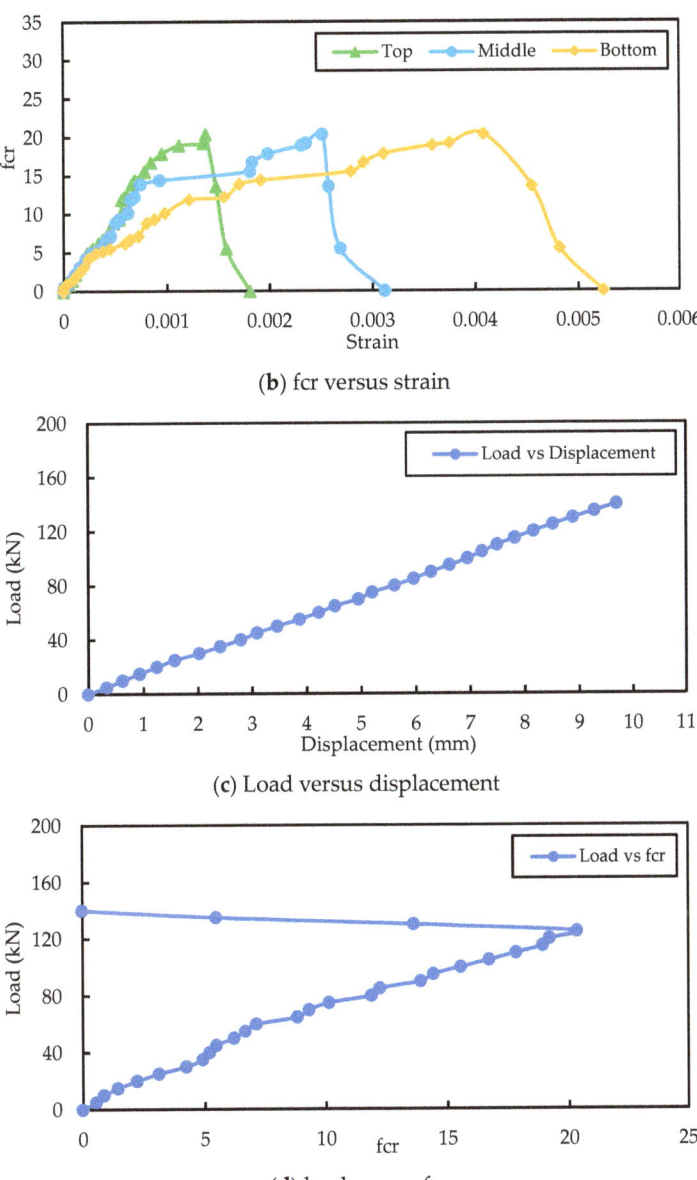

Figure 8. Graphical representation of (**a**) stress vs. strain, (**b**) fcr vs. strain, (**c**) load vs. displacement, and (**d**) load versus fcr behaviour of the brass fibre added fresh mortar layer cast on an existing concrete beam of length 1.5 m.

3.3. Brass Fibre Added Mortar Layer Pasted on Existing Beam of Length 1.5 m with Araldite Paste

The final load of the brass fiber-added mortar layer cast and placed on the existing beam with the help of Araldite paste was 155 kN. The behaviour of the beam in terms of displacement, strain, and stress was identical to a fresh brass fibre mortar layer cast on a fresh beam. The displacement was reduced by 8.71% when brass fibre was added to the mortar layer placed on the existing beam with the help of Araldite paste as compared to a fresh mortar layer placed on a fresh beam. The electrical resistance improved by 15.43% more when brass fibre was added to the mortar layer placed on the existing beam with the help of Araldite paste as compared to a fresh mortar layer cast on a fresh beam. The addition of a smart mortar layer to the existing RCC beam with Araldite paste improved the piezoelectricity of the beam greatly. This is due to the improved bonding between the beam and the smart mortar layer due to the Araldite paste. Hence, among all three methods used to implement the smart mortar on the reinforced concrete fibre-added mortar layer cast on the existing beam with the help of Araldite paste, one was observed to be more effective.

The development and spread of flexural cracks on the beam caused the self-sensing effect to be disconnected, which resulted in the failure of the smart mortar layer. The maximum bottom strain at failure was observed as 0.00234 for the brass fibre-added mortar layer pasted on the existing concrete beam of length 1.5 m with Araldite paste. In Figure 9, the relationships between stress and strain, fcr and strain, load and displacement, and load and fcr are represented graphically.

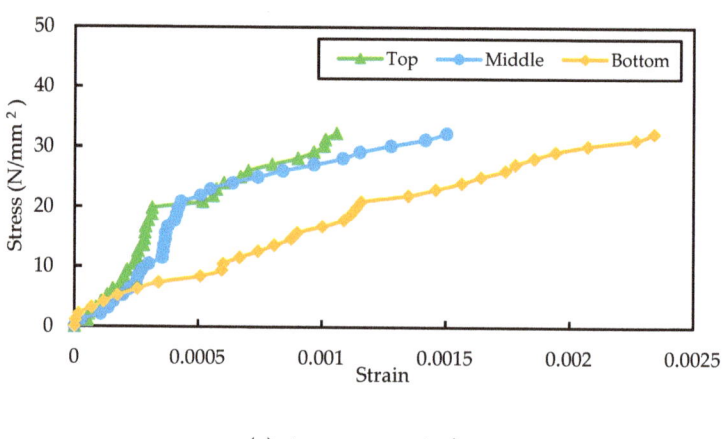

(a) stress versus strain

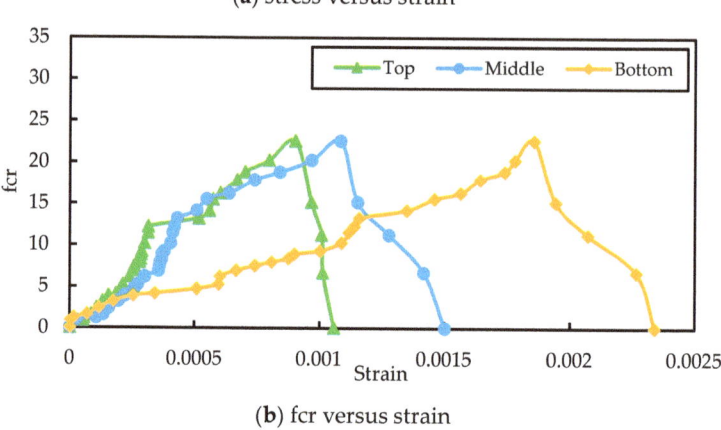

(b) fcr versus strain

Figure 9. Cont.

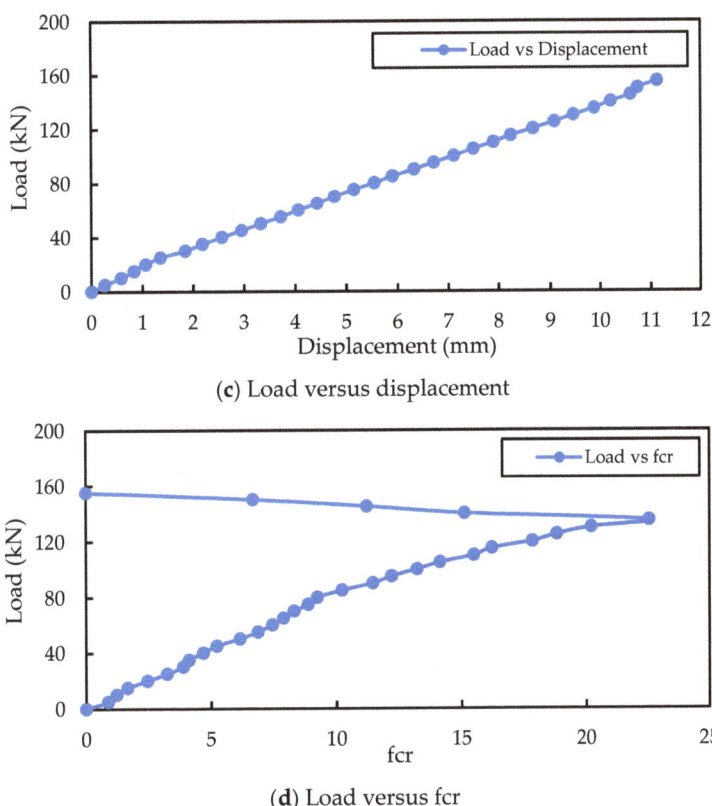

Figure 9. Graphical representation of (**a**) stress vs. strain, (**b**) fcr vs. strain, (**c**) load vs. displacement and (**d**) load versus fcr behaviour of brass fibre added mortar layer pasted on existing concrete beam with Araldite paste of length 1.5 m.

3.4. Hybrid Brass Carbon Fresh Mortar Layer Cast on a Fresh Concrete Beam of Length 1.5 m

The ultimate load of the mortar layer added with fresh hybrid brass-carbon fibre cast on a fresh beam was measured at 140 kN. In comparison to the centre and top of the beam, there was more strain towards the bottom of the beam. The reinforced beam experienced very little strain during the early stages of loading. When the loads were increased, the bottom strain progressively developed until failure, and the first crack was observed. The mid-span displacement of the beam was also found to significantly increase from the point of initial loading until failure.

The fcr values were calculated using voltage and current readings from a multimeter connected to the beam via the four-probe winding method. It was discovered that, initially, the fcr values increased continuously with the increase in loading. In contrast, a sharp change in the fcr values was seen as the loading approached the failure load. This was carried on by the formation and growth of further cracks in the layer of mortar.

The fcr values also decreased as the applied load approached the failure load and were at their lowest at the maximum load. This happened as a result of the cracks in the smart mortar layer that developed at the bottom of the beam. The fracturing of the smart mortar layer interrupted the electrical circuit established by the four probes, which reduced the fcr values. In Figure 10, the relationships between stress and strain, fcr and strain, load and displacement, and load and fcr are represented graphically.

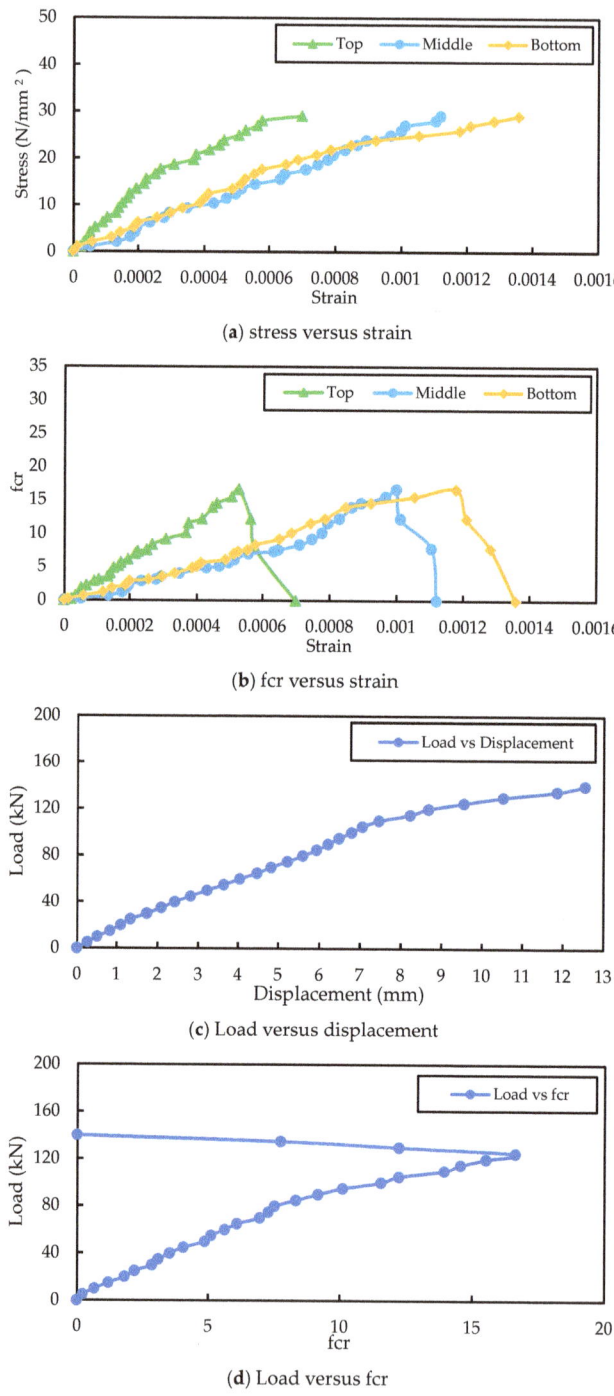

Figure 10. Graphical representation of (**a**) stress vs. strain, (**b**) fcr vs. strain, (**c**) load vs. displacement, and (**d**) load versus fcr behaviour of hybrid brass carbon fibre added fresh mortar layer cast on a fresh concrete beam of length 1.5 m.

When compared to a fresh brass fibre added mortar layer cast on a fresh beam, the displacement in a fresh hybrid brass–carbon fibre-added mortar layer cast on a fresh beam increased by 3.04%. Similarly, when compared to a fresh brass fibre added mortar layer cast on a fresh beam, the fractional change in electrical resistivity decreased by 14.76% in a fresh hybrid brass and carbon fibre added mortar layer cast on a fresh beam. The maximum bottom strain at failure was observed as 0.00136 for the hybrid brass carbon fibre-added fresh mortar layer cast on a fresh concrete beam of length 1.5 m.

3.5. Hybrid Brass–Carbon Fibre Added Fresh Mortar Layer Cast on an Existing Concrete Beam of Length 1.5 m

The peak load of the fresh hybrid brass–carbon fibre-added mortar layer cast on an existing beam was 130 kN. The beam behaved similarly to a fresh hybrid brass–carbon fibre-added mortar layer cast on a fresh beam in terms of displacement, strain, and stress. When fresh hybrid brass with a carbon fibre added mortar layer cast on an existing beam was compared to fresh hybrid brass with a carbon fibre added mortar layer cast on a new beam, displacement was reduced by 15.94%. When comparing fresh hybrid brass with a carbon fibre-added mortar layer cast on a fresh beam to fresh hybrid brass with a carbon fibre-added mortar layer cast on an existing beam, the electrical resistivity increased by 6.62%.

In comparison to a fresh brass fibre-added mortar layer cast on an existing beam, the displacement in a fresh hybrid brass–carbon fibre-added mortar layer cast on an existing beam increased by 8.77%. Similarly, when compared to a fresh brass fibre-added mortar layer cast on an existing beam, the fcr of a fresh hybrid brass and carbon fibre added mortar layer cast on an existing beam decreased by 12.83%. The maximum bottom strain at failure was observed as 0.00455 for the hybrid brass–carbon fibre-added fresh mortar layer cast on an existing concrete beam of length 1.5 m. The graphical representation of stress versus strain, fcr versus strain, load versus displacement, and load fcr is shown in Figure 11.

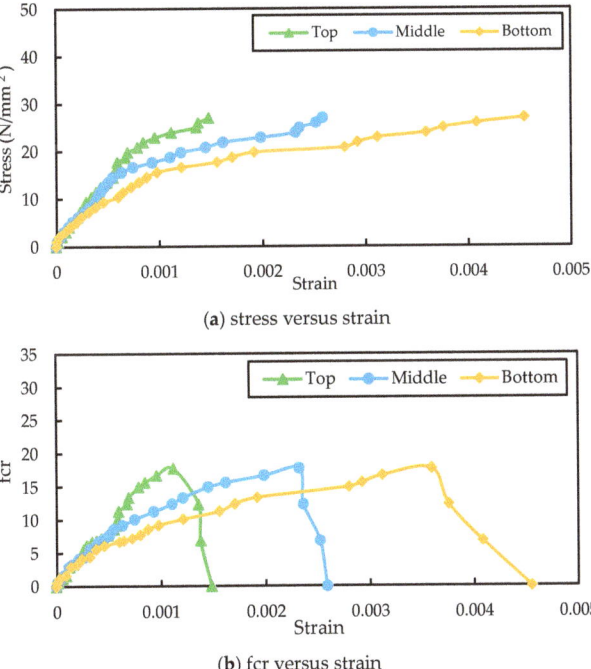

(a) stress versus strain

(b) fcr versus strain

Figure 11. Cont.

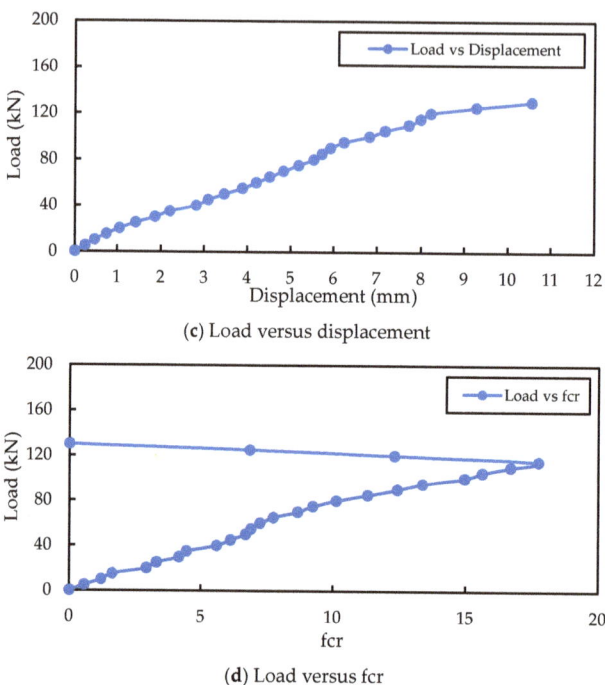

Figure 11. Graphical representation of (**a**) stress vs. strain, (**b**) fcr vs. strain, (**c**) load vs. displacement, and (**d**) load versus fcr behaviour of hybrid brass–carbon fibre-added fresh mortar layer cast on an existing concrete beam of length 1.5 m.

3.6. Hybrid Brass–Carbon Fibre-Added Mortar Layer Pasted on Existing Concrete Beam with Araldite Paste of Length 1.5 m

The ultimate load of the hybrid brass-carbon fibre mortar layer placed on the existing beam with the help of Araldite paste was 150 kN. The beam behaved in the same manner as the fresh hybrid brass–carbon fibre-added mortar layer cast on a fresh beam in terms of displacement, strain, and stress. In comparison to fresh hybrid brass with carbon fibre added mortar applied to a new beam, displacement was reduced by 11.16% in the hybrid brass, with carbon fibre-added mortar layer applied to an existing beam using Araldite paste. In comparison to a hybrid brass–carbon fibre-added fresh mortar layer cast on a fresh beam, the electrical resistance increased by 15.51% when the hybrid brass–carbon fibre-added fresh mortar layer was placed on an existing beam with the aid of Araldite paste.

When compared to the brass fibre added mortar layer placed on the existing beam with the help of Araldite paste, the displacement increased by 0.27% in the hybrid brass and carbon fibre added mortar layer placed on the existing beam with the help of Araldite paste. Similarly, in the brass fibre-added mortar layer placed on an existing beam with the help of Araldite paste, the f_{cr} decreased by 14.70% in the hybrid brass–carbon fibre-added mortar layer placed on an existing beam with the aid of Araldite paste. The maximum bottom strain at failure was observed as 0.00156 for the hybrid brass–carbonfibre-added mortar layer pasted on an existing concrete beam of length 1.5 m with Araldite paste. The graphical representation of stress versus strain, f_{cr} versus strain, load versus displacement, and load versus f_{cr} is shown in Figure 12.

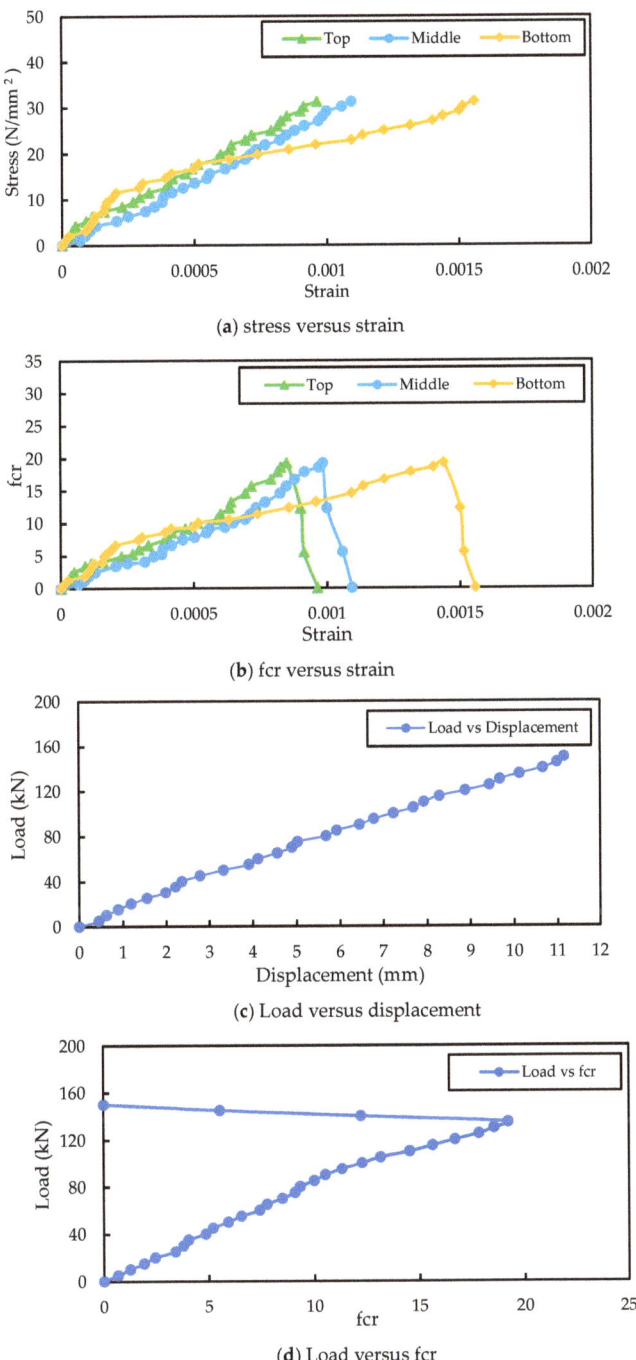

Figure 12. Graphical representation of (**a**) stress vs. strain, (**b**) fcr vs. strain, (**c**) load vs. displacement, and (**d**) load versus fcr behaviour of hybrid brass–carbon fibre-added mortar layer pasted on existing concrete beam with Araldite paste of length 1.5 m.

The load was applied with a 2 kN increment in UTM. A strain gauge and digital multimeter were used for testing. The fractional change in electrical resistance (fcr) was determined using the resistivity value for each 2 kN increment in load.

4. Discussion of Results

From the experimental results, it was found that the fractional change in the electrical resistivity values was found to steadily increase with the increase in loading initially. In contrast, a drastic reduction in the fractional change in electrical resistivity values was observed when the loading approached the failure load. Both brass and carbon fibres performed well as conductive fillers in the production of cement composites that helped to measure strains on the surface of a structural member, irrespective of the local stresses being in tension or in compression, which was similar to the experimental observations of Baeza et al. [27]. The most sensitive dosage of brass fibres in the mortar strip was 0.25% (by volume of the mortar). Similarly, the optimum dosages of the hybrid brass-carbon fibre were 95% for brass fibres and 5% for carbon fibres, respectively (by volume of the mortar) [30].

The ultimate load at failure was also observed in a separately cast, fibre-added mortar layer pasted to the existing beam with the use of Araldite paste. Compared to the hybrid brass–carbon fibre-added mortar layer, the brass fibre-added mortar layer's fractional change in electrical resistivity performance improved by 14–18%. Similarly, the ultimate load at failure was improved by 3–8% when the brass fibre was added to the mortar layer as compared to the hybrid brass-carbon fibre-added mortar layer.

From the experimental results presented in this paper, it was found that the addition of brass fibre and hybrid brass–carbon fibre made mortar more self-sensing. With the newly developed smart mortar, the strain level in any flexural member can be assessed. Additionally, the load level on the beam or on the slab can be assessed by measuring the electrical resistance of the member. The structural elements can maintain their load level within a safe limit. The developed smart mortar can be used for structural applications where temperature changes are severe, such as chemical storage facilities, pavements, dams, bridges, and for tall buildings. A quick, affordable, and less time-consuming technique of monitoring the structural health of a building is possible by the development of this smart mortar layer that can be applied to an existing structural member to monitor its performance throughout its lifecycle.

5. Conclusions

The following findings were obtained from experiments on reinforced concrete beams with smart mortar layers containing either pure brass fibre or a hybrid of brass and carbon fibre. By adding a smart mortar layer to the bottom of a reinforced concrete beam, the electrical resistivity values were able to detect a failure in the beam before it actually occurred, making the beam element more sustainable. The application of Araldite paste, which improves bonding and piezoelectricity, is the most effective method for implementing the smart mortar layer in the reinforced concrete beam. The brass fibre-added smart mortar performed better when compared to the hybrid brass–carbon fibre-added smart mortar in terms of piezoresistive effect. The behaviour of the beam in terms of displacement, strain, and the ultimate load was more or less similar, since all the beams were prepared with the same materials and specifications. The formation and progression of cracks in the smart mortar layer before the failure of the beam resulted in prior intimation through a sudden drop in a fractional change in electrical resistivity values. The ultimate load at failure was observed in a separately cast fibre-added mortar layer pasted to the existing beam with the use of Araldite paste. Compared to the hybrid brass–carbon fibre-added mortar layer, the brass fibre-added mortar layer showed an increase in the electrical resistivity values by 14–18%. Similarly, the ultimate load at failure was improved by 3–8% in the case of brass fibre-added mortar layer when compared to the case of hybrid brass–carbon fibre-added mortar layer.

As a result, structural members can be coated with either carbon or brass fibres in thin mortar layers to detect damage to the member. It can also be used to detect damage in members before they fail and collapse, making the structure more sustainable. Further research can be carried out with different conductive fillers in the mortar matrix. Studies on the applications of such smart mortars in high-rise buildings, pavements, and dams can also be carried out. The incorporation of smart mortars into other structural elements subjected to cyclic loading can be performed in the future.

Author Contributions: R.D.: Conducted the experiments; T.V. and S.K.S.: Supervised the research as well as the Validation of results; R.D., T.V. and S.K.S.: Introduced the idea of self-sensing in this project, wrote, reviewed, and submitted the paper, and collaborated in and coordinated the research; T.V., S.K.S., B.G.A.G. and K.R.: Suggested and chose the journal for submission; R.D., T.V., S.K.S., B.G.A.G. and K.R.: Participated in the manuscript revision phase. All authors have read and agreed to the published version of the manuscript.

Funding: This research received no external funding.

Data Availability Statement: The data presented in this study are available on request from the corresponding author.

Acknowledgments: The authors acknowledge SRMIST, Kattankulathur for their support in conducting this research study.

Conflicts of Interest: This manuscript has not been submitted to, nor is it under review by, another journal or other publishing venue. The authors have no affiliation with any organization with a direct or indirect financial interest in the subject matter discussed in the manuscript. The authors declare no conflict of interest.

References

1. Arya, C.; Clarke, J.L.; Kay, E.A.; O'Regan, P.D. Design guidance for strengthening concrete structures using fiber composite materials: A review. *Eng. Struct.* **2001**, *24*, 889–900. [CrossRef]
2. Gao, B.; Kim, J.K.; Leung, C.K.Y. Experimental investigation of taper ended FRP strips in FRP strengthened RC beams. In Proceedings of the 4th International Conference on Advanced Composite Materials in Bridges and Structures, Calgary, AB, Canada, 20–23 July 2004.
3. Khalifa, A.; Nanni, A. Rehabilitation of rectangular simply supported RC beams with shear deficiencies using CFRP composites. *Constr. Build. Mater.* **2002**, *16*, 135–146. [CrossRef]
4. Pesic, N.; Pilakoutas, K. Concrete beams with externally bonded flexural FRP-reinforcement: Analytical investigation of debonding failure. *Compos. Part B Eng.* **2003**, *34*, 327–338. [CrossRef]
5. Mohammad Noh, N. Effectiveness of RC Beams Strengthened with Near Surface Mounted (NSM) Fiber Reinforced Polymer (FRP) Bars and Externally Bonded (EB) FRP Sheets. Master's Thesis, Universiti Teknologi MARA, Shah Alam, Malaysia, 2003.
6. Gao, B.; Kim, J.K.; Leung, C.K.Y. Optimization of tapered end design for FRP strips bonded to RC beams. *Compos. Sci. Technol.* **2005**, *66*, 1266–1273. [CrossRef]
7. Ceroni, F.; Pecce, M.; Mattys, S.; Taerwe, L. Debonding strength and anchorage devices for reinforced concrete elements strengthened with FRP sheets. *Compos. Part B Eng.* **2007**, *30*, 429–441. [CrossRef]
8. Xiong, G.J.; Jiang, X.; Liu, J.W.; Chen, L. A way for preventing tension delamination of concrete cover in midspan of FRP strengthened beams. *Constr. Build. Mater.* **2005**, *21*, 402–408. [CrossRef]
9. Al-Saidy, A.H.; Al-Harthy, A.S.; Al-Jabri, K.S.; Abdul-Halim, M.; Al Shidi, N.M. Structural performance of corroded RC beams repaired with CFRP sheets. *Compos. Struct.* **2010**, *92*, 1931–1938. [CrossRef]
10. El-Ghandour, A.A. Experimental and analytical of CFRP flexural and shear strengthening efficiencies of RC beams. *Constr. Build. Mater.* **2011**, *25*, 1419–1429. [CrossRef]
11. Buyle-Bodin, F. Use of carbon fiber textile to control premature failure of reinforced concrete beams strengthened with bonded CFRP plates. *J. Ind. Text.* **2010**, *33*, 145–157. [CrossRef]
12. Arafah, A.M. Factors affecting the reliability of reinforced concrete beams. *WIT Trans. Ecol. Environ.* **2000**, *45*, 379–387.
13. Khalel, R.I. Torsional behavior of high-strength reinforced concrete beams. *J. Eng. Sustain. Dev.* **2013**, *17*, 317–332.
14. Jagana, R.; Kumar, C.V. High strength concrete. *Int. J. Eng. Sci. Res. Technol.* **2017**, *6*, 394–407.
15. Rashid, M.A.; Mansur, M.A. Reinforced high-strength concrete beams in flexure. *ACI Struct. J.* **2005**, *102*, 462.
16. Fu, X.; Ma, E.; Chung, D.D.L.; Anderson, W.A. Self-monitoring in carbon fiber reinforced mortar by reactance measurement. *Cem. Concr. Res.* **1997**, *27*, 845–852. [CrossRef]
17. Fu, X.; Chung, D.D.L. Effect of curing age on the self-monitoring behavior of carbon fiber reinforced mortar. *Cem. Concr. Res.* **1997**, *27*, 1313–1318. [CrossRef]

18. Chung, D.D.L. Self-monitoring structural materials. *Mater. Sci. Eng.* **1998**, *22*, 57–78. [CrossRef]
19. Teomete, E.; Kocyigit, O.I. Tensile strain sensitivity of steel fiber reinforced cement matrix composites tested by split tensile test. *Constr. Build. Mater.* **2013**, *47*, 962–968. [CrossRef]
20. Teomete, E. Transverse strain sensitivity of steel fiber reinforced cement composites tested by compression and split tensile tests. *Constr. Build. Mater.* **2014**, *55*, 136–145. [CrossRef]
21. Teomete, E. Measurement of crack length sensitivity and strain gage factor of carbon fiber reinforced cement matrix composites. *Measurement* **2015**, *74*, 21–30. [CrossRef]
22. Teomete, E. The effect of temperature and moisture on electrical resistance, strain sensitivity and crack sensitivity of steel fiber reinforced smart cement composite. *Smart Mater. Struct.* **2016**, *25*, 075024. [CrossRef]
23. Han, B.; Zhang, K.; Yu, X.; Kwon, E.; Jinping, O.U. Nickel particle based self-sensing pavement for vehicle detection. *Measurement* **2011**, *44*, 1645–1650. [CrossRef]
24. Han, B.; Yu, Y.; Han, B.Z.; Ou, J.P. Development of a wireless stress/strain measurement system integrated with pressure-sensitive nickel powder-filled cement-based sensors. *Sens. Actuators A* **2008**, *147*, 536–543. [CrossRef]
25. Han, B.; Han, B.Z.; Ou, J.P. Experimental study on use of nickel powder-filled Portland cement-based composite for fabrication of piezoresistive sensors with high sensitivity. *Sens. Actuators A* **2009**, *149*, 51–55. [CrossRef]
26. Sun, M.; Liew, R.J.Y.; Zhang, M.H.; Li, W. Development of cement-based strain sensor for health monitoring of ultra-high strength concrete. *Constr. Build. Mater.* **2014**, *65*, 630–637. [CrossRef]
27. Baeza, F.J.; Galao, O.; Zornoza, E.; Garcés, P. Multifunctional cement composites strain and damage sensors applied on reinforced concrete (RC) structural elements. *Materials* **2013**, *6*, 841–855. [CrossRef]
28. Xiao, H.; Li, H.; Ou, J.P. Strain sensing properties of cement-based sensors embedded at various stress zones in a bending concrete beam. *Sens. Actuators A* **2011**, *167*, 581–587. [CrossRef]
29. Kumar, D.R.; Thirumurugan, V.; Satyanarayanan, K.S. Experimental study on optimization of smart mortar with the addition of brass fibres. *Mater. Today Proc.* **2021**, *50*, 388–393. [CrossRef]
30. Durairaj, R.; Varatharajan, T.; Srinivasan, S.K.; Gurupatham, B.G.A.; Roy, K. An Experimental Study on Electrical Properties of Self-Sensing Mortar. *J. Compos. Sci.* **2022**, *6*, 208. [CrossRef]
31. Sevim, O.; Jiang, Z.; Ozbulut, O.E. Effects of graphene nanoplatelets type on self-sensing properties of cement mortar composites. *Constr. Build. Mater.* **2022**, *359*, 129488. [CrossRef]
32. Erman, D.; Egemen, T. Cross tension and compression loading and large-scale testing of strain and damage sensing smart concrete. *Constr. Build. Mater.* **2022**, *316*, 125784.
33. Cheng, A.; Lin, W.-T. Electrical resistance and self-sensing properties of pressure-sensitive materials with graphite filler in Kuralon fiber concrete. *Mater. Sci.* **2022**, *40*, 223–239. [CrossRef]
34. Dong, W.; Li, W. Application of intrinsic self-sensing cement-based sensor for traffic detection of human motion and vehicle speed. *Constr. Build. Mater.* **2022**, *355*, 129130. [CrossRef]
35. Bellic, A.; Mobili, A. Commercial and recycled carbon/steel fibers for fiber-reinforced cement mortars with high electrical conductivity overlay panel. *Cem. Concr. Compos.* **2020**, *109*, 103569. [CrossRef]
36. Zhao, J.; Wang, K.; Wang, S.; Wang, Z.; Yang, Z.; Shumuye, E.D.; Gong, X. Effect of Elevated Temperature on Mechanical Properties of High-Volume Fly Ash-Based Geopolymer Concrete, Mortar and Paste Cured at Room Temperature. *Polymers* **2021**, *13*, 1473. [CrossRef]
37. Tucci, F.; Vedernikov, A. Design Criteria for Pultruded Structural Elements. *Encycl. Mater. Compos.* **2021**, *3*, 51–68.
38. Vedernikov, A.; Safonov, A.; Tucci, F.; Carlone, P.; Akhatov, I. Analysis of Spring-in Deformation in L-shaped Profiles Pultruded at Different Pulling Speeds: Mathematical Simulation and Experimental Results. In Proceedings of the ESAFORM 2021, 4th International ESAFORM Conference on Material Forming, Liège, Belgium, 14–16 April 2021.
39. Minchenkov, K.; Vedernikov, A. Effects of the quality of pre-consolidated materials on the mechanical properties and morphology of thermoplastic pultruded flat laminates. *Compos. Commun.* **2022**, *35*, 101281. [CrossRef]
40. Zhou, P.; Li, C. Durability study on the interlaminar shear behavior of glass-fibre reinforced polypropylene (GFRPP) bars for marine applications. *Constr. Build. Mater.* **2022**, *349*, 128694. [CrossRef]
41. Madenci, E.; Özkılıç, Y.O.; Aksoylu, C.; Safonov, A. The Effects of Eccentric Web Openings on the Compressive Performance of Pultruded GFRP Boxes Wrapped with GFRP and CFRP Sheets. *Polymers* **2022**, *14*, 4567. [CrossRef]
42. Ozkilic, Y.O.; Gemi, L.; Madenci, E.; Aksoylu, C.; Kalkan, İ. Effect of the GFRP wrapping on the shear and bending Behavior of RC beams with GFRP encasement. *Steel Compos. Struct.* **2022**, *45*, 193–204.
43. *IS 456-2000*; Indian Standard Plain and Reinforced Concrete Code of Practice. Bureau of Indian Standards: New Delhi, India, 2000.
44. *IS 2250-1981*; Indian Standard Code of Practice for Preparation and Use of Masonry Mortars. Bureau of Indian Standards: New Delhi, India, 1981.

Disclaimer/Publisher's Note: The statements, opinions and data contained in all publications are solely those of the individual author(s) and contributor(s) and not of MDPI and/or the editor(s). MDPI and/or the editor(s) disclaim responsibility for any injury to people or property resulting from any ideas, methods, instructions or products referred to in the content.

Article

Fabrication and Experimental Analysis of Bricks Using Recycled Plastics and Bitumen

Naveen Kumar Koppula *, Jens Schuster and Yousuf Pasha Shaik

Institute for Polymer Technology West-Palatinate, University of Applied Sciences Kaiserslautern, 66953 Pirmasens, Germany
* Correspondence: nako1001@stud.hs-kl.de

Abstract: Plastic is being used increasingly in daily life. Most of it is not recyclable, and the remaining plastic cannot be used or decomposed. This causes increased plastic waste, contributing to global warming due to thermal recycling. The major objective of this research was to utilise the maximum plastic waste possible to manufacture bricks that compete with the properties of conventional bricks without affecting the environment and the ecological balance. A balanced mixture of high-density polyethylene (HDPE), quartz sand, and some additive materials, such as bitumen, was used to produce these bricks. Various tests were performed to assess the bricks' quality, such as compression, water absorption, and efflorescence tests. These bricks had a compression strength of 37.5 MPa, which is exceptionally strong compared to conventional bricks. The efflorescence and water absorption tests showed that the bricks were nearly devoid of alkalis and absorbed almost no water. The obtained bricks were light in weight and cost-effective compared to conventional bricks.

Keywords: HDPE; quartz sand; bitumen; compression testing; water absorption test; efflorescence test

---

Citation: Koppula, N.K.; Schuster, J.; Shaik, Y.P. Fabrication and Experimental Analysis of Bricks Using Recycled Plastics and Bitumen. *J. Compos. Sci.* **2023**, *7*, 111. https://doi.org/10.3390/jcs7030111

Academic Editors: G. Beulah Gnana Ananthi and Krishanu Roy

Received: 13 February 2023
Revised: 5 March 2023
Accepted: 8 March 2023
Published: 10 March 2023

Copyright: © 2023 by the authors. Licensee MDPI, Basel, Switzerland. This article is an open access article distributed under the terms and conditions of the Creative Commons Attribution (CC BY) license (https://creativecommons.org/licenses/by/4.0/).

1. Introduction

Plastic, a synthetic material made from various organic compounds with a high molecular mass, was first introduced in 1907 by Leo Baekeland [1]. Since then, it has changed numerous industries. During World War II, there was rapid growth in the plastic industry, as manufacturers could use plastic to replace products previously manufactured using natural resources. The best example was nylon, which replaced silk and was used for parachutes, helmets, and body armour; another was plexiglass, used as a substitute for glass in aircraft [2].

Over time, plastics replaced the use of traditional materials. It has become one of the essential materials in day-to-day life. Properties such as high strength, corrosion resistance, easily mouldability, waterproofness, and its ductile nature make it fit a wide range of applications. It is used in almost every sector, such as in electrical and electronic applications, packaging, logistics, and industrial machinery. More than 50% of plastic was produced after the year 2000. Most plastic is used for packaging [3].

Its production involves the polymerisation or polycondensation of natural materials, such as cellulose, coal, natural gas, salt, crude oil, minerals, and plants. It can be moulded into various shapes, sizes, and forms. It also possesses remarkable properties, such as lightness, durability, flexibility, and affordability, and is thus a popular choice over traditional materials, such as wood, natural fibres, rubber, and paper [4].

Despite having advantages, such as cheaper cost and ease of production, it affects the environment if not utilised properly. According to EEA, plastics have many other effects, including contributions to climatic change, and the report says that there is no control over the production and consumption of plastics [5]. There is a great need to develop a circular trend by inventing various recycling techniques. However, according to the Ellen MacArthur Foundation, only 14% of the plastic produced is recycled [6]. Every year, countries worldwide discard millions of tons of plastic waste. Less than 20% of waste plastic is recycled to make new plastics, while the remaining plastics are either disposed of in landfills or burnt or dumped [7]. The only way to minimise its impact is to stop using it; however, in the modern world, this is impossible due to excessive reliance on plastic items. Thus, one of the valuable solutions is to convert the available waste plastic into raw material for various new plastic goods, speeding up the recycling process.

Technically, all polymers are recyclable and can be used once or more to create an identical product [8]. However, it must be carried out in controlled conditions and requires high-end machinery and technology. This is not a feasible solution, because it harms the environment and makes industries unprofitable. Therefore, it is best to sort waste plastic according to its categorisation and qualities to extract recyclable plastic and use it to produce plastic products in the future. Because of the process-induced breakdown of the polymer chain, the properties of such recycled materials may differ from those of the original. These properties can be regained by adding the appropriate additives and strength-enhancing materials [9].

Researchers conducted experiments to investigate the optimal combination of raw materials, plastic, and sand to achieve maximum brick strength. These experiments involved varying proportions of the raw materials. One specific experiment conducted by Wahid et al. [10] utilised a mixture of sand, sand dust, and cement in a ratio of 9:9:4. The mixture was then mixed with plastic waste in weights of 0%, 5%, 10%, and 15%. The highest compression strength observed was 12.4 N/mm^2 when 0% plastic was used, and the strength decreased as the percentage of plastic increased. The reduction in strength was attributed to poor adhesion between the plastic waste and cement. Additionally, a longer curing time was required due to the significant volume of cement and sand. These findings suggest that plastic waste in brick production may not necessarily improve the strength of bricks, and further research is needed to optimise the proportions of the raw materials and curing process.

In an extension of the research of Wahid et al. [10], another researcher, Agyeman et al. [11], mixed plastic: quarry dust: sand ratios of 1:1:2 and 2:1:2 by weight. The resulting samples were then cured for seven days by being sprinkled with water. The bricks produced from these mixtures were tested for their compressive strengths and water absorption rates. It was found that the bricks produced from the 1:1:2 mixture had a compressive strength of 6.07 N/mm^2 and a water absorption rate of 4.9%, while the bricks produced from the 2:1:2 mixture had a higher compressive strength of 8.53 N/mm^2 and a lower water absorption rate of 0.5%.

As per Circular Action Hub (CAH) [12], mixing recyclable plastics with sawdust, concrete, mud, or sand can replace conventional building materials. High-density polyethylene is one of those recyclable plastics. Typically, HDPE produces storage containers for milk, shampoo, oil, and chemicals. It is comparatively more stable and emits fewer hazardous gases, and is thus acceptable for recycling under certain controlled circumstances. Globally, about 40 million tonnes of HDPE waste are generated [13].

High mechanical strength, transparency, non-toxicity, no effect on taste, and permeability that may be disregarded for carbon dioxide are all characteristics of HDPE plastic. In addition to being transparent, processable, colourable, and thermally stable, HDPE plastic is chemically resistant, tensile, and impact-resistant [14].

The polyethylene grade with the most stiffness and least flexibility is HDPE. It works well for various uses, such as garbage cans and everyday home items, including miniature bottles and clothespins. This non-toxic, lightweight material can replace less eco-friendly materials because it is readily recyclable [15].

An analysis conducted by Sahani et al. [16] found that recycling plastic waste in manufacturing bricks and tiles is an effective way to reduce waste. The study suggested that when plastic and sand are combined in the proper ratio, the resulting bricks have a higher compression strength than traditional clay bricks. The highest compression strength was achieved with a plastic-to-sand ratio of 1:4, valued at 12.28 N/mm^2. However, as the amount of plastic in the mixture decreased, the compression strength also decreased.

The research of Kulkarni et al. [17] shows that HDPE plastics have a higher compression strength of 14.6% than conventional bricks, with a brick wall's ultimate load-carrying capacity being 197.5 KN. This means they can be used to build structures supporting higher loads. According to Maneeth et al. [18], bitumen acts as a binder and enhances the strength of the bricks because of its stability and density. However, as the bitumen percentage increases, the compression strength declines from 10 N/mm^2 to 2.04 N/mm^2, with the ideal bitumen percentage being 2%. According to Benny T.K. et al. [19], bitumen-added bricks have a hydrophobic property that prevents water from infiltrating them. They also have a higher compression strength than standard bricks, with a compression loading of 120 KN. In this way, large amounts of waste can be used without affecting the environment and reducing the cost of construction.

This publication aims to manufacture plastic sand bricks using bitumen and find the optimum percentages of plastic and sand for producing bricks with higher compression strength and a lower water absorption rate than conventional bricks.

2. Materials and Methods

2.1. Materials Used

2.1.1. HDPE Plastics

High-density polyethylene (HDPE) is usually derived from petroleum. HDPE is a low-cost thermoplastic material that performs well for low- and medium-technical applications. It is used to produce various products, such as pipelines, milk jars, cutting boards, and plastic bottles. HDPE does not crack under stress due to its very ductile behaviour. Table 1 shows the properties of HDPE.

Table 1. Properties of HDPE [20].

Property	Units	HDPE
Yield stress	MPa	18
Youngs modulus	MPa	960–1000
Density	Kg/m^3	941–967
Melting point	°C	130–133
Coefficient of thermal elongation	%	20–100
Impact resistance	J/m	27–160

2.1.2. Sand

High-quality, naturally occurring quartz stone that has been carefully chosen and finely processed is used to make the high-purity quartz sand (SiO$_2$ ≥ 99.5–99.9%, Fe$_2$O$_3$ ≤ 0.001%). High-purity quartz sand produces glass, refractories, ferrosilicon flux, ceramics, grinding materials, and casting-moulding quartz sand. It has significant anti-acid medium-erosion properties for making concrete that is resistant to acid [21].

The particle size range of 0.1–0.3 mm was selected for the experiment. Quartz sand chemically resistant, and because it has a higher melting temperature than metals, it is used as a foundry sand and can be used for manufacturing bricks. Table 2 shows the properties of quartz sand. Quartz sand was purchased from OBI Markt, Kaiserslautern, Germany.

Table 2. Properties of quartz sand [22].

Property	Units	Value
Specific gravity	-	2.45
Water absorption	%	1.9
Fineness modulus	-	2.2

2.1.3. Bitumen

Bitumen is mainly used for construction because of its higher binding characteristics; it is less costly compared to other binding materials. Bituminous materials have adhesive properties, are soluble in carbon disulphide, and are mostly made of high-molecular-weight hydrocarbons produced by distilling petroleum or asphalt [23]. Bitumen is also known for its binding properties. In this experiment, different percentages of bitumen (1%, 2%, and 3%) were considered. Table 3 shows the properties of bitumen. Bitumen was purchased from BAUHAUS, Ludwigshafen, Germany.

Table 3. Properties of bitumen [24].

Property	Units	Value
Specific gravity	-	22.4
Softening point	°C	35–70
Ductility	m	0.0264

2.2. Methodology
2.2.1. Mould Design

A rectangular aluminium mould with (230 × 120 × 150) mm dimensions was manufactured. There were four side plates, a base plate, and a top plate on the brick mould. Figure 1 illustrates the cutting of four plates with a thickness of 10 mm and dimensions of 230 mm × 150 mm. Since it was positioned on a flat surface, the base plate's thickness was reduced to 5 mm, significantly reducing the possibility of deflection. In order to make it simple to apply pressure to the molten mixture and facilitate the easy removal of the brick, the top plate was fastened with a knob-shaped component. An Allen key of 8 mm was used to assemble the plates after they were drilled with 8 mm holes.

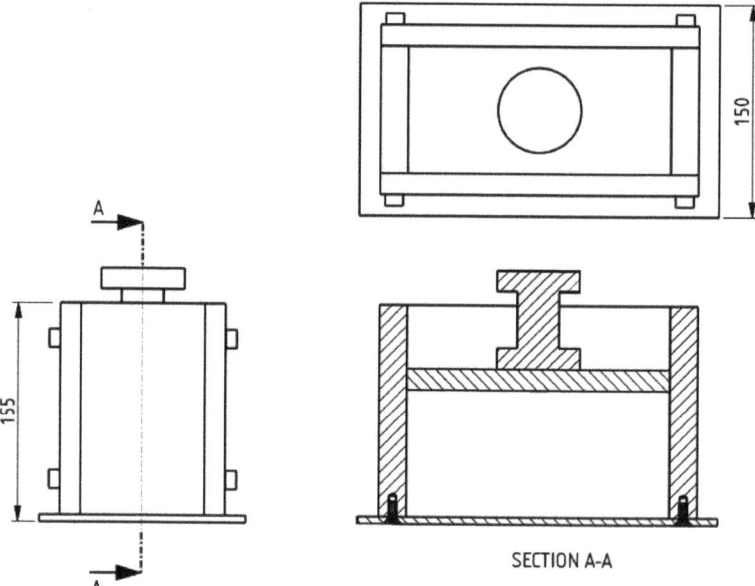

Figure 1. Mould design.

2.2.2. Plastics Collection

Initially, used HDPE containers/bottles were collected and cleaned with warm water. Next, the HDPE waste was dried to enable further processing without any moisture content. These plastics were then broken down into smaller pieces using a shredding machine containing a series of rotating blades that cut the plastic containers/bottles into small pieces. The sizes of these pieces ranges from 1 mm to 5 mm. To produce the desired brick, these plastics were then melted with sand.

2.2.3. Quartz Sand and Bitumen

A small quantity of bitumen was added as a binding agent to increase the strength of the brick, induce better bonding of granules and cover any voids. Different amounts of bitumen were added to the sample to determine the maximum strength (0%, 1%, 2%, and 3%). The mixing ratios of plastic and sand were varied (3:1, 3:2, 1:1, 2:3, and 1:3). The mixture was then heated using a kneader to firmly bond the plastic and sand.

2.2.4. Shredding and Mixing

The mixture was heated in the kneader until the plastic melted and firmly bonded with the sand. The kneader was operated at a temperature of 180 °C and a rotational speed of 40 rpm. The resulting lump immediately solidified and hardened, making melting in a hot press difficult. These lumps were subsequently granulated into tiny particles using a grinder. The mixture was then poured into a mould of the appropriate size and set in the hot press. It was left for about an hour at a temperature of 300 °C, which was higher than the melting point of the plastic. As the plastic granules began to melt, pressure was applied (about 15 bar) to the top plate of the brick to obtain the desired shape and thickness. The specimens were subjected to a compressed force of 1 kN to make the material more compact and reduce the number of voids. The pressure was applied during the cooling down of the press tool to ambient temperature. As a result of the rapid cooling, the bending of the brick could be observed after demoulding.

2.3. Test Equipment and Test Parameters

2.3.1. Compression Tests

The compression strength of the plastic bricks was tested using the DIN EN ISO 604. Twenty bricks were tested in total. Because of their ductile nature, the bricks were cut into 10 mm × 10 mm × 10 mm pieces and tested on a compression machine. The load was applied until the brick broke or showed deformation. The test was performed on a Universal Testing Machine with a maximum force of 100 kN at a speed of 1 mm/min. The ultimate stress at which the brick deformed or broke was noted, and the compressive strength was calculated using the formula shown in Equation (1).

$$\text{Compression strength} = \frac{P}{A} \tag{1}$$

with

P as the maximum load [kN] and
A representing the area of the specimen [mm^2].

2.3.2. Water Absorption Tests

The test was used to determine the amount of water absorbed by the brick. The quality of the brick was determined by its water absorption rate, with a lower water absorption rate considered the best. The bricks were heated to remove any moisture content. The weight of the dry brick was noted as W_1. The brick was then fully immersed in water and undisturbed for 24 h. After 24 h, the brick was removed from the water and gently wiped with a cloth. The weight of the wet brick was noted as W_2. The percentage of water absorbed was calculated using the formula shown in Equation (2).

$$\text{Water absorption} = \frac{W - W_1}{W_1} * 100\%, \qquad (2)$$

with

W being the weight of the dry brick [kg]
W_1 being the weight of the wet brick [kg]

2.3.3. Efflorescence Test

The test was used to determine the presence of hazardous alkalis in bricks. A circular vessel was used, and enough water was added for testing. The immersion depth was 25 mm. The bricks were then soaked in distilled water for 24 h. After 24 h, the bricks were removed from the vessel and dried for the same period. A high-quality brick should not possess any alkalis on the bricks and should be free of soluble salts. If alkalis were present on the bricks, efflorescence might occur, resulting in a layer forming. Table 4 shows the percentage of alkali presence.

Table 4. Percentage of alkali presence after efflorescence test.

Efflorescence Test	Extent of Deposits
Nil	0%
Low	≤10%
Medium	10% to 50%
Heavy	More than 50% without powdered flakes
Serious	More than 50% with powdered flakes

2.3.4. Test to Determine the Relative Rise in Temperature

The test aimed to determine how much the temperature would increase on one side of a brick when the other side was in direct contact with a heat source. A simple setup was created, with the brick as a dividing wall. An electric induction plate was used and heated to a temperature of 410 °C. One side of the brick was placed next to the plate, while the other was exposed to room temperature. The brick was heated for three minutes, resulting in an increase in the temperature. After three minutes, the temperature on both sides of the brick was measured, and the relative rise in temperature was calculated.

3. Results

3.1. Compression Test

The test was carried out at a speed of 1 mm/min using a Universal Testing Machine (UTM). A total of five brick samples were taken, consisting of 20 bricks in total. The following results were obtained. The stress values were recorded when the strain value was 0.2, which is acceptable in most applications.

3.1.1. Brick Sample 1 (Plastic:Sand—3:1)

Figure 2a shows that the bricks with more plastic (i.e., a plastic-to-sand ratio of 3:1) had higher compression strengths. When 0% bitumen was added, the strength of the brick was 22.08 MPa; when bitumen was added, the values varied between 26.6 MPa and 33.46 MPa, with 2% (9 g bitumen) being the optimal amount; and as bitumen was removed, the value fell to 31.8 MPa. The highest strength was attained at an optimal bitumen percentage of 2% (i.e., 9 g bitumen).

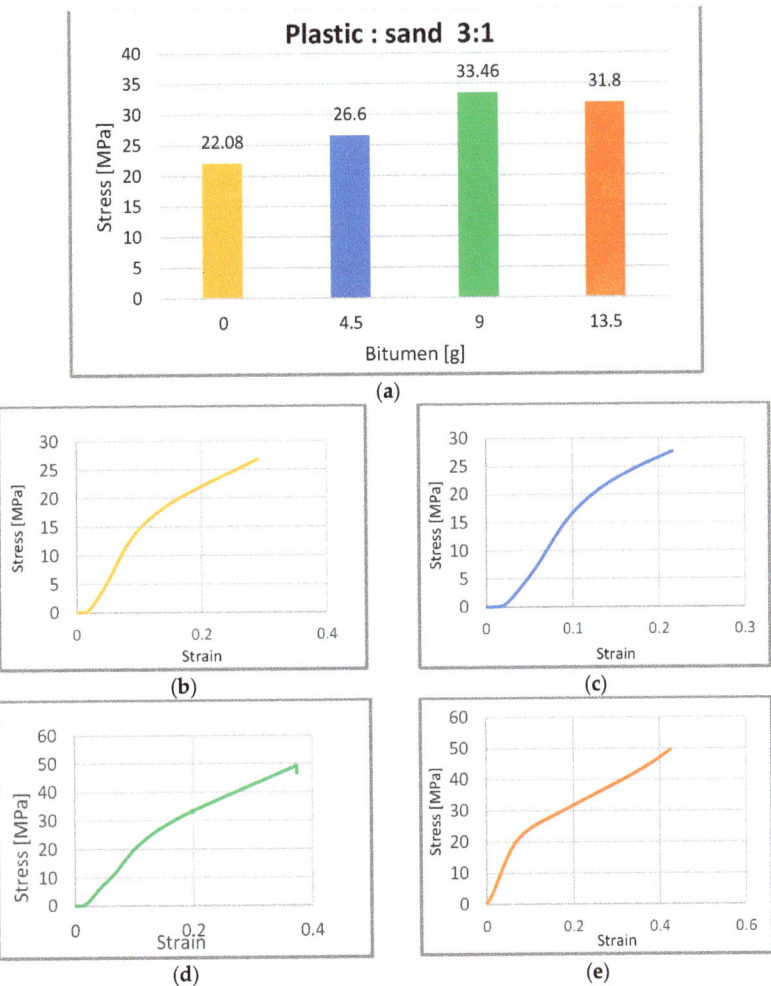

Figure 2. Results of brick sample 1. (**a**) Compression strength values of brick plastic to sand 3:1. (**b**) Stress-strain curve with 0 g bitumen. (**c**) Stress-strain curve with 4.5 g bitumen. (**d**) Stress-strain curve with 9 g bitumen. (**e**) Stress-strain curve with 13.5 g bitumen.

3.1.2. Brick Sample 2 (Plastic:Sand—3:2)

The bricks with a plastic-to-sand ratio of 3:2 produced the best results among all other bricks, with each brick having a compression strength that was almost higher than that of other bricks, as seen in Figure 3a. The value of the compression strength was 22.7 MPa if no bitumen was applied, and it grew further when bitumen was deposited, rising to 28.7 MPa and reaching a maximum of 37.5 MPa, before decreasing to 33.9 MPa as additional bitumen was added, with the optimum being 2% (i.e., 9 g bitumen).

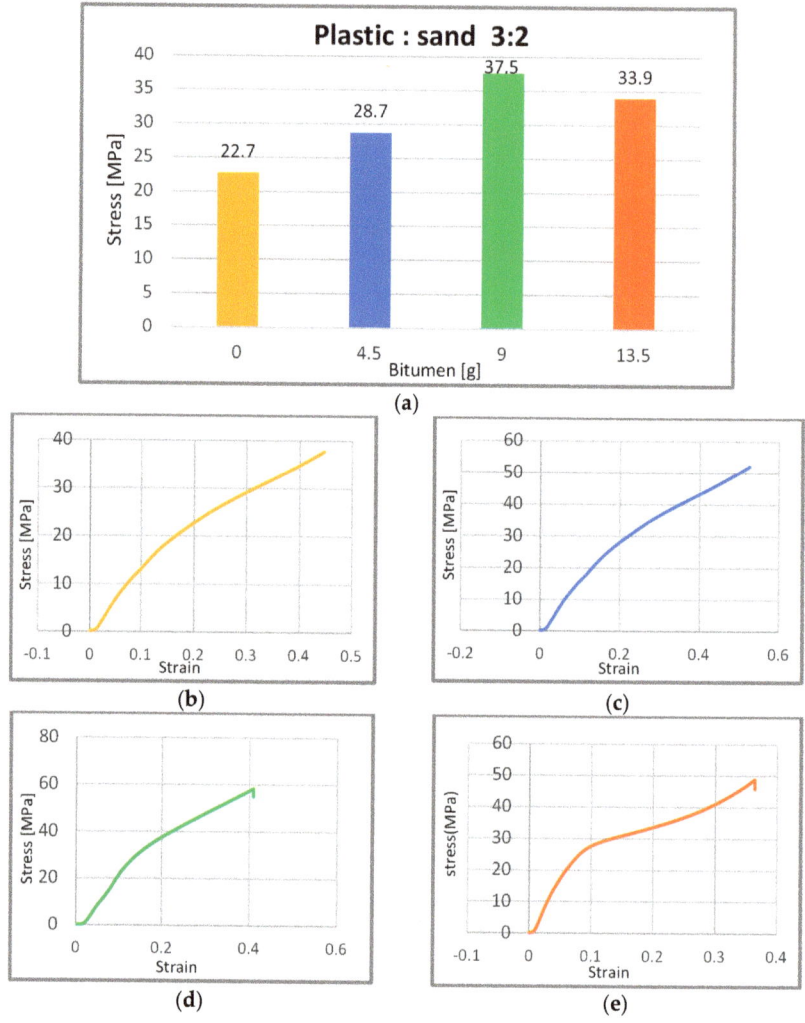

Figure 3. Results of brick sample 2. (**a**) Compression strength values of brick plastic to sand 3:2. (**b**) Stress-strain curve with 0 g bitumen. (**c**) Stress-strain curve with 4.5 g bitumen. (**d**) Stress-strain curve with 9 g bitumen. (**e**) Stress-strain curve with 13.5 g bitumen.

3.1.3. Brick Sample 3 (Plastic:Sand—1:1)

The bricks with the same percentage of plastics and sand showed average results compared to those with more plastics. Figure 4a indicates that the brick without bitumen had a compression strength of 19.39 MPa. Adding bitumen by 4.5 g and 9 g increased the value to 24.28 MPa and 29.18 MPa. The strength again dropped to 27.82 MPa as the bitumen content was increased to 13.5 g.

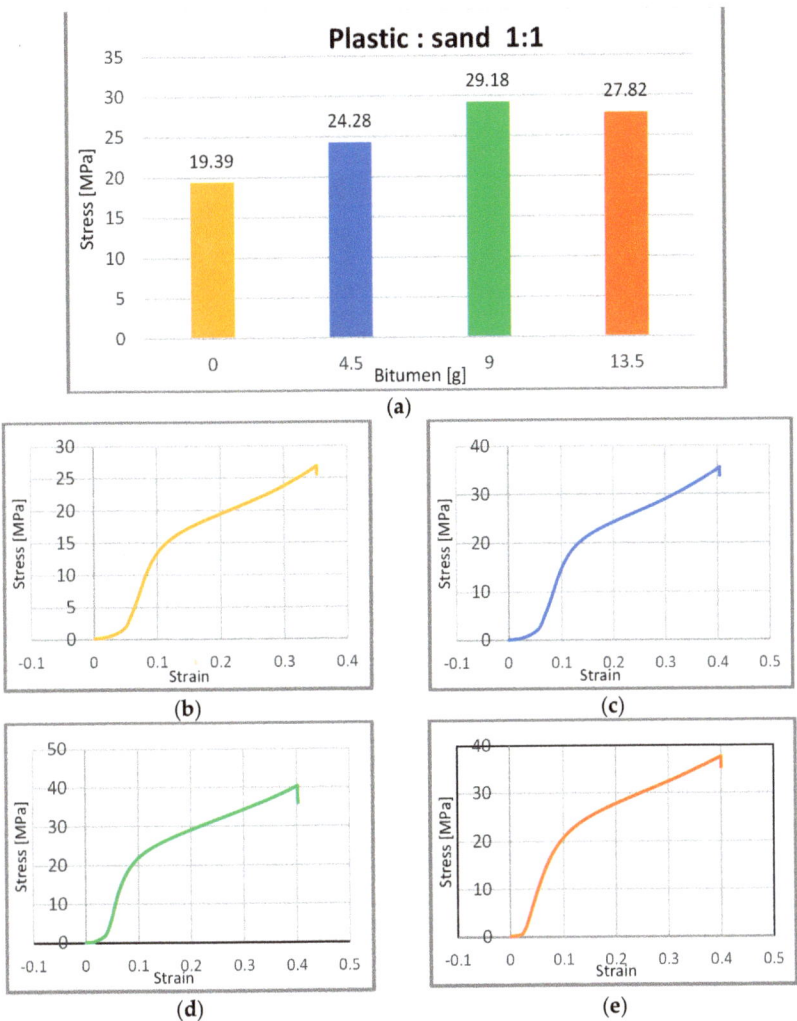

Figure 4. Results of brick sample 3. (**a**) Compression strength values of brick plastic to sand 1:1. (**b**) Stress-strain curve with 0 g bitumen. (**c**) Stress-strain curve with 4.5 g bitumen. (**d**) Stress-strain curve with 9 g bitumen. (**e**) Stress-strain curve with 13.5 g bitumen.

3.1.4. Brick Sample 4 (Plastic:Sand—2:3)

From Figure 5a, the results of the bricks with a lower plastic percentage (i.e., plastic: sand—2:3) showed less strength compared to the bricks with an equal plastic–sand ratio and the bricks with more plastic, with the highest strength being 26.49 MPa for 9 g bitumen and decreased to 25.6 MPa when the bitumen was increased again to 13.5 g.

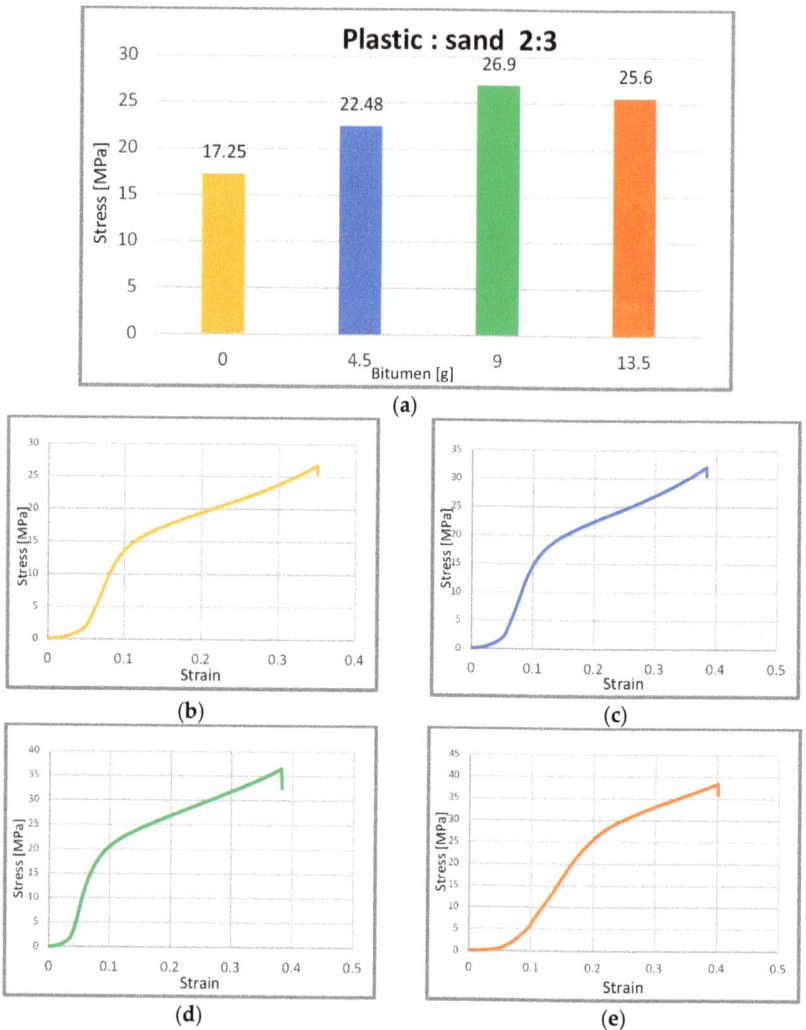

Figure 5. Results of brick sample 4. (**a**) Compression strength values of brick plastic-to-sand 2:3. (**b**) Stress-strain curve with 0 g bitumen. (**c**) Stress-strain curve with 4.5 g bitumen. (**d**) Stress-strain curve with 9 g bitumen. (**e**) Stress-strain curve with 13.5 g bitumen.

3.1.5. Brick Sample 5 (Plastic: Sand—1:3)

Out of all the bricks, the bricks with the lowest percentage of plastic (i.e., 25%) exhibited very low compression strength even when bitumen was added, which had a strength of 25.36 MPa for 9 g bitumen and the lowest being 16.8 MPa with no bitumen, as seen in Figure 6a. Even in this case, the strength declined as bitumen content increased.

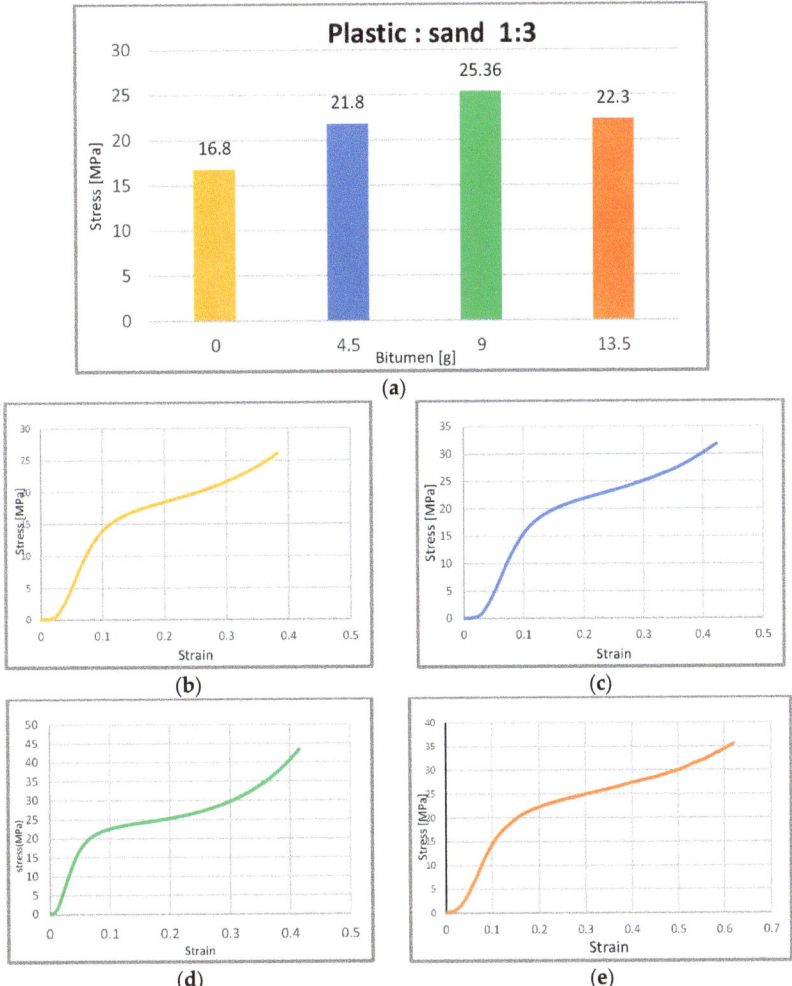

Figure 6. Results of brick sample 5. (**a**) Compression strength values of brick plastic-to-sand 1:3. (**b**) Stress-strain curve with 0 g bitumen. (**c**) Stress-strain curve with 4.5 g bitumen. (**d**) Stress-strain curve with 9 g bitumen. (**e**) Stress-strain curve with 13.5 g bitumen.

Compression strength was found to be greater in samples with 60 percent plastic, so the mean stress of these samples was calculated. The bricks were initially tested for compression, and five observations were recorded. The stress values were added and divided by the number of observations to determine the mean stress. The standard deviation was also calculated to understand the range and distribution of compression stress values. Table 5 shows the mean value of compression stress. Figure 7 shows the standard deviation values.

Table 5. Mean value of compression stress.

Plastic%	Units	0 g Bitumen	4.5 g Bitumen	9 g Bitumen	13.5 g Bitumen
60	MPa	19.644	24.772	30.48	28.284

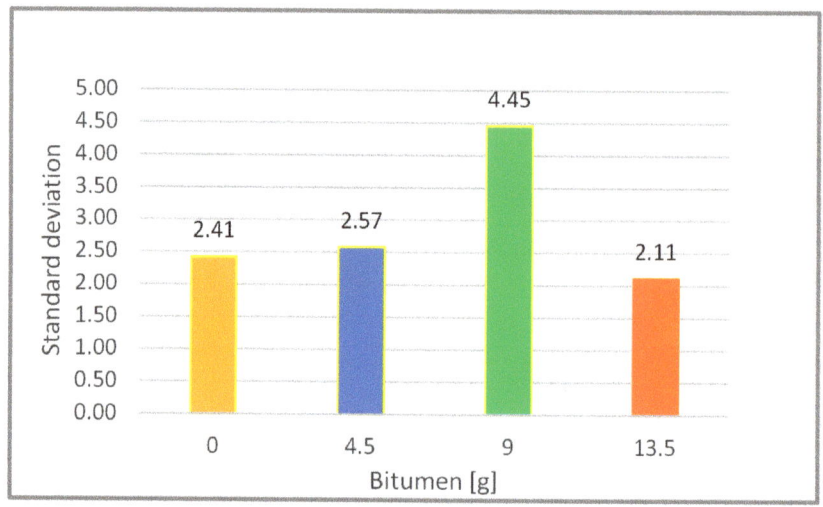

Figure 7. Standard deviation value of compression stress.

3.2. Water Absorption Test

The test was carried out by the weighing of all 20 bricks in their dry state. The weights were measured, and the bricks were submerged in water for 24 h, as shown in Figure 8a. The weights of all 20 bricks were noted after 24 h, as shown in Figure 8b. The water absorption was calculated using Formula (2). It was evident from the graph that plastic bricks had a lower water-absorption rate. Bricks with more plastics (i.e., a ratio of 3:1 plastic to sand) tend to absorb less water than bricks with fewer plastics (i.e., a ratio of 1:3). When the plastic content was higher in the bricks; there was less chance of water molecules filling the voids in the plastic–sand mixture. All plastic bricks had an overall water absorption rate of less than 1%. Figure 9 shows the water absorption rate on different composition of plastic bricks.

(a)

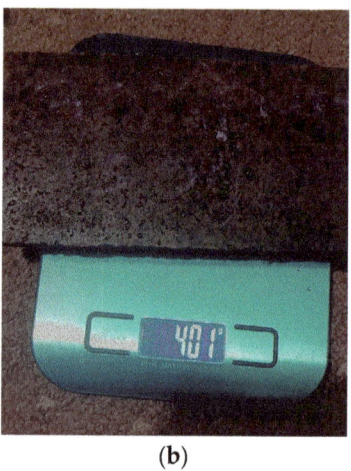
(b)

Figure 8. Performing the water absorption test. (a) Bricks immersed in water. (b) Weight check after 24 h.

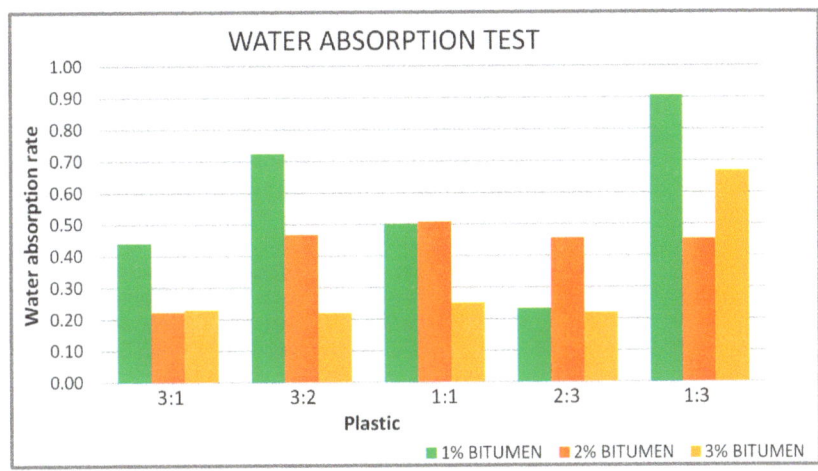

Figure 9. Water absorption rate on different composition of plastic bricks.

3.3. Efflorescence Test

The efflorescence test showed impressive results on plastic bricks with zero presence of alkalis on almost all of the bricks. Bricks with a lower percentage of plastics had fewer alkalis, whereas bricks with a higher percentage had no significant alkalis. For bricks made of 1:3 plastic and sand, all of the bricks exhibited a small number of alkalis. However, bricks made of 3:1 and 3:2 plastic showed excellent results, with no alkalis or other white particles. Table 6 shows the severity of the alkali presence on the bricks.

Table 6. Efflorescence test on different composition of plastic bricks.

Plastic: Sand	Bitumen			
	0 g	4.5 g	9 g	13.5 g
3:1	Nil	Nil	Nil	Nil
3:2	Nil	Nil	Nil	Nil
1:1	Nil	Nil	Nil	Nil
2:3	Slight	Slight	Nil	Nil
1:3	Slight	Slight	Slight	Slight

3.4. Test to Determine the Relative Rise in Temperature

This test was used to determine the relative rise in temperature by placing the bricks on a heat source. The temperature differential between the two faces of the bricks with different plastic compositions was depicted in the table. The temperature difference increased as the amount of plastic increased, indicating that heat conduction decreased. This was a result of plastics having a lower heat conductivity than sand. The study involved testing samples that did not contain bitumen to measure the temperature that rose on one side of the brick when the other side was exposed to heat. This test aimed to evaluate the thermal conductivity of the brick material. The bitumen in plastic sand bricks can make them soft and malleable at elevated temperatures. However, this was not a factor in this test, since the samples being tested did not contain bitumen. Additionally, the bitumen used in these plastic sand bricks was typically relatively small (i.e., 1%, 2%, 3%) and did not significantly contribute to temperature rise. The test was conducted over a relatively short period (i.e., 3 min). As a result, the temperature difference observed between the various samples was comparatively small. If the testing time is prolonged, the brick will eventually melt, leading to an uneven surface area at the source. This can result in inaccurate temperature measurements on the other side of the brick.

4. Discussion

The idea of using HDPE bricks instead of standard bricks was to turn waste plastic into a usable product. Using a novel manufacturing process, the primary goal of the study was to determine the ideal plastic-to-sand ratio. Brick samples can be prepared by varying the proportion of plastic in the mixture. The proportions considered were 75%, 60%, 50%, 40%, and 25% by weight. In samples where 75% plastic was used, bitumen was added in proportions of 1%, 2%, and 3% by weight. It was observed that the compression stress of the bricks increased gradually as the amount of bitumen was increased until it reached a maximum stress of 33.46 MPa. However, if the amount of bitumen added exceeded 2%, the bricks became softer. This can be observed in Figure 2a, where the stress value is decreased to 31.8 MPa. Due to its good adhesive property, bitumen was used to increase the strength of the bricks, even though the plastic in the mixture was sufficient to bind the sand. However, excessive amounts of bitumen had a negative impact on the quality of the bricks.

It was observed that the strength of bricks could be improved by varying the proportion of plastic used in the mixture. Specifically, when the amount of plastic was reduced by 15%, the stress values increased from 21.2 MPa to 22.6 MPa. The stress values decreased to 16.25 MPa when the amount of plastic was further reduced to 25%. Moreover, when 9 g of bitumen was added to a sample containing 60% plastic, the stress values reached a maximum of 37.5 MPa.

The increase in stress values with a reduction in the amount of plastic or the addition of bitumen was likely due to the changes in the physical properties of the brick. With a reduction in the amount of plastic or adding bitumen, the interlocking of the sand particles can be improved, leading to increased strength and stress values. However, adding too much bitumen may have adverse effects, as it can make the bricks softer and weaker. The mixture of 60% plastic and 9 g bitumen in sand produced the strongest results. The average stress of all the samples was found to be 25 MPa, and it was noted that the majority of stress readings deviated by an average of 2.54 MPa from the average stress. In order to find the average compression stress, a set of five observations of compression stress values were taken and mean stress was calculated. Additionally, the standard deviation was calculated to find the spread or variation in the compression values. This helped in the assessment of risk and of how much the stress values differed from the mean stress.

Bricks should absorb the least amount of water possible. As seen in Figure 9, the water absorption test demonstrated that no bricks absorbed more water than 1% after being soaked for 24 h. This is especially helpful in the construction sector, because when bricks absorb more water, damage to the building arose. This test is also valuable in areas where water leakage is a primary concern. The bricks with the highest water absorption percentage were 0.9%. The decrease in plastic content in bricks increased their water absorption rate, with most bricks having a water absorption rate of around 0.2%. However, as plastic content decreased, the water absorption rate increased, potentially reaching a rate of 0.9%. The permissible range was 1% to 2%, while zero was the best value for a brick. The 0.9% water content was caused by the tiny gaps between the granules and the quick cooling of the bricks in the hot press.

The efflorescence test (Table 6) demonstrated that there were no soluble salts or alkalis present on the bricks with higher plastic–sand ratio (i.e., 3:1, 3:2, and 1:1), but the bricks with less plastic (i.e., 2:3) showed a slight alkali presence. When bitumen percentage increased, the alkalis were reduced, and for the ratio of 1:3, all bricks showed alkalis, which demonstrated that with more plastics, the presence of alkalis decreased. The relative temperature rise shown in Table 7 demonstrated that the temperature difference was greatest for bricks that contained the most plastic and bitumen. As a result, the temperature transfer from one side of the brick to the other took longer, which could be advantageous in a fire accident. The results showed that the initial temperatures of the bricks were all around 19.2 °C. When the samples were placed on a heat source maintained at 400 °C, the temperatures at the left and right faces of the bricks after 3 min ranged from 325 °C to

338 °C and 42 °C to 62 °C, respectively. It was observed that the temperature difference increased as the amount of plastic in the bricks increased, indicating that heat conduction was decreasing.

Table 7. Temperature difference of different bricks.

Plastic %	Initial Temperature [°C]	After 3 min, Temperature at Left Face [°C]	After 3 min, Temperature at Right Face [°C]	Temperature Difference [°C]
75	19.2	325	42	283
60	19.2	328	48	280
50	19.2	327	51	276
40	19.2	330	56	274
25	19.2	338	62	276

As in Kulkarni's [17] study, bitumen was added after the HDPE and quartz sand were ground into granules. This caused the bitumen to be distributed equally, which strengthened bonds. It was discovered through several tests, including the efflorescence, water absorption, and compression tests, that HDPE bricks performed better than conventional bricks. The combination of high-density polyethylene, sand, and bitumen in this study increased the strength of the brick. The bricks' compression strength gradually improved for the plastic sand ratio, rising from 22.08 MPa with 0% bitumen to a peak of 33.46 Mpa with 2% bitumen and falling just short of 32 MPa for 3% bitumen. These results suggested that bitumen aided in fusing sand and plastic, increasing the strength of the brick. During the initial point of the compression test, no stress was observed, indicating that the specimen may have been able to undergo further compression and that a more compact specimen could be made with less thickness.

5. Conclusions

This research work inferred that recycling plastic waste for building projects was the way of the future, as it would help reduce plastic waste and the price of bricks in the construction industry. HDPE, quartz sand, and bitumen were used in this research work to make bricks, and it was found that this combination led to an improvement in the strength of the bricks while reducing their weight. Additionally, the cost of these bricks was almost 50% less than conventional bricks. The cost of one conventional brick is around 0.20 to 0.50 euros. However, the cost of one plastic sand brick is around 0.092 euros, including the raw materials cost (i.e., for 1 kg of plastics, it costs 0.33 euros; for 1 kg of quartz sand, it costs 0.248 euros; and for 2% bitumen by weight (9 g), it costs 0.057 euros). HDPE and bitumen made the bricks more water-resistant by reducing their permeability, and quartz sand improved the binding between the plastics due to its strong binding properties.

This research indicates that adding bitumen to bricks can significantly enhance their tensile strength, with a maximum value of 37.5 MPa observed. However, it is important to note that an increase in the proportion of bitumen led to a decrease in the strength of the bricks. Therefore, it is recommended to use bitumen in moderation (i.e., around 9 g to 13.5 g) to achieve optimal results. Additionally, the combination of HDPE and quartz sand ensured that the bricks were void-free and free of alkalis, making them a suitable choice for the construction industry.

Overall, bricks with a bitumen content of 2% (i.e., 9 g of bitumen) and a plastic–sand ratio of 3:2 showed better properties than other bricks. Every brick had a compression strength that was better than typical clay bricks. For use in building, homes, and pallets, bricks with a higher percentage of plastics were preferable to those with a lower percentage of plastic (i.e., 2:3 and 1:3) since the bricks with fewer plastics possessed a significantly lower compression strength and a higher presence of alkalis. Plastic bricks are the best option for construction, parking chairs, and pathway pallets since they are very light, have a good load bearing capacity, have less water absorption, are inexpensive, and can gradually reduce plastic waste.

Although this is a useful method for recycling used plastic, plastic has the potential to emit greenhouse gases when used in excessively high temperatures. Conducting this experiment in a vacuum is always advised so that the carbon atoms released during heating cannot produce carbon dioxide and carbon monoxide, which affects the environment.

Further, this formula can be developed into high-strength tiles and hallow interlocking bricks that could replace the current materials without compromising strength. It was also found that the surfaces of the bricks were smoother. Hence, this process can be used to manufacture parts that need surface lubrication to reduce friction.

Author Contributions: N.K.K. contributed with the original manuscript draft writing, data synthesis, and data gathering. J.S. aided with the report's review and research progress. The methodology was devised by Y.P.S. and N.K.K. Reviewing the report was aided by the supervision of Y.P.S. All authors have read and agreed to the published version of the manuscript.

Funding: The APC was sponsored by Hochschule Kaiserslautern, whereas no external funding was provided for this research.

Institutional Review Board Statement: Not applicable.

Data Availability Statement: Not applicable.

Acknowledgments: I thank Hochschule Kaiserslautern Pirmasens campus for financially supporting this research. I would also like to thank Hochschule Kaiserslautern, Kammgarn campus, for providing me with the resources for testing my bricks.

Conflicts of Interest: The authors declare no conflict of interest.

References

1. Gottlieb, H.M. Filling the Gaps in the Global Governance of Marine Plastic Pollution. *Nat. Resour. Environ.* **2021**, *35*, 1–5.
2. Nayeema, M. Prospects of Biodegradable Polymer (Polyhydroxyalkanoate) Production with Wastewater. Ph.D. Thesis, Brac University, Dhaka, Bangladesh, 2019.
3. Plastic Soup Foundation. Available online: https://www.plasticsoupfoundation.org/en/plastic-facts-andfigures/#:~:text=The%20amount%20of%20plastic%20that,production%20is%20a%20lot%20higher (accessed on 6 February 2023).
4. Andrady, A.L. *Plastics and Environmental Sustainability*; John Wiley & Sons: Hoboken, NJ, USA, 2015.
5. European Environment Agency. Available online: https://www.eea.europa.eu/highlights/plastics-environmental-concern (accessed on 6 February 2023).
6. Ellen Macarthur Foundation. Available online: https://ellenmacarthurfoundation.org/the-new-plastics-economy-catalysing-action (accessed on 6 February 2023).
7. Future Planet BBC. Available online: https://www.bbc.com/future/article/20210510-how-to-recycle-any-plastic (accessed on 6 February 2023).
8. Alliance to End Plastic Waste. Available online: https://endplasticwaste.org/en/our-stories/all-plastics-are-technically-recyclable-so-whats-holding-us-back (accessed on 6 February 2023).
9. Sulyman, M.; Haponiuk, J.; Formela, K. Utilization of recycled polyethylene terephthalate (PET) in engineering materials: A review. *Int. J. Environ. Sci. Dev.* **2016**, *7*, 100. [CrossRef]
10. Wahid, S.A.; Rawi, S.M.; Desa, N.M. Utilization of plastic bottle waste in sand bricks. *J. Basic Appl. Sci. Res.* **2015**, *5*, 35–44.
11. Agyeman, S.; Obeng-Ahenkora, N.K.; Assiamah, S.; Twumasi, G. Exploiting recycled plastic waste as an alternative binder for paving blocks production. *Case Stud. Constr. Mater.* **2019**, *11*, e00246. [CrossRef]
12. Circular Action Hub. Available online: https://www.circularactionhub.org/plastics-could-help-build-a-sustainable-future-heres-how/?lang=id (accessed on 5 January 2023).
13. Plastic Waste. Our World in Data. Available online: https://ourworldindata.org/grapher/plastic-waste-polymer (accessed on 6 February 2023).
14. Silviyati, I.; Zubaidah, N.; Amin, J.M.; Supraptiah, E.; Utami, R.D.; Ramadhan, I. The Effect of Addition of High-Density Polyethylene (HDPE) as Binder on Hebel Light Brick (celcon). *J. Phys. Conf. Ser.* **2020**, *1500*, 012083. [CrossRef]
15. Wani, T.; Pasha, S.A.Q.; Poddar, S.; Balaji, H.V. A Review on the use of High Density Polyethylene (HDPE) in Concrete Mixture. *Int. J. Eng. Res. Technol.* **2022**, *9*, 861–864.
16. Sahani, K.; Joshi, B.R.; Khatri, K.; Magar, A.T.; Chapagain, S.; Karmacharya, N. Mechanical Properties of Plastic Sand Brick Containing Plastic Waste. *Adv. Civ. Eng.* **2022**, *2022*, 8305670. [CrossRef]
17. Kulkarni, P.; Ravekar, V.; Rao, P.R.; Waigokar, S.; Hingankar, S. Recycling of waste HDPE and PP plastic in preparation of plastic brick and its mechanical properties. *Clean. Mater.* **2022**, *5*, 100113. [CrossRef]
18. Maneeth, P.D.; Pramod, K.; Kumar, K.; Shetty, S. Utilization of waste plastic in manufacturing of plastic-soil bricks. *Int. J. Eng. Res. Technol.* **2014**, *3*, 530–536.

19. Benny, T.K.; Wilson, A.J.; Krishn, B.D.; Nair, A.M.; Rasheed, J.K.J. Asphalt Plasto A Brick from Waste Plastic. *Int. J. Eng. Res. Technol. (IJERT)* **2019**, *8*, 222–227.
20. Favaro, S.L.; Pereira, A.G.B.; Fernandes, J.R.; Baron, O.; da Silva, C.T.P.; Moisés, M.P.; Radovanovic, E. Outstanding impact resistance of post-consumer HDPE/Multilayer packaging composites. *Mater. Sci. Appl.* **2016**, *8*, 15–25. [CrossRef]
21. Beidoou. Available online: https://www.beidoou.com/materials/what-is-quartz-sand.html (accessed on 6 February 2023).
22. Soumya, G.; Karthiga, S. Study on mechanical properties of concrete using silica fume and quartz sand as replacements. *Int. J. Pure Appl. Math.* **2018**, *119*, 151–157.
23. Civil Work Study. Available online: https://www.civilworkstudy.com/2020/05/bitumen-properties-and-its-detail-why.html (accessed on 6 February 2023).
24. Sarkar, D.; Pal, M. A laboratory study involving use of brick aggregate along with plastic modified bitumen in preparation of bituminous concrete for the roads of Tripura. *ARPN J. Eng. Appl. Sci.* **2016**, *11*, 570–576.

Disclaimer/Publisher's Note: The statements, opinions and data contained in all publications are solely those of the individual author(s) and contributor(s) and not of MDPI and/or the editor(s). MDPI and/or the editor(s) disclaim responsibility for any injury to people or property resulting from any ideas, methods, instructions or products referred to in the content.

Article

Influence of Steel Fiber and Carbon Fiber Mesh on Plastic Hinge Length of RCC Beams under Monotonic Loading

Pradeep Sivanantham [1,*], Deepak Pugazhlendi [1], Beulah Gnana Ananthi Gurupatham [2,*] and Krishanu Roy [3,*]

[1] Department of Civil Engineering, SRM Institute of Science and Technology, Chennai 603203, India
[2] Division of Structural Engineering, College of Engineering Guindy Campus, Anna University, Chennai 600025, India
[3] School of Engineering, The University of Waikato, Hamilton 3216, New Zealand
* Correspondence: pradeeps@srmist.edu.in (P.S.); beulah28@annauniv.edu (B.G.A.G.); krishanu.roy@waikato.ac.nz (K.R.)

Abstract: The most susceptible area of a structural member, where the most inelastic rotation would take place, is the plastic hinge. At this stage, flexural elements in particular achieve their maximal bending flexibility. This study uses finite element analysis (FEA) and experimental inquiry to analyze and test the effects of carbon fiber mesh jacketing and steel fiber reinforcement at the concrete beam's plastic hinge length subjected to a vertical monotonic load. The compressive strength, split tensile strength, and flexural strength tests are used to evaluate the mechanical qualities, such as compressive strength and tensile strength, of M25 grade concrete that is used to cast specimens. While conducting this analysis, seven different parameters are taken into account. After the conventional concrete beam has been cast, the steel-fiber reinforced beam is cast. Several empirical formulas drawn from Baker, Sawyer, Corley, Mattock, Paulay, Priestley, and Park's methods were used to calculate the length of the beam's plastic hinge. Finally, the steel fiber was inserted independently at 150 mm into the concrete beam's plastic hinge length mechanism using the techniques described by Paulay and Priestley. The analytical and experimental results are compared. The results obtained from the investigations by applying monotonic loads to the beam show that fibers used at specific plastic hinge lengths show a 41 kN ultimate load with 11.63 mm displacement, which is similar to that of conventional beam displacement, and performance. Meanwhile, the carbon fiber mesh wrapped throughout the beam behaves better than other members, showing an ultimate load of 64 kN with a 15.95 mm deflection. The fibers provided at the plastic hinge length of the beam perform similarly to those of a conventional beam; eventually, they become economical without sacrificing strength.

Keywords: steel fiber reinforced concrete beam; carbon fiber mesh; jacketing; monotonic loading; plastic hinge length

Citation: Sivanantham, P.; Pugazhlendi, D.; Gurupatham, B.G.A.; Roy, K. Influence of Steel Fiber and Carbon Fiber Mesh on Plastic Hinge Length of RCC Beams under Monotonic Loading. *J. Compos. Sci.* **2022**, *6*, 374. https://doi.org/10.3390/jcs6120374

Academic Editor: Francesco Tornabene

Received: 12 October 2022
Accepted: 30 November 2022
Published: 6 December 2022

Publisher's Note: MDPI stays neutral with regard to jurisdictional claims in published maps and institutional affiliations.

Copyright: © 2022 by the authors. Licensee MDPI, Basel, Switzerland. This article is an open access article distributed under the terms and conditions of the Creative Commons Attribution (CC BY) license (https://creativecommons.org/licenses/by/4.0/).

1. Introduction

The strengthening and restoration of existing structures is a hot issue these days, attracting the attention of both researchers and engineers. The possibilities presented by new materials and technologies ideal for granting the best performance of existing structures have given a new pulse to the research of matrix composites such as fiber-reinforced polymers (FRP) and fiber-reinforced concrete (FRC) in this context. The most widely used construction material in the world is concrete. It is a composite material mainly composed of materials like cement, fine aggregate, coarse aggregate, and water. These materials are mixed to get a fluid material that is cast into different shapes according to need. The materials used in concrete have different physical properties. Under compression, concrete behaves well, whereas it is weak under tension. To improve the behavior of concrete under tension, reinforcements are provided to the concrete member. To enhance the quality of concrete, different additives and admixtures are added to the concrete depending upon the need. Enhanced durability and strength of concrete can be achieved by changing ingredients like cement and aggregate. There are different grades of concrete depending upon the mixed proportions of the materials used and also its strength. The concrete is prepared according to the mix designs of different grades. They are cast into different shapes according to their needs. The concrete gets hardened after which the concrete is allowed for the curing process for different time durations like 3 days, 1 week, 2 weeks, and 4 weeks. RCC is a type of concrete where reinforcement is provided to the cement-aggregate matrix to increase its tensile strength and durability. Steel reinforcing bars, otherwise called rebars, are used mostly as reinforcement in concrete to improve its behavior in the tension zone of concrete. Different members of a structure, like slabs, beams, columns, and foundations, are reinforced with rebar to counteract different forces like axial, compression, shear, flexure, and torsion. Meanwhile, fiber mesh strengthening also influences the behavior of members by influencing load and displacement. Excellent strength and modulus along with the strength-to-weight ratio of carbon fiber mesh take advantage of other mesh such as glass fiber mesh, basalt fiber mesh, and kevlar mesh in strengthening characteristics.

Carbon fiber mesh is widely used in many industries because of its high strength, low density, and thinness. It also does not increase the self-weight of reinforced components or cross-section size. Jacketing is the technique of strengthening the reinforced cement concrete (RCC) columns that have weakened over a period of time as a consequence of poor maintenance or weather-related conditions. Other issues that arise during the building phase include design flaws, poor concrete manufacturing, and sloppy execution procedures. For this reason, the study is devoted to the definition of a "ductile" FRC material, characterized by limited softening, or better yet, by plastic behavior by achieving ultimate strain. The composite material is then applied as a thin layer on reinforced concrete (RC) beams, and the strength and ductility characteristics of the structure are evaluated with analytical and experimental models.

The mechanical behavior of concrete, such as concrete strength, split tensile strength, and flexural strength, can be enhanced by adding different types of fibers to the concrete mix. The tensile strength of the mix can be enhanced by the usage of fibers. It also resists crack growth in the concrete members. The major classification of fibers is natural and synthetic. The most effectively used synthetic fibers are steel fiber, glass fiber, polypropylene fibers, etc. Steel fiber comes in a variety of shapes and sizes, including hooked edge, flat edge, and hooked flat edge. Hooked edge and hooked flat edge fibers offer greater benefits than the others. Furthermore, fibers with a high aspect ratio have a greater capability for energy absorption. Steel fiber reinforced concrete increases the ductility and mechanical qualities of the concrete by reducing its brittleness. The kind and volume of steel fiber used in the concrete determine the mechanical qualities of steel fiber reinforced concrete (SFRC). The fiber dose ranges from 0% to 2% of the total volume of concrete. After modelling, the first phase of the experiment examined the mechanical properties of both conventional and SFRC concrete [1,2]. The length of a plastic hinge can be calculated using a variety of expressions. In this investigation, Priestley and Park's expressions are employed. Instead of employing steel fiber across the slab, the fiber can be used solely along the length of the plastic hinge, as it offers similar resistance. It will cut down on the use of steel fiber. Findings of the tests says, giving steel fibers along the plastic hinge length (PHL) is more effective and cost-effective for practical application. The largest bending moment would occur at the plastic hinge length, which is thought to be the seriously damaged section of the RCC member and will suffer increased inelastic deflection in the member [3,4]. The length of plastic hinges has drawn more interest from researchers in recent years. Experimental research on plastic hinge length is used to determine how well it can bend while supporting a load [5]. When the RCC element is loaded, it deforms in an inelastic manner, creating zones where the bending moment is higher than in other locations. As the applied stress grows, the zones have a propensity to rotate until the ultimate hinge is formed, which would result in the structural component collapsing. The length of the rebar-yielding zone is thought to represent the upper bound of the three physical inelastic deformation zones since it has a value that has never exceeded twice the effective depth of the cross-section in any of the conditions evaluated in this study. The diameter of the rebar under stress has a diminished influence on the member's plastic hinge length and flexural capacity as a result of its effect on bond strength. The genuine plastic hinge zone is much smaller than the member's yielding zone, which accounts for the majority of the plastic rotation. None of the empirical models for forecasting plastic hinge length that are currently available have taken into account all of the important variables that affect the result [6,7].

Numerous examinations have already been conducted on various RC member kinds (beams, columns, and walls). Following that, member geometry and mechanical properties were used to represent the deformations of RC members at yielding or failure under cyclic loading. In general, although with a sizable dispersion, the yield and final curvature calculations based on the plane-section assumption agree well with the test results. Additionally, it has been demonstrated that models with curved surfaces, ultimate drift, and chord-rotation capability are accurate. The rotational capacity, the load–displacement behavior, and the reinforcing strain and stress distributions were all taken into account in a computer model that was developed and validated using available experimental data. The concrete crush zone, curvature localization zone, and genuine plastic hinge length are all studied using the calibrated FEM model (PHL). The plastic hinge length is examined using parametric studies in relation to reinforcement, concrete mechanical characteristics, element size, and flexural modulus. This FE model can be used to perform a simulation analysis of the plastic hinge zone in RC members, from which plastic hinge length can be calculated. It is concluded that the plastic hinge length should be designed properly with reinforcement to make the structural elements more ductile.

According to this article, large bond stresses exist in zero shear zones, and overall shear and local bond pressures are directly related. Deformed bars without hooks of the kind and embedment utilized in this series produced bond strengths high enough to cause bar fracture rather than bond failure in all specimens [8]. Mattock's formulas overestimate hinge lengths by a large amount. In many beams, Sawyer's calculations overstate the hinge lengths [9]. Although in some formulas, the lengths of plastic hinges are overestimated, the forecasts are more accurate than the experimental measurements. For beams with axial loads, Park formulas provide reasonable estimates. The hinge span of all flexural member are underestimated by the Baker and Amerakone formulas. Several expressions have been offered for estimating the corresponding plastic hinge lengths of flexural member and axial member, with Corley, Mattock, Paulay, Priestley, Panagiotakos, and Fradis also proposing some of the expressions [10]. The results of Paulay and Priestley, Panagiotakos, and Fradis indicate a similar trend to those of Finite Element Modelling. Mattock's expression was created with monotonic loading in mind, whereas the others were created with cyclic loading in mind [11].

The experimental examination on the mechanical properties of high-strength concrete used various combinations of steel fibers together with steel bar reinforcement. According to our experimental examination, the kind and volume of steel fiber employed in the mix design affects the rise in compressive strength, tensile strength, shear, flexural toughness, and resistance [12]. The behavior of slabs in terms of flexure due to the addition of fibers was studied. The ratio of steel fibers used in this experimental investigation is 0.5%, 1%, and 1.5%, respectively. This project concludes that the fibers with a high aspect ratio provide more energy absorption capacity, similar to the studies conducted by previous researchers [13–16]. The length of the plastic hinge zone is a significant design parameter that should be closely regulated in order to maximize the flexibility of the member and make it capable of withstanding severe events like earthquakes.

The behavior of plastic hinges is highly complicated because of the materials' significant nonlinearity, contact, relative movement between the component materials, and strain localization. As a result, the majority of researchers used experimental testing to investigate the problem. Using experimental and computational investigations, it is determined how fiber clustering affects the fatigue behavior of steel-fiber-reinforced concrete (SFRC) beams with reinforcements. In order to comprehend the underlying process better, an experimental research of the fiber dispersion in the concrete cross-section was carried out. The findings indicated that the fatigue life of the beam rose as the percentage of fiber volume grew. The shear strength of steel fiber-reinforced concrete beams has been predicted using a mechanics-based mathematical model that considers the effects of all shear-resisting mechanisms (without transverse reinforcement). When the suggested model's efficacy was evaluated using a range of datasets, it shown strong correlations with experimental results, with a mean, standard deviation, and coefficient of variation of 0.94, 0.22, and 22.99 percent, respectively. It has also been postulated how each shear-resisting mechanism contributes. The application of SFRC improves the flexural and cyclic responses of reinforced concrete bridge deck slabs [17]. In other cases, cyclic loads were applied to a plain concrete slab, two SFRC slabs reinforced with mill-cut steel fibres and corrugated steel fibres, respectively [18]. Concrete used in building is made with regular Portland cement (OPC). Due to its low cost and easy availability of raw materials, it is the most widely used construction material. OPC manufacturing, however, necessitates calcareous and argillaceous ingredients and is energy-intensive. Calcination and the burning of fossil fuels are the primary causes of greenhouse gas emissions during the manufacturing of OPC [19,20].

A solid 3D shell element 190 with finite strain was utilized in a finite element study for the composite material IM7/8552 to assess the validity of the finite element method. How pre- and post-failure material nonlinearity in composite materials functions has been discussed. It was discovered that IM7/8552 failed as a result of the orthotropic features of its material nonlinearity. The solution showed optimal and precise convergence, and the finite element analysis findings were successfully confirmed [21]. Finite element software ABAQUS 6.13 is used to run complementary FE simulations using RVE in order to rationalize the full set of micromechanical models that were described in the previous section. In the micromechanics-based technique for the efficient determination of the elastic characteristics of fiber-reinforced polymer matrix composites, the outcomes of FE simulations are compared with experimental observation [22]. The results of 15 push-off tests provided a precise evaluation of the longitudinal splitting characteristics of a concrete slab in a distinctive steel-concrete composite beam with headed shear stud connections [23]. Concrete's strength is decreased when PEG-600% content rises with concentration. As a consequence, 0.5% to 1% PEG-600 employed as an internal curing agent in concrete increases its effectiveness [24–26].

Strain, load-deflection responses, cycles of deformation, and fracture growth was examined. The results show that adding SFRC to deck slabs improves cyclic deformation behavior, decreases residual strain in the slab section, and improves crack behavior by lowering residual fracture breadth and raising cracking stiffness. The mechanical characteristics of conventional and fiber-reinforced concrete, as well as the calculation of plastic hinge length, are the only topics covered by the prior study. The utilization of steel fiber-reinforced concrete and fiber mesh along the length of the beam's plastic hinge has not been studied. This work addresses this problem by applying fiber and mesh internally and externally on the specific plastic hinge length obtained from various expressions which improve the performance of the Beam under bending.

2. Plastic Hinge Length (PHL)

When the load is applied to a structural member, the member undergoes bending. When the applied load is increased further, the structure changes from elastic behavior to plastic behavior at a particular moment value called the plastic moment. When a plastic moment is reached in the member, the plastic hinge produced in the structural element. This plastic hinge allows large inelastic rotations to occur in the structure. These rotations make the structure change into a mechanism and make it fail without any warning. The inelastic rotation occurs at a particular length during the application of load, and it is called the plastic hinge length. Table 1 shows that the empirical formula derived by Baker, Sawyer,

Coreley, Mattock, Park, Pristley, Paulay, Fardis, and Panagiotakos developed expressions that may be used to calculate the non-elastic hinge span of members.

Table 1. Empirical formula to derive PHL.

Description	Empirical Formula
Baker—1956	$k(z/d)^{1/4}d$
Herbert and Sawyer—1964	$0.25d + 0.075z$
Corley—1966	$0.5d + 0.2\sqrt{d}(z/d)$
Priestley and Park—1987	$0.08z + 6d_b$
Paulay and Priestley—1992	$0.08z + 0.022d_b\, f_y$

The following notation should be taken into consideration: d = effective depth of beam or column; db = diameter of longitudinal reinforcement; fy = yielding stress of reinforcement; and z = distance from critical section to point of contra flexure

The Paulay and Priestley formula ($0.08z + 0.022d_b\, f_y$) was used to calculate the plastic hinge lengths of beams. Paulay and Priestley equations were chosen since they give peak plastic hinge span and consider both the span and size of the reinforcement of the beam. The plastic hinge span was determined to be 150 mm. (Paulay and Priestley) [3].

3. Materials and Specifications

3.1. General

The experimental inquiry makes use of the following materials:

i. 53 graded Ordinary Portland cement is used in this study with a specific gravity of 3.14 according to the Indian standard IS 12269 (1987) [27] for conventional concrete.

ii. According to IS 383:1970 [28], M-sand is added as the fine aggregate of zone II for conventional concrete that passes through 4.75 mm and has a specific gravity of 2.6.

iii. Coarse aggregate, 20 mm in diameter in dry condition with a specific gravity of 2.69 was used for casting concrete.

iv. Water is a major component in concrete because it is responsible for the workability of the concrete. Portable water meeting requirements as per IS 456-2000 [29] is used for casting and curing.

v. A fabric-reinforced bidirectional carbon mesh is designed to be field installed with a cementitious matrix to create an FRCM as a composite system for structural reinforcement applications. In particular, it makes beams and columns more flexible structurally. Epoxy (Ly556) and hardener (HY951) were employed in a ratio of 1 (hardener) to 4 to bond carbon fiber mesh to concrete (epoxy). Figure 1 depicts the carbon fiber mesh that was utilized in this work, and Tables 2–4 provide material parameters for concrete, carbon fiber mesh, and steel fibers that will be used in ABAQUS [30]. Table 5 displays the mix ratio and the material proportions.

vi. In this investigation, steel fiber with hooked ends was employed. The steel fiber with the hooked end was chosen among the many fiber kinds because, as was already explained, it is utilized to increase strength and offer additional anchoring in the concrete. The steel fiber utilized had a diameter of 0.50 mm and a length of 30 mm. The steel fiber utilized has a 60 aspect ratio. The steel fiber utilized in this investigation is seen in Figure 2.

Figure 1. Carbon fiber mesh.

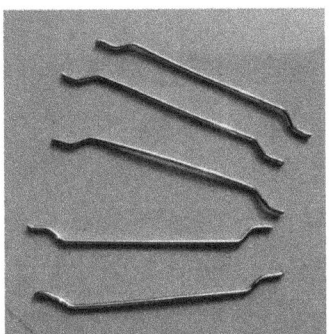

Figure 2. Steel fiber.

Table 2. Concrete properties.

Properties	Conventional M25	SFRC M25
Density	2.3×10^{-9}	2.3×10^{-9}
Youngs Modulus (MPa)	25,000	29,025
Poisson Ratio	0.2	0.2

Table 3. Carbon fiber mesh properties.

Description	Properties
Density	1750 kg/m^3
Young's modulus of elasticity	230,000 Mpa
Poisson's Ratio	0.3
Fiber weight	200 g/m^2
Tensile Strength	2500 Mpa
Thickness	0.048 mm

Table 4. Steel fibers properties.

Description	Properties
Density	7.85×10^{-9}
Young's modulus of elasticity	200,000 Mpa
Poisson's Ratio	0.3
Yield Stress	450
Plastic Strain	0.3
Geometry	linear

Table 4. Cont.

Description	Properties
Length	30 mm
Diameter	0.5 mm
Length/Diameter	1/60

Table 5. Mix design.

Component	Quantity (kg/m³)
Cement	338.18
Fine Aggregate	723.96
Coarse Aggregate	1141.09
Water/cement	0.55
Mix ratio	1:2.14:3.37

3.2. Mix Proportions

For the target strength M25, the mix proportions chosen for the nominal mix with cement shown in Table 5. For each mix, 9 cubes of 150 mm are casted for the determination of compressive strength. Similarly, 9 cylinders of dimension 150 × 300 mm are casted for the determination of split tensile strength. Vibrators were used to mix, pour, and compress the concrete. The specimen was then taken out of the mold after 24 h and allowed to cure for 7, 14, and 28 days in a curing tank.

3.3. Specimen Details

A cube specimen of proportion 150 mm was made for conventional concrete and SFRC with a 1.5% dosage of the weight of concrete. The sample of the cube specimen cast is represented in Figure 3a. A cylinder specimen of diameter 150 × 300 mm was cast for conventional concrete and SFRC with a 1.5% fiber dosage of the weight of concrete. The sample of the cylinder specimen cast is represented in Figure 3b. A flexure beam specimen of size 500 × 100 × 100 mm was cast for both conventional and SFRC with a 1.5% fiber dosage weight of concrete. The sample the of beam specimen cast is represented in Figure 3c.

Figure 3. Test specimens (**a**) Cube specimens (**b**) Cylinder specimens (**c**) Flexure beam specimen.

3.4. Beam Specimens

A reinforced concrete beam of size 1500 × 150 ×× 150 mm was cast according to IS code (IS 456:2000) with reinforcements. Seven different beams are cast with varying parameters like plastic hinge length and carbon fiber mesh jacketing. Based on the mechanical properties of the conventional concrete and SFRC, a differentiation study was made. The steel fiber reinforced concrete with a fiber dosage of 1.5% of the total weight of the concrete shows increased mechanical properties than other fiber percentages. Hence, beams are cast with a 1.5% fiber dosage to the total weight of concrete. Researchers such as Baker, Sawyer, Corley, Mattock, Park, Pristley, Paulay, fradis, and panagiotakos expressions developed by them are used to calculate the lengths of plastic hinges on beams. Seven different beams are cast by varying their parameters as shown in Figure 4a conventional beam, Figure 4b carbon fiber mesh at the plastic hinge length of a conventional beam, Figure 4c carbon fiber mesh jacketing for the total length of a control beam, Figure 4d steel fiber reinforced concrete SFRC beam, Figure 4e SFRC only at the plastic hinge length (PHL) of a beam, Figure 4f SFRC and carbon fiber mesh jacketing at the plastic hinge length of a beam, and Figure 4g carbon fiber mesh (CFM) jacketing for total length with SFRC at the plastic hinge length of a beam, respectively.

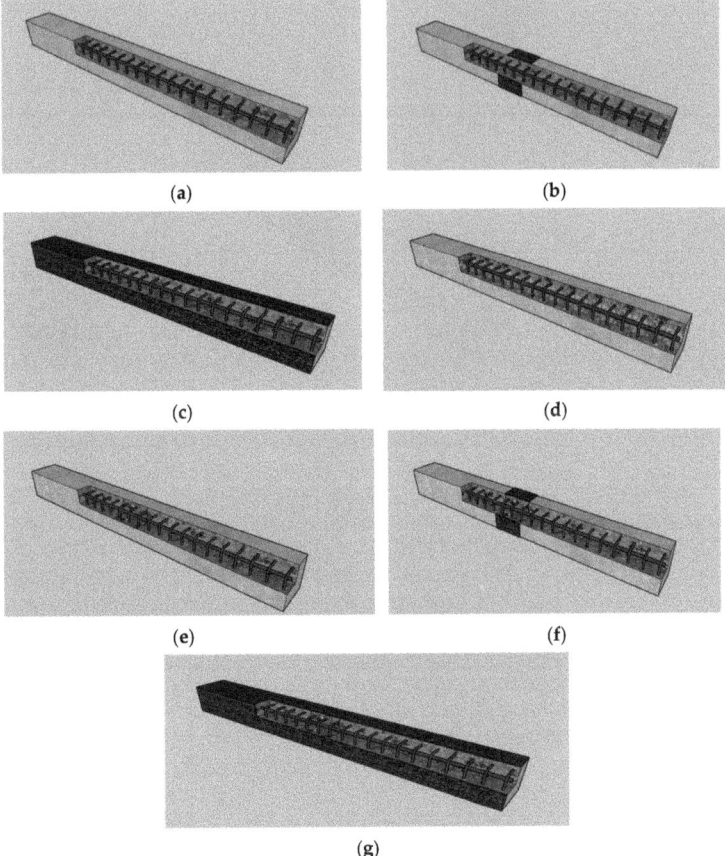

Figure 4. Schematic representation of beam specimens. (**a**) Conventional beam. (**b**) Conventional beam with CFM at PHL. (**c**) Conventional beam with CFM for full length. (**d**) Steel fiber reinforced concrete (SFRC). (**e**) SFRC only at PHL. (**f**) SFRC and CFM at PHL. (**g**) Carbon fiber mesh (CFM) jacketing for total length with SFRC at plastic hinge length.

4. Analytical Investigation

Figure 5 depicts the design of a 1500 mm span, 150 mm wide, and 150 mm deep beam using the standard finite element (FE) software ABAQUS, version 2020 [30]. The material properties of M25 for concrete and Fe415 for steel were applied. A carbon fiber jacketing technique was applied for the total length of a beam. The carbon fiber mesh model was created as a shell type. For the longitudinal reinforcement, 12 mm in diameter and stirrups of 8 mm in diameter were used. The reinforcement was placed with a cover provided with 25 mm on each side of a beam as shown in Figure 5. The concrete, reinforcement, and stirrups are given constraints of the embedded region for interaction with each other. The point load of 30 kN concentrated force and the boundary condition with both ends fixed are applied.

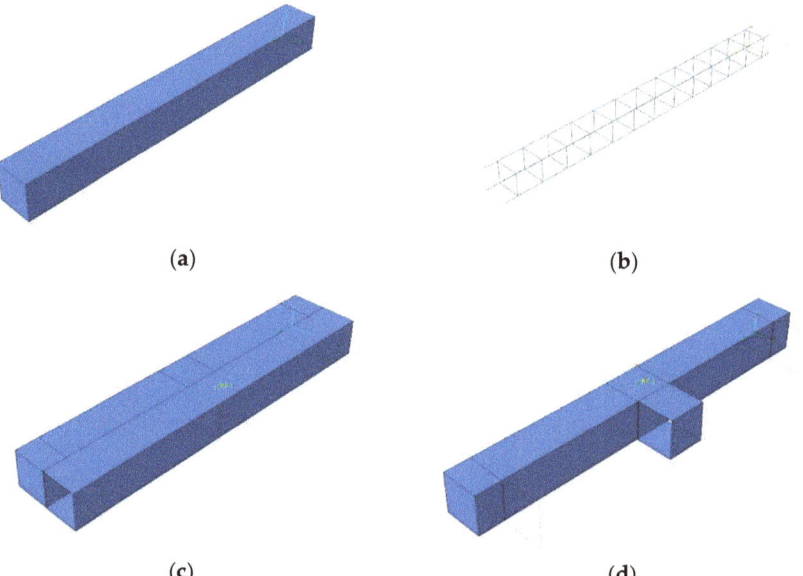

Figure 5. Details of the finite element model for RC beams. (**a**) Conventional beam. (**b**) Reinforcement details. (**c**) Beam with carbon fiber mesh jacketing for the total length. (**d**) Beam with carbon fiber mesh at plastic hinge length.

Convergence Study

The convergence study was analyzed by applying different mesh sizes with the same load. The mesh sizes of 200 mm, 150 mm, 100 mm, 75 mm, 50 mm, 40 mm, 30 mm, 20 mm, and 10 mm were applied and analyzed for convergence. By taking the results of load and displacement for each size, the convergence graph is drawn as shown in Figure 6. In the graph, when the linear pattern of the line moves, the value is noted. The linear pattern of line forms from 75 mm to 20 mm mesh size and the convergence value of 75 mm are chosen for the study.

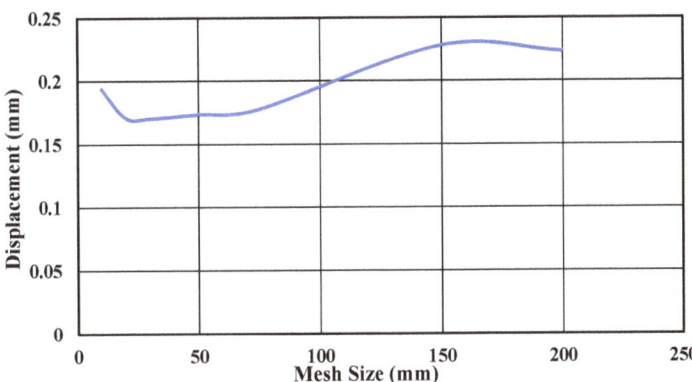

Figure 6. Results of the convergence study.

5. Experimental Investigation

5.1. General

Concrete has the ability to bear compressive forces on its own. In this research work, cube specimens of 150 mm are cast to test crushing strength. The casting was performed for all of the combinations with varying steel fiber dosages. The cubes are cured for the period of seven, fourteen, and twenty-eight days, respectively. The cubes were examined in compression testing equipment after curing (CTM), as shown in Figure 7a. The split tensile strength of ordinary concrete was measured using cylinders measuring 150 × 300 mm. The cylinder was cured for 7, 14, and 28 days. The cylinders are examined in compression testing equipment after curing (CTM) as shown in Figure 7b. For determining the flexural strength of conventional concrete and SFRC, beams of size 500 × 100 × 100 mm were cast. The testing of the beam is done in the compression testing machine (CTM) after the curing period, as shown in Figure 7c.

(a)   (b)

Figure 7. *Cont.*

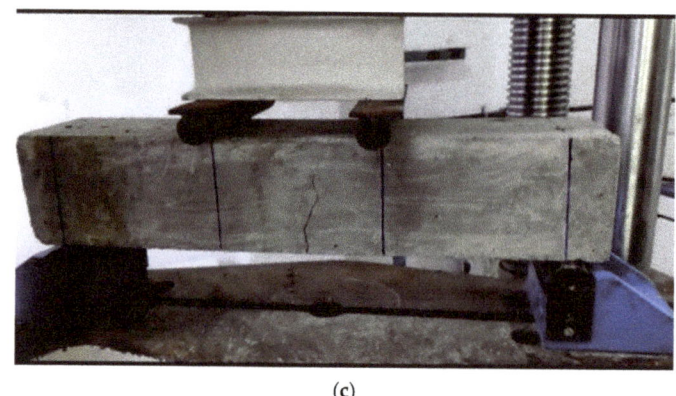

(c)

Figure 7. Testing of specimens. (**a**) Compressive strength test. (**b**) Tensile strength test. (**c**) Flexural strength test.

5.2. Modulus of Elasticity

The modulus of elasticity was calculated by applying uniaxial compression to the cylinder specimen and measuring the deformations with a dial gauge positioned between the 200 mm gauge length, as illustrated in Figure 8. The test was carried out using a compressometer in line with IS 516-1959 [31]. The cylinder specimens were installed on a compression testing equipment, and a consistent load was applied until the cylinder collapsed. The target load and deflection are taken into consideration.

Figure 8. Compressometer test for lateral strain and linear strain measurements.

The deflection values were determined as the strain value based on the length change. The strain is calculated by dividing the applied load by the cylinder's cross-sectional area, and the stress is calculated by multiplying the dial gauge readings by the gauge length. The deformation of various loads was measured in order to compute the Young's modulus of concrete for both the conventional and SFRC specimens. The findings are graphically displayed versus the tension. The computed modulus of elasticity for conventional concrete with SFRC is 25.47 N/mm^2 and 29.025 N/mm^2, as indicated in Table 6. Analytical modelling inputs are based on the outcomes of the test.

Table 6. Modulus of elasticity of specimens.

S.No	Specimen	Modulus of Elasticity (MPa)
1	Conventional Plain Cement Concrete	25.47
2	Steel—Fiber Reinforced concrete	29.03

5.3. Monotonic Loading on Beams

The 200 kN load cell is fastened to the self-straining testing frame by a hydraulic jack, and the beam is hinged on both sides. As seen in Figure 9, the repeated loading is delivered at the middle of the beam's top. To measure the deflection of the slab under loading until the final failure cracks occur, a mechanical dial gauge is mounted below the loading region of the beam.

Figure 9. Experimental setup for testing of beam.

6. Result and Discussion

6.1. Stress and Deflection

The point load of a 30 kN concentrated force and with simply supported boundary conditions were applied. The mesh size of 75 mm was considered in the convergence study. A peak load of 40 kN with a deflection of 12 mm in the loading direction was achieved. From the analysis, the stress diagram is derived as mentioned in Figure 10. The maximum stress concentration is visible near the support and loading areas.

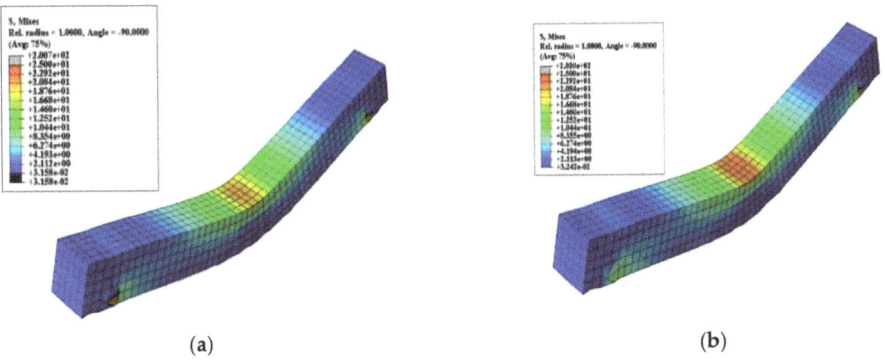

(a)　　　　　　　　　　　　　　　(b)

Figure 10. *Cont.*

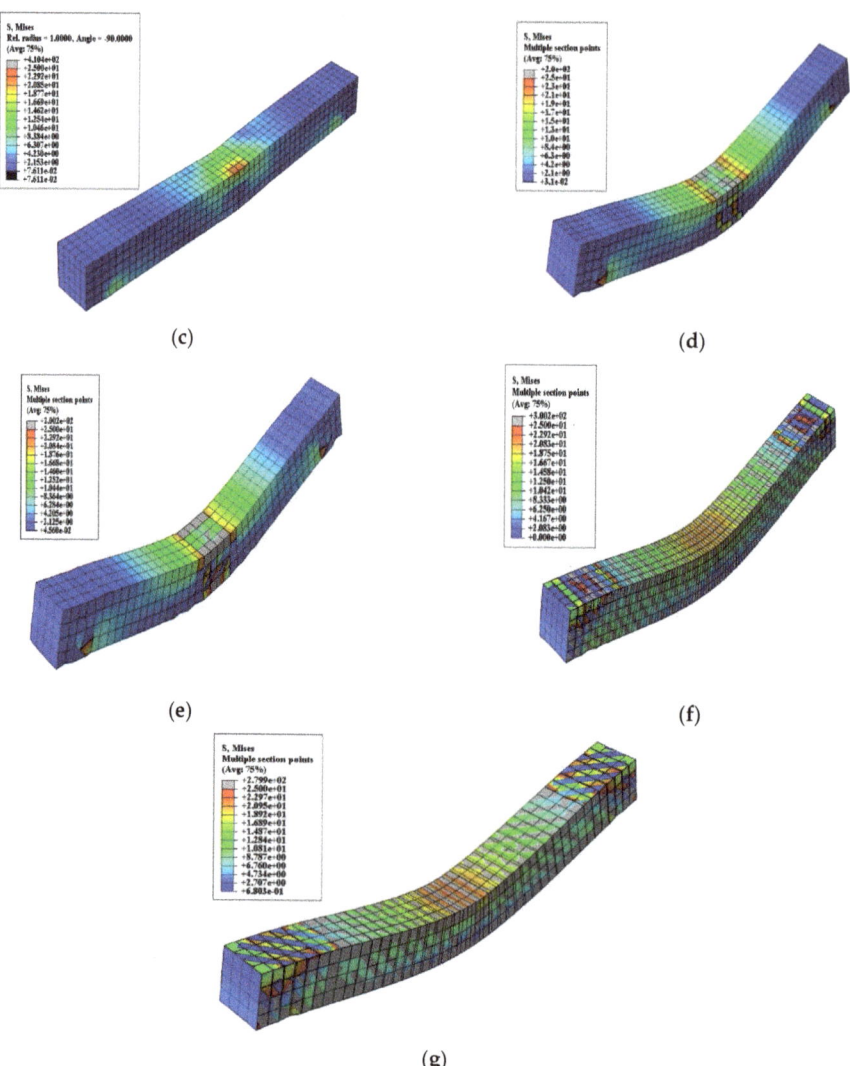

Figure 10. Stress Pattern- (**a**) Conventional Beam, (**b**) SFRC beam, (**c**) SFRC only at plastic hinge length of a beam, (**d**) Carbon fiber mesh jacketing at Plastic hinge length of a conventional beam, (**e**) SFRC and Carbon fiber mesh jacketing only at a plastic hinge length of a beam, (**f**) Carbon fiber mesh jacketing for a total length of a conventional beam, (**g**) Carbon fiber mesh jacketing for total length with SFRC at plastic hinge length.

Carbon fiber mesh jacketing for the total length of the beam achieved a peak load of 90 kN with a deflection of 18 mm. Meanwhile, a peak load of 65 kN with a deflection of 18 mm was attained by the beam strengthened with carbon fiber mesh at the plastic hinge span location alone. Where the stress distribution evenly takes place throughout the span of the beam, jacketing is done throughout the span. However, the stress concentration is converging towards the mid-span.

The load was applied as displacement rotation and the boundary condition with simple support was applied. The mesh size of 30 mm was considered in the convergence study. From the analysis, the load-deflection graph is derived as shown in Figure 11.

The maximum load of 57 kN with a deflection of 18 mm was achieved for a beam reinforced with steel fiber throughout the span. Steel fiber reinforcement at the plastic hinge length of the beam achieved the maximum load of 45 kN with a deflection of 18 mm. At the same time, another set of beams are cast with SFRC and CFM together at the plastic hinge length of the beam, which shows the maximum load of 60 kN, and the last beam jacketed throughout the span with SFRC at the plastic hinge length achieved 92 kN. Hence, the beam jacketed throughout the span with carbon fiber mesh withstand the ultimate load of 64 kN due to the overall combined performance in reducing bending. Conventional RCC Beam shows 50 kN ultimate load and 45 kN load achieved by SFRC at plastic hinge length. Steel fiber at plastic hinge length with overall wrapping of carbon fiber mesh reached 11.58 mm the least deflection due to resistance to deflection which shows drastic change in stiffness. Conventional beam shows the maximum deflection if 17.8 mm due to the ultimate.

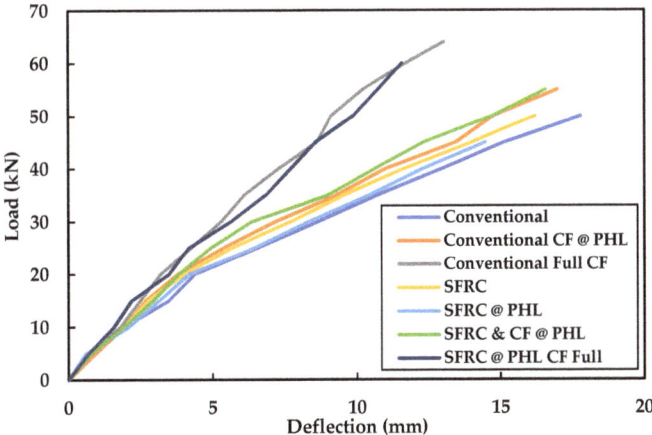

Figure 11. Analytical load-deflection curve.

6.2. Experimental Results

The compression strengths of conventional concrete and SFRC for 7, 14, and 28 days are shown in Table 7. Where steel fiber reinforced concrete on 28 of days curing shows the maximum compressive strength of 31.2 N/mm².

Table 7. Compressive strength.

Concrete Type	7 Days (MPa)	14 Days (MPa)	28 Days (MPa)
Conventional	16.3	20.5	26.59
SFRC 1.5%	21.58	25.86	31.2

The split tensile strength of conventional concrete and SFRC for 7, 14, and 28 days is shown in Table 8. The tensile strength of concrete improved by 4.76 N/mm² after 1.5% weight of steel fibers was added to the conventional concrete. Meanwhile, the flexural strength of concrete also improved to 5.1 N/mm² on 28 days curing as mentioned in Table 9.

Table 8. Split tensile strength.

Concrete Type	7 Days (MPa)	14 Days (MPa)	28 Days (MPa)
Conventional	2.54	3.61	3.9
SFRC 1.5%	3.02	4.45	4.76

The Flexural strength of conventional concrete and SFRC for 7, 14, 28 days are shown in Table 9.

Table 9. Flexure strength.

Concrete Type	7 Days (MPa)	14 Days (MPa)	28 Days (MPa)
Conventional	2.1	3.2	4.12
SFRC 1.5%	2.92	3.9	5.1

6.3. Ultimate Load and Deflection of Beams

The ultimate load and defection of the beam are determined by testing the beam with one-point loading as shown in Figure 12, and the ultimate load and deflection of the RCC beam are shown in Table 10. The ultimate load of 64 kN was attained by a beam where carbon fiber mesh jacketing was done for the full length. In the same way, the SFRC only at plastic hinge length and carbon fiber mesh jacketing for the total length of beam reached 60 kN. Conventional RCC beam and SFRC and carbon fiber mesh jacketing only in the zone of plastic hinge of a beam are in the same range as 38 kN. The deflection of the beam seems to be similar for SFRC only at plastic hinge length and SFRC and carbon fiber mesh jacketing only at plastic hinge length. Beams where carbon fiber mesh jacketing is used for strengthening show almost 15 mm of deflection. The comparison of all the above load and deflection data is mentioned in Figure 13 and Table 10.

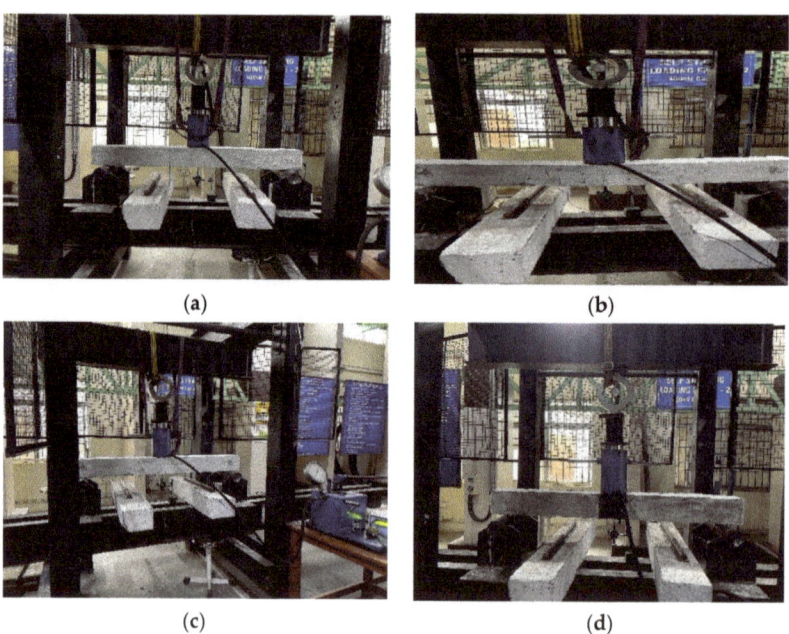

(a) (b) (c) (d)

Figure 12. Cont.

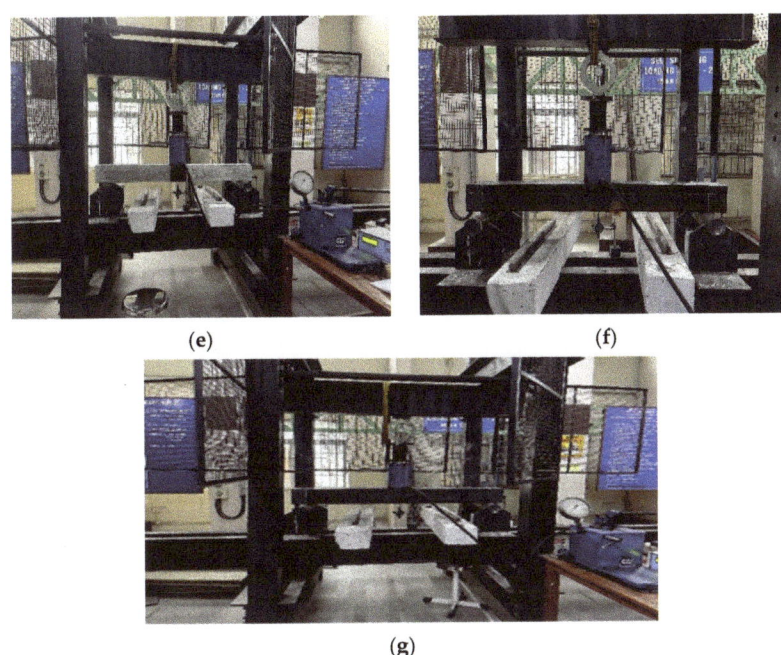

Figure 12. Experimental setup (**a**) Conventional Beam, (**b**) SFRC beam, (**c**) SFRC only at plastic hinge length of a beam, (**d**) Carbon fiber mesh jacketing at plastic hinge length of a conventional beam, (**e**) SFRC and carbon fiber mesh jacketing only at a plastic hinge length of a beam, (**f**) Carbon fiber mesh jacketing for a total length of a conventional beam, (**g**) Carbon fiber mesh jacketing for total length with SFRC at plastic hinge length of a beam.

Table 10. Ultimate load and defection of beams.

Beam Description	Ultimate Load (KN)	Deflection (mm)
Conventional beam	38	8.21
Carbon fiber mesh jacketing at plastic hinge length of conventional beam	48	15.43
Carbon fiber mesh jacketing for the full length of conventional beam	64	15.95
SFRC	43	13.78
SFRC only at the plastic hinge length of a beam	41	11.9
SFRC and carbon fiber mesh jacketing only at the plastic hinge length of a beam	38	11.63
SFRC only at plastic hinge length and carbon fiber mesh jacketing for a total length of a beam	60	15.67

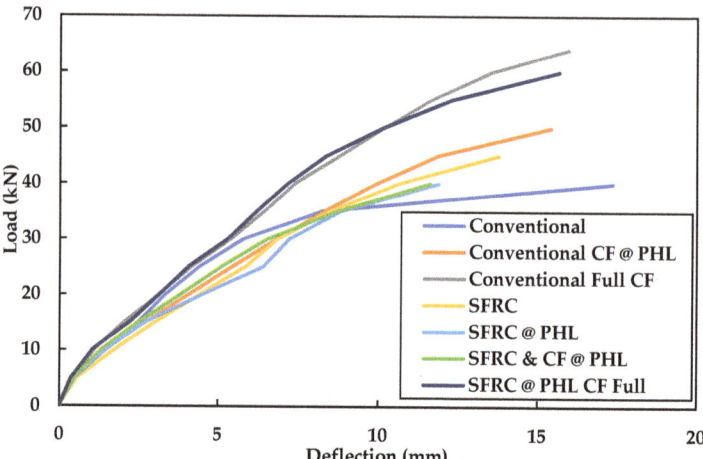

Figure 13. Load-deflection graphs from experiments.

6.4. Comparative Study on Analytical and Experimental Investigation

In this work, the displacement of the conventional beam, carbon fiber mesh jacketing for a full length, and carbon fiber mesh jacketing on the plastic hinge length of a beam for conventional, SFRC, and SFRC only at PHL are compared as shown in Table 11.

Table 11. Comparative study of displacements predicted from the experiments against analytical predictions.

Model	Experimental Displacement	Analytical Displacement
Conventional Beam	5.8	8.6
Carbon fiber mesh jacketing for full length	5.4	5.46
Carbon fiber mesh jacketing on plastic Hinge Length	6.85	7.04
SFRC	6.86	7.7
SFRC only at plastic hinge length	7.2	8.32
SFRC and Carbon fiber mesh only at plastic hinge length	6.54	6.39
SFRC only at plastic hinge length and carbon fiber mesh jacketing for full length	5.31	5.5

The experimental and analytical results were compared, which shows the deviation in results is nominal and in most cases, it's almost similar.

6.5. Stiffness

The theoretical calculation of stiffness calculated by using l/250 of the beam as per IS 456:2000 is 6 mm. Figure 14 shows the difference in stiffness of different types of beam specimens under ultimate loading conditions, which helps to understand the role of the carbon fiber mesh and SFRC in providing strength to the beams from the experimental investigation.

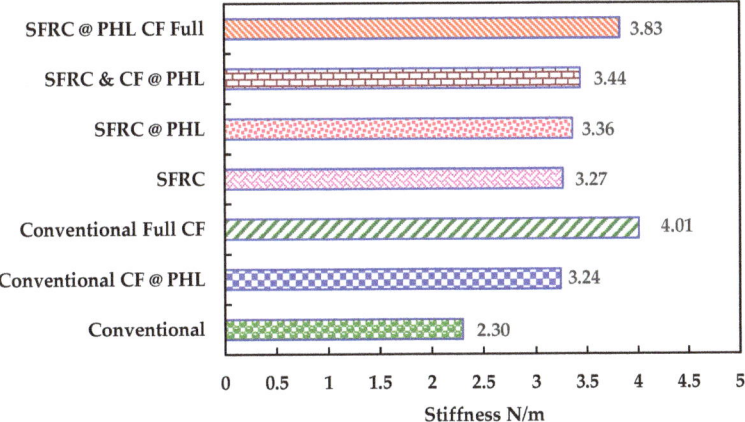

Figure 14. Stiffness of different RC beams.

The stiffness is calculated by the force-displacement relation and the stiffness is shown in Table 12. From the above results, SFRC only at PHL is stiffer when compared to conventional and SFRC. From the graph, it is obvious that among all the evaluated specimens, the steel fiber reinforcement at the plastic hinge length and the carbon fiber mesh jacketing along the complete span acquired the maximum stiffness.

Table 12. Stiffness of different RC beams.

Model	Stiffness (N/m)
Conventional beam	2.30×10^6
Carbon fiber mesh jacketing for full length	4.01×10^6
Carbon fiber mesh jacketing on PHL	3.24×10^6
SFRC	3.27×10^6
SFRC only at plastic hinge length	3.36×10^6
SFRC and carbon fiber mesh only at plastic hinge length	3.44×10^6
SFRC only at plastic hinge length and carbon fiber mesh jacketing for full length	3.83×10^6

7. Conclusions

The performance of a beam with steel fiber reinforcement along the length of the plastic hinge is presented in this analytical and experimental examination. After being cast and loaded monotonously, seven different beams' mechanical characteristics are examined. These are the findings that were drawn from them.

1. Among different plastic hinge length expressions from Baker, Sawyer, Coreley, Mattock, Park, Pristley, Paulay, Fardis, and Panagiotakos, the Paulay and Pristley expressions have been considered for plastic hinge length calculations in experimental investigation.
2. Steel fiber reinforced concrete with a 1.5% content performs better in terms of compressive strength and split tensile strength when compared to regular concrete.
3. The split tensile strength of steel fiber-reinforced concrete seems to be 1.5 times larger than that of regular concrete due to the dispersion of steel fiber in the concrete, which influences the bonding and promotes tensile strength. The cylinder specimen was compressed using uniaxial compression in order to calculate the modulus of elasticity, and the results showed that the SFRC specimen with 1.5% steel fiber outperformed the conventional concrete specimen by a factor of 1.14. The same was thus used to casting beam specimens. The behavior under bending is clearly shown by the flexural

strength test, which shows that the flexible beam with steel fiber exhibits a 1.39 times larger performance gain than that of a normal beam.
4. The results of conventional concrete, CFM jacketed on total length, CFM jacketed on PHL, SFRC, and SFRC at PHL, SFRC & CF at PHL, and SFRC at PHL & CFM jacketed on the total length of a beam were compared.
5. Steel fiber reinforcement in concrete has superior tensile strength and can withstand more severe loads than regular concrete when exposed to monotonic stress. Similar types of fracture patterns may be seen in steel fiber reinforced concrete with a 150 mm plastic hinge length and steel fiber dosed over the beam span. The failure happened simultaneously because of the material's higher ductility. Cracks are appearing far from the zone of greatest deflection because of the steel fiber at the hinge length.
6. When compared to the RCC beam, the SFRC beam exhibits comparatively less deflection. The inclusion of SFRC at simply the length of the plastic hinge of a beam led to a similar discovery since steel fiber boosts the beam's strength. Steel fiber reinforcement offers the same ductility and resilience to the load as SFRC and SFRC solely at PHL at a 150 mm plastic hinge length. When the traditional beam is reinforced across its span and steel fibres are used at the length of the plastic hinge, the total stiffness of the beam is fairly high. The ultimate load-bearing capacity and deflection due to delay in failure are both increased by carbon fiber mesh jacketing for the complete span with steel fiber dosage at the plastic hinge length.
7. Hence, from the above discussion two things are very clear, one is instead of providing steel fiber throughout the span of the beam, provide it at plastic hinge length alone as both of them provide the same performance under monotonic loading. This will reduce the number of fibers used for construction and which will be economical as well. Meanwhile, the next one is carbon fiber jacketing done for the whole beam span with fiber placed at plastic hinge length shows the best performance when compared to that of other techniques.

8. Scope for Future Work

Experiments with various types of fibers and with different percentages can be carried out to extend this work.

Author Contributions: P.S. Conceived the model, developed the empirical theorem, and conducted experiments. B.G.A.G., K.R. and P.S. I oversaw the research and the outcomes analysis. D.P.: Introduced monotonic loading as a concept in this project, constructed the slab, and produced the article. reviewed and submitted the work, worked on and oversaw the research. K.R. and B.G.A.G. The journal was recommended by K.R. and selected for submission. K.R., B.G.A.G. and K.R. Took part in the editing stage of the manuscript. All authors have read and agreed to the published version of the manuscript.

Funding: This research received no external funding.

Data Availability Statement: The data presented in this study are available on request from the corresponding author.

Conflicts of Interest: This manuscript has not been submitted to, nor is it under review by, another journal or other publishing venue. The authors have no affiliation with any organization with a direct or indirect financial interest in the subject matter discussed in the manuscript. The authors declare no conflict of interest.

References

1. Barros, J.A.O.; Figueiras, J.A. Flexural Behavior of Steel Fiber Reinforced Concrete: Testing and Modelling. *J. Mater. Civ. Eng.* **1999**, *11*, 331–352. [CrossRef]
2. Neto, B.N.M.; Barros, J.A.; Melo, G.S. A model for the prediction of the punching resistance of steel fibre reinforced concrete slabs centrically loaded. *Constr. Build. Mater.* **2013**, *46*, 211–223. [CrossRef]
3. Sivanantham, P.; Gurupatham, B.G.A.; Roy, K.; Rajendiran, K.; Pugazhlendi, D. Plastic Hinge Length Mechanism of Steel-Fiber-Reinforced Concrete Slab under Repeated Loading. *J. Compos. Sci.* **2022**, *6*, 164. [CrossRef]

4. Pradeep, S.; Vengai, V.E.; More, D.F. Experimental Investigation on the Usage of Steel Fibres and Carbon Fibre Mesh at Plastic Hinge Length of Slab. *Mater. Today Proc.* **2019**, *14*, 248–256. [CrossRef]
5. Gopinath, A.; Nambiyanna, B.; Nakul, R.; Prabhakara, R. Parametric study on rotation and plastic hinge formation in RC beams. *J. Civ. Eng. Technol. Res.* **2014**, *2*, 393–401.
6. Zhao, X.; Wu, Y.-F.; Leung, A.; Lam, H.F. Plastic hinge length in reinforced concrete flexural members. *Procedia Eng.* **2011**, *14*, 1266–1274. [CrossRef]
7. Zhao, X.-M.; Wu, Y.-F.; Leung, A. Analyses of plastic hinge regions in reinforced concrete beams under monotonic loading. *Eng. Struct.* **2012**, *34*, 466–482. [CrossRef]
8. Mainst, R.M. Measurement of the Distribution of Tensile and Bond Stresses Along Reinforcing Bars. *ACI J. Proc.* **1951**, *48*, 225–252. [CrossRef]
9. Mattock, A.H. Rotational Capacity of Hinging Regions in Reinforced Concrete Beams, Flexural Mechanics of Reinforced Concrete. In Proceedings of the ASCE-ACI International Symposium, Miami, FL, USA, 10–12 November 1964; pp. 143–180.
10. Sümer, Y. Determination of Plastic Hinge Length for RC Beams Designed with Different Failure Modes under Static Load. *Acta Phys. Pol. A* **2019**, *135*, 955–960. [CrossRef]
11. Paulay, T.; Priestley, M.N. *Seismic Design of Reinforced Concrete and Masonry Buildings*; Wiley: New York, NY, USA, 1992; Volume 768.
12. Holschemacher, K.; Mueller, T.; Ribakov, Y. Effect of steel fibres on mechanical properties of high-strength concrete. *Mater. Des.* **2010**, *31*, 2604–2615. [CrossRef]
13. Khaloo, A.R.; Afshari, M. Flexural behaviour of small steel fibre reinforced concrete slabs. *Cem. Concr. Compos.* **2005**, *27*, 141–149. [CrossRef]
14. Nguyen, N.T.; Bui, T.-T.; Bui, Q.-B. Fiber reinforced concrete for slabs without steel rebar reinforcement: Assessing the feasibility for 3D-printed individual houses. *Case Stud. Constr. Mater.* **2022**, *16*, e00950. [CrossRef]
15. Xiang, D.; Liu, Y.; Gu, M.; Zou, X.; Xu, X. Flexural fatigue mechanism of steel-SFRC composite deck slabs subjected to hogging moments. *Eng. Struct.* **2022**, *256*, 114008. [CrossRef]
16. McMahon, J.A.; Birely, A.C. Service performance of steel fiber reinforced concrete (SFRC) slabs. *Eng. Struct.* **2018**, *168*, 58–68. [CrossRef]
17. Xiang, D.; Liu, S.; Li, Y.; Liu, Y. Improvement of flexural and cyclic performance of bridge deck slabs by utilizing steel fiber reinforced concrete (SFRC). *Constr. Build. Mater.* **2022**, *329*, 127184. [CrossRef]
18. Paramasivam, P.; Ong, K.; Ong, B.; Lee, S. Performance of Repaired Reinforced Concrete Slabs under Static and Cyclic Loadings. *Cem. Concr. Compos.* **1995**, *17*, 37–45. [CrossRef]
19. Laila, L.R.; Gurupatham, B.G.A.; Roy, K.; Lim, J.B.P. Effect of super absorbent polymer on microstructural and mechanical properties of concrete blends using granite pulver. *Struct. Concr.* **2020**, *22*, 1–18. [CrossRef]
20. Laila, L.R.; Gurupatham, B.G.A.; Roy, K.; Lim, J.B.P. Influence of super absorbent polymer on mechanical, rheological, durability, and microstructural properties of self-compacting concrete using non-biodegradable granite pulver. *Struct. Concr.* **2020**, *22*, 1–24. [CrossRef]
21. Kathavate, V.; Amudha, K.; Ramesh, N.; Ramadass, G. Failure Analysis of Composite Material under External Hydrostatic Pressure: A Nonlinear Approach. *Mater. Today Proc.* **2018**, *5*, 24299–24312. [CrossRef]
22. Kathavate, V.S.; Pawar, D.N.; Adkine, A.S. Micromechanics-based approach for the effective estimation of the elastic properties of fiber-reinforced polymer matrix composite. *J. Micromech. Mol. Phys.* **2019**, *4*, 1950005. [CrossRef]
23. Lowe, D.; Roy, K.; Das, R.; Clifton, C.G.; Lim, J.B. Full scale experiments on splitting behaviour of concrete slabs in steel concrete composite beams with shear stud connection. *Structures* **2020**, *23*, 126–138. [CrossRef]
24. Madan, C.S.; Munuswamy, S.; Joanna, P.S.; Gurupatham, B.G.A.; Roy, K. Comparison of the Flexural Behavior of High-Volume Fly AshBased Concrete Slab Reinforced with GFRP Bars and Steel Bars. *J. Compos. Sci.* **2022**, *6*, 157. [CrossRef]
25. Madan, C.S.; Panchapakesan, K.; Reddy, P.V.A.; Joanna, P.S.; Rooby, J.; Gurupatham, B.G.A.; Roy, K. Influence on the Flexural Behaviour of High-Volume Fly-Ash-Based Concrete Slab Reinforced with Sustainable Glass-Fibre-Reinforced Polymer Sheets. *J. Compos. Sci.* **2022**, *6*, 169. [CrossRef]
26. Chinnasamy, Y.; Joanna, P.S.; Kothanda, K.; Gurupatham, B.G.A.; Roy, K. Behavior of Pultruded Glass-Fiber-Reinforced Polymer Beam-Columns Infilled with Engineered Cementitious Composites under Cyclic Loading. *J. Compos. Sci.* **2022**, *6*, 169. [CrossRef]
27. IS 12269; Ordinary Portland Cement, 53 Grade-Specification. Bureau of Indian Standards: New Delhi, India, 1987; pp. 1–17.
28. IS: 383; Specification for Coarse and Fine Aggregates from Natural Sources for Concrete. Bureau of Indian Standards: New Delhi, India, 1970; pp. 1–24.
29. IS 456; Plain Concrete and Reinforced. Bureau of Indian Standards: New Delhi, India, 2000; pp. 1–114.
30. SIMULIA. *ABAQUS Standard User's Manual, Version 6.14*; Dassault Systèmes Simulia Corp.: Johnston, RI, USA, 2013.
31. IS 516; Method of Tests for Strength of Concrete. Bureau of Indian Standards: New Delhi, India, 1959; pp. 1–30.

Article

Behavior of Pultruded Glass-Fiber-Reinforced Polymer Beam-Columns Infilled with Engineered Cementitious Composites under Cyclic Loading

Yoganantham Chinnasamy [1], Philip Saratha Joanna [2,*], Karthikeyan Kothanda [3], Beulah Gnana Ananthi Gurupatham [4] and Krishanu Roy [5,*]

1. School of Planning, Architecture and Design Excellence, Hindustan Institute of Technology and Science, Padur, Chennai 603103, India
2. Department of Civil Engineering, Hindustan Institute of Technology and Science, Padur, Chennai 603103, India
3. School of Civil Engineering, Vellore Institute of Technology, Chennai 600127, India
4. Department of Civil Engineering, College of Engineering Guindy Campus, Anna University, Chennai 600025, India
5. School of Engineering, The University of Waikato, Hamilton 3216, New Zealand
* Correspondence: joanna@hindustanuniv.ac.in (P.S.J.); krishanu.roy@waikato.ac.nz (K.R.)

Abstract: Glass-fiber-reinforced polymer (GFRP) is an advanced material that has superior corrosion resistance, a high strength-to-weight ratio, low thermal conductivity, high stiffness, high fatigue strength, and the ability to resist chemical and microbiological compounds. Despite their many advantages compared with traditional materials, GFRP sections exhibit brittle behavior when subjected to severe loading conditions such as earthquakes, which could be overcome by infilling the GFRP sections with concrete. This paper presents the results of an experimental investigation carried out on the cyclic response of a GFRP beam-column infilled with high-volume fly ash engineered cementitious composites (HVFA-ECC) consisting of 60%, 70%, and 80% fly ash as a replacement for cement. Finite element analysis was also conducted using robot structural analysis software, and the results were compared with the experimental results. The mechanical properties of GFRP sections presented are the compressive strength of ECC, the direct tensile strength of ECC determined using a dog-bone-shaped ECC specimen, the hysteresis behavior of the beam-column, and the energy dissipation characteristics. The lateral load-carrying capacity of beam-column GFRP infilled with HVFA-ECC consisting of 60%, 70%, and 80% fly ash was found to be, respectively, 43%, 31%, and 20% higher than the capacity of GFRP beam-columns without any infill. Hence the GFRP sections infilled with HVFA-ECC could be used as lightweight structural components in buildings to be constructed in earthquake-prone areas. Also in the structural components, as 70% of cement could be replaced with fly ash, it can potentially lead to sustainable construction.

Keywords: pultruded glass-fiber-reinforced polymer; engineered cementitious composite; high-volume fly ash; finite element analysis; cyclic loading

1. Introduction

Composites have started replacing traditional materials in most engineering fields, such as the aerospace, marine, automobile, electrical, chemical, and construction industries where high strength and stiffness-to-weight ratios are required. Fiber-reinforced polymer (FRP) is one of the composite materials in which fibers such as glass, aramid, and carbon are embedded in the matrix material. As FRP composites have high stiffness, light weight, corrosion resistance, and chemical resistance, they are being utilized in the construction industry. In FRP sections, fibers carry the load and the matrix protects the fibers from the atmosphere and also helps in transferring load to the fibers. Carbon and graphite fibers are light in weight and stronger than other fibers. However, the cost of carbon fibers is higher than that of glass or aramid fibers. Glass fibers are more ductile and cheaper than carbon fibers, which leads to their utility in the construction industry [1].

Pultruded glass-fiber-reinforced polymers (GFRP) are made with 12–35 m diameter glass fibers and are embedded with vinyl ester/polyester/epoxy resins through a continuous rowing method called the pultrusion process. The process consists of pulling impregnated filaments together with a mat or fabric through a heated die. The fiber composition and stiffness of GFRP sections vary with the manufacturer, and hence, it is very important to find the mechanical properties of GFRP sections with suitable test methods [2,3]. The load-carrying capacity and stiffness of the GFRP sections increase when infilled with concrete. Concrete-infilled GFRP tubes have a higher flexural strength than conventional reinforced concrete [4–7]. The GFRP tube confines the entire cross-section of the concrete, and longitudinal fibers act as reinforcement in the longitudinal direction of the beam [8]. Concrete-infilled glass-/carbon-fiber-reinforced polymer tubes could fail sequentially and progressively and exhibit pseudo-ductile behavior [9–14]. The use of high-grade concrete as an infill in GFRP sections exhibits only a 20% increase in load-carrying capacity compared to low-strength concrete as an infill [15–17].

Engineered cementitious composite (ECC) is a special type of concrete that exhibits increased tensile strength and strain-hardening behavior compared to conventional concrete. In ECC, strain-hardening takes place after the first cracking like a ductile metal and exhibits 3% to 5% tensile strain capacity, which is 300 to 500 times higher than that of normal concrete. ECC typically has a tensile strain capacity of more than 3% of compressive strength due to the interaction between fibers and matrix and exhibits closely spaced cracks, resulting in decreased water permeability or chloride ion penetration into the mixture [18]. ECC consists of cement, fine aggregates of a maximum size of 200 m, water, and high-range water-reducing (HRWR) admixture to increase workability and less than 2% volume of fibers. Different varieties of ECC with fibers such as polyvinyl alcohol (PVA), polypropylene (PP), and polyethylene (PE) fibers have been developed, and their properties have been investigated. ECC with PVA fibers exhibits higher tensile strength, toughness, and flexural strength than that of PP fibers. Further, the cost of PVA fiber is eight times less than that of PE fiber [19]. The elimination of coarse aggregate (CA) from ECC results in relatively higher cement content in the mixture and also leads to higher costs and environmental pollution. The production of one ton of cement emits 0.94 tons of CO_2 into the atmosphere, along with other glasshouse gases such as nitrogen oxide and sulfur dioxide. However, industrial solid waste materials such as fly ash (FA), ground granulated blast furnace slag (GGBS), silica fume (SF), and inert limestone powder could be added to ECC, which acts as a filler material and results in good workability, lower cement content, and a reduction in embodied carbon in ECC. The shear strength and ductility of concrete with cement replaced with FA are higher than those of cement replaced with GGBS and SF [20–24]. The use of FA in ECC results in substantial energy savings and decreases greenhouse gas emissions [25]. Further, the use of fly ash reduces the requirement for a large land area for its disposal, thus creating significant benefits for the environment. A feasible design approach was carried out in the development of ECC with FA of various quantities based on simple flow testing as a guideline [26–31]. There is a decrease in the strength of the ECC following an increase in fly ash content. Cracks with smaller widths were noticed when the fly ash content increased [32–34]. The use of fly ash in ECC contributes to the self-compactability of the fresh ECC and also helps in achieving the strain-hardening behavior of hardened ECC, which in turn leads to sustainability [35–38].

The durability of the concrete is greatly influenced by curing since it has a significant effect on the hydration process. Negligence in curing will hurt the strength and durability of concrete [39]. On the 7th and 28th days, the compressive and split tensile strength of concrete with 0.5, 1, and 1.5 percent polyethylene glycol (PEG-600) as an internal curing agent were tested and compared to conventional concrete [40]. PEG-600 in concrete not only helps with self-curing but also helps with better cement hydration and increases compressive strength by trapping the moisture within the concrete, preventing it from evaporating [41]. An increase in the amount of PEG-600% in concrete decreases the strength of the concrete. As a result, adding 0.5% to 1% PEG-600 as an internal curing agent to concrete improves its effectiveness [42–44]. The literature available on ECC with HVFA is rather limited, and no literature is available on the effect of an internal curing agent on ECC with manufactured sand (M-sand).

In the construction industry, FRP composites have been used because of their high stiffness, lightweight, corrosion resistance, and easy installation. However, its utilization has a limitation because of its brittle failure. Hence, in this invention, the GFRP beam-column was infilled with the eco-friendly high-volume fly ash engineered cementitious composite. The present study is to investigate the mechanical properties of pultruded GFRP sections and to develop the ECC with HVFA, M-sand, and self-curing agent which is to be used as an infill in pultruded GFRP square sections, as well as to investigate the hysteretic behavior of beam-columns made of pultruded GFRP sections infilled with HVFA-ECC. Numerical investigations were carried out using Autodesk robot structural analysis (RSA) professional software to compare with the experimental investigation.

2. Materials and Methods

2.1. GFRP Sections

In this research, pultruded GFRP sections of size 100 mm × 100 mm, 5 mm thick, were used. The mechanical properties of the GFRP sections were carried out on coupons extruded from GFRP sections. Tensile, compressive, flexural, and shear strength tests were carried out on coupons extruded from GFRP sections as per ASTM D3039 [45], ASTM D3410 [46], ASTM D790 [47], and ASTM D2344 [48], respectively, as shown in Figure 1.

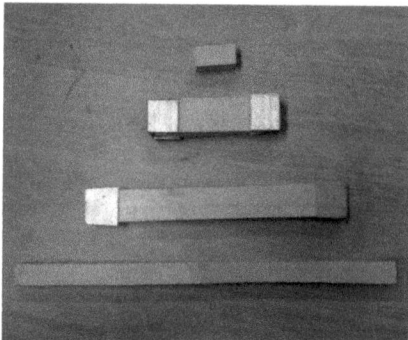

Figure 1. GFRP coupons.

All tests were carried out on a servo-controlled universal testing machine (UTM) of 100 kN capacity. Three specimens were prepared to conduct the test, and the details of the coupons extruded to find the mechanical properties of GFRP sections are given in Table 1.

Table 1. Details of the GFRP coupons.

Name of the Test	Coupon Size (mm)
Tensile strength	250 × 25 × 5
Compressive strength	125 × 25 × 5
Flexural strength	360 × 15 × 5
Interlaminar shear strength	50 × 15 × 5

The average ultimate tensile strength, compressive strength, elastic modulus, flexural strength, flexural modulus, shear strength, and shear modulus are 387.5 MPa, 150 MPa, 17.2 Mpa, 215 MPa, 1.1 GPa, 29 MPa, and 3 GPa, respectively.

2.2. Engineered Cementitious Composite with High-Volume Fly Ash (ECC-HVFA)

ECC contains ordinary Portland cement (OPC)-53 grade, manufactured sand (M-sand) having a size of 150–300 m, "Class F" fly ash, PVA fiber of 12 mm length, CONXL PCE RHEOPLUS 2635, a high-range water-reducing (HRWR) agent, and PEG 600, an internal curing agent. The properties of PVA fibers and PEG600 are given in Tables 2 and 3. In ECC, cement was replaced with fly ash ranging from 60% to 80%.

Table 2. Properties of PVA fibers.

Fibre	Density	Initial Modulus	Specification	Oil Agent Content
PVA	1.29	280 cN/dtex	12 mm	0.2%

Table 3. Properties of PEG-600 agent.

Sl. No	PEG600	
1	Solubility	Soluble in water
2	Density	1.126 kg/m^3
3	Odor	Mild odor
4	Mean molecular weight	570–630 kg/m^3
5	Appearance	Clear liquid

In a pan mixer, cement, fly ash, and M-sand were mixed for 5 to 6 min, and then HRWR and PEG 600 mixed with water were added gradually. The mixing continued for 10 to 15 min. After ensuring the minimum spread value using a mini-slump flow test, fibers were added to the mix, and the pan mixer was continuously rotated to avoid the formation of lumps in the mix. All the mixes were designed to have spread values of between 450 mm and 500 mm as stipulated in the standard slump flow test. A mini-slump cone test was carried out before the addition of fibers in ECC using a 60 mm high mini-slump cone. All the mixes were designed to ensure the achievement of a spread value of between 270 mm and 300 mm. A workability test for the final mix with fiber was carried out on a standard slump cone of 300 mm in height to find the flowability of ECC. All the mixtures were designed to ensure the achievement of spread values ranging from 450 mm to 500 mm. The workability test is shown in Figure 2. The mix proportion details of ECC are shown in Table 4.

(a)

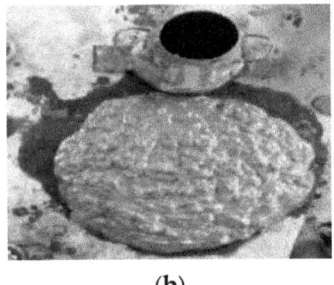
(b)

Figure 2. Workability test on ECC. (a) Mini-slump test; (b) standard slump test.

Table 4. Strength of concrete mixes.

Mix	Description	Cement to Binder (B)	Fly Ash to B	M-Sand to B	Water to B	HRWR to B	Fiber to B	PEG to B
1	ECC-0	1	0	0.6	0.35	0.005	0.01	0
2	ECC-60P	0.4	0.6	0.6	0.37	0.005	0.01	0.02
3	ECC-70P	0.3	0.7	0.6	0.4	0.005	0.01	0.02
4	ECC-80P	0.2	0.8	0.6	0.44	0.005	0.01	0.02

2.2.1. Mechanical Properties of HVFA-ECC

Compressive Strength

The compressive strength test on four mixes of ECC was carried out in the compression testing machine. The three cubes in each mix were tested at the ends of 7 days, 28 days, and 56 days after casting, and the average values were taken as the compressive strength of the mixes, as shown in Table 5.

Table 5. Compressive strength for different concrete mixes.

Mix	Mix Description	7 Days (MPa)	28 Days (MPa)	56 Days (MPa)
1	ECC-0	15.6	33.5	36.2
2	ECC-60P	12.9	28.7	34.5
3	ECC-70P	9.36	26.9	31.2
4	ECC-80P	6.01	23.7	28.4

The compressive strength of self-cured ECC-60P, ECC-70P, and ECC-80P was 20%, 66%, and 15% less than ECC-0 at 7 days. The compressive strength of self-cured ECC-60P, ECC-70P, and ECC-80P was, respectively, 16%, 25%, and 41% less than that of ECC-0 at 28 days, and 4%, 16%, and 27% less than that of ECC-0 at 56 days.

Direct Tensile Strength

The direct tensile strength of ECC was determined using a dog-bone-shaped specimen having an 80 mm gauge length with a 36 mm × 20 mm cross-section. Three dog-bone-shaped specimens were cast from the same batch of ECC. The specimens made of ECC-0 were cured with water, and the specimens of ECC-60P, ECC-70P, and ECC-80P added with self-curing agents were cured under shade. The detailed and direct tensile strength tests were carried out on a dog-bone-shaped specimen as shown in Figure 3 in UTM of 100 kN capacity as per ASTM C1273 [49].

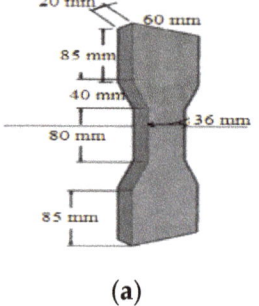

(a)

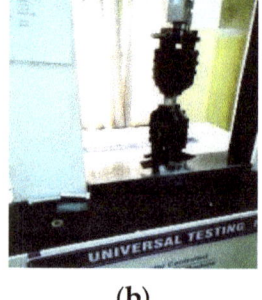

(b)

Figure 3. Direct tensile strength test setup. (a) Dog-bone-shaped specimen; (b) direct tensile strength test setup on ECC.

The rate of displacement of the crosshead was kept at 0.1 mm/min. The tensile strengths of three dog-bone-shaped specimens in each series of mixes were tested after

28 days of casting, and the average values were taken as the tensile strengths of ECC mixes. The tensile strength test results of all the mixes are shown in Table 6.

Table 6. Direct tensile test results.

Mix Description	Mix	Tensile Stress (MPa)		Tensile Strain (%)
		At Initial Crack	At Ultimate Level	Ultimate Level
ECC-0	2	4.40	4.80	1.22
ECC-60P	4	4.10	4.30	0.97
ECC-70P	7	4.05	4.15	0.97
ECC-80P	10	3.90	3.98	0.97

ECC with and without fly ash exhibited a fluctuation in the stress–strain curve due to the propagation of cracks during the time of loading. The ultimate tensile strength of ECC-60P, ECC-70P, and ECC-80P was, respectively, 7%, 9%, and 11.5% less than that of ECC-0, and the ultimate tensile strains of ECC-60P, ECC-70P, and ECC-80P were 20% less than that of ECC-0. The stress–strain curve obtained from a direct tensile strength test is shown in Figure 4.

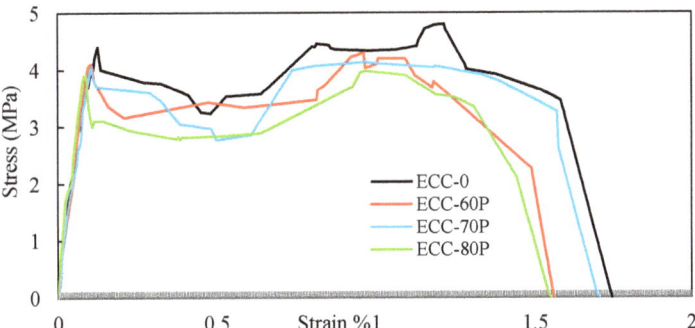

Figure 4. Stress–strain curves obtained from the direct tensile strength tests.

2.3. Beam-Column Specimens

Eight GFRP beam-column specimens having a beam of 1.5 m length and a column of 1.1 m height were connected using steel angle plates of size 200 mm × 100 mm × 6 mm and four numbers of 10 mm diameter bolts. The specimens were subjected to lateral loading to obtain the hysteresis curve, peak load–deflection, pseudo-ductile behavior, and energy dissipation. Base plates were used to avoid punching shear. Two specimens in each series of GFRP beam-column sections infilled with ECC-60P, ECC-70P, and ECC-80P were cast and tested at the end of 28 days. The results of GFRP beam-column sections infilled with HVFA-ECC were compared with those of pultruded GFRP beam-columns without infill. The details of the beam-column specimens are given in Table 7. The preparation of beam-column specimens is shown in Figure 5.

Table 7. Details of the GFRP beam-column specimens.

Sl. No.	Beam-Column ID	No of Specimens	Outer Material	Infill Material
1	BCG-H	2	GFRP Section	-
2	BCG-E60P	2	GFRP Section	ECC-60P
3	BCG-E70P	2	GFRP Section	ECC-70P
4	BCG-E80P	2	GFRP Section	ECC-80P

Figure 5. The casting of GFRP beam-column specimens.

2.4. Experimental Investigation

Lateral Loading on Pultruded GFRP Beam-Column with and without HVFA-ECC

GFRP beam-columns with and without HVFA-ECC infill consist of a 1.5 m long beam and a 1.1 m high column connected using steel angle plates and bolts. A schematic diagram of the experimental setup for beam-columns is shown in Figure 6.

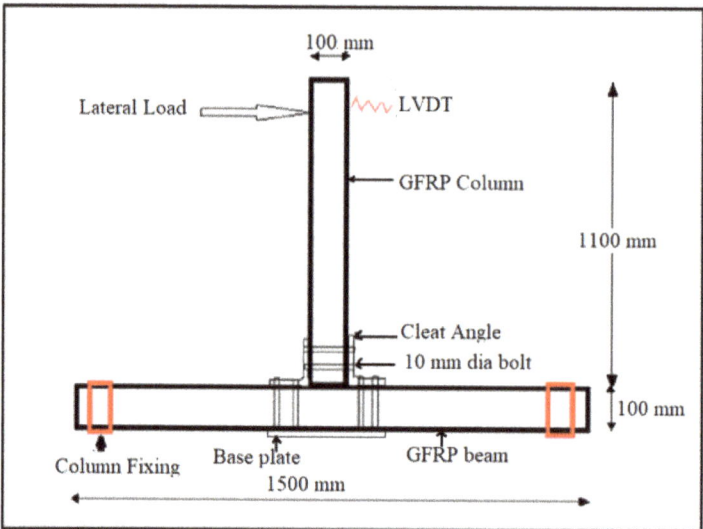

Figure 6. A schematic diagram of the experimental setup for beam-column testing.

The test was conducted on a reaction frame of 200 kN capacity having a stroke length of 100 mm. A hydraulic jack attached to the load cell was used for the measurement of the applied load on the column, and an LVDT was used for the measurement of lateral displacement at the top of the column, as shown in Figure 7. The lateral load and the corresponding displacement readings were obtained from the data logger connected to a computer, which captured the values until the completion of the test. Cyclic loading was applied on the top of the column until the specimen failed.

Figure 7. Experimental setup for lateral loading on beam-columns.

The test was conducted on a reaction frame of 200 kN capacity having a stroke length of ±100 mm. A hydraulic jack attached to the load cell was used for the measurement of the applied load on the column, and an LVDT was used for the measurement of lateral displacement at the top of the column, as shown in Figure 7. The lateral load and the corresponding displacement readings were obtained from the data logger connected to a computer, which captured the values until the completion of the test. Cyclic loading was applied on the top of the column until the specimen failed.

3. Results and Discussion
3.1. Lateral Load–Deformation Behavior

The failure of the GFRP beam-column without HVFA-ECC infill was sudden, but the failure of GFRP sections infilled with HVFA-ECC was in a sequential manner and exhibited a larger load-carrying capacity due to the confinement effect provided by ECC with the GFRP section. The failure pattern of the beam-column specimens is shown in Figure 8.

Figure 8. Failure pattern of beam-columns manufactured from GFRP infilled with and without HVFA-ECC.

Figure 9 shows the hysteretic curve of BCG-H specimens. The average ultimate load and maximum deflection of BCG-H specimens were 13.6 kN and 29.6 mm, respectively. When subjected to forward lateral loading, the BCG-E60P specimens recorded an average ultimate lateral load of 19.15 kN with a maximum lateral deflection of 51.2 mm. Figure 10 shows the hysteretic curve of BCG-E60P specimens. The BCG-E70P exhibited an ultimate load of 17.82 kN with a maximum lateral deflection of 45.80 mm. Figure 11 shows the hysteretic curve of BCG-E70P specimens. The BCG-E80P failed with an average lateral load capacity of 16.35 kN with an average displacement of 38.35 mm. Figure 12 shows the hysteretic curve of BCG-E80P specimens.

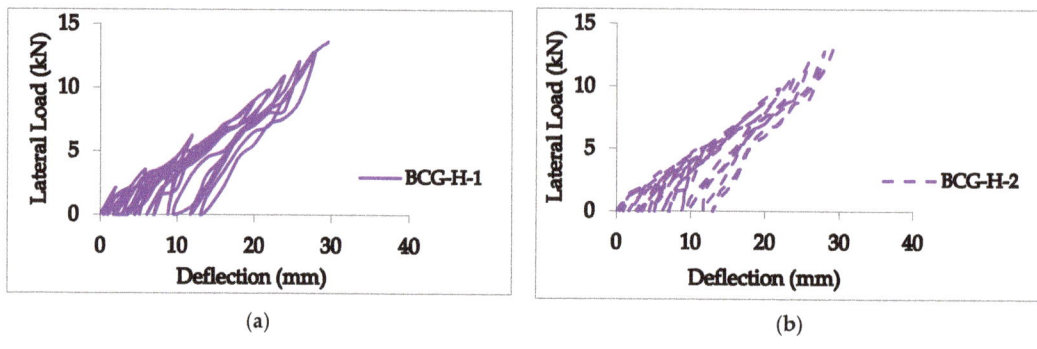

Figure 9. Hysteretic curves of hollow GFRP beam-columns. (**a**) Specimen 1; (**b**) specimen 2.

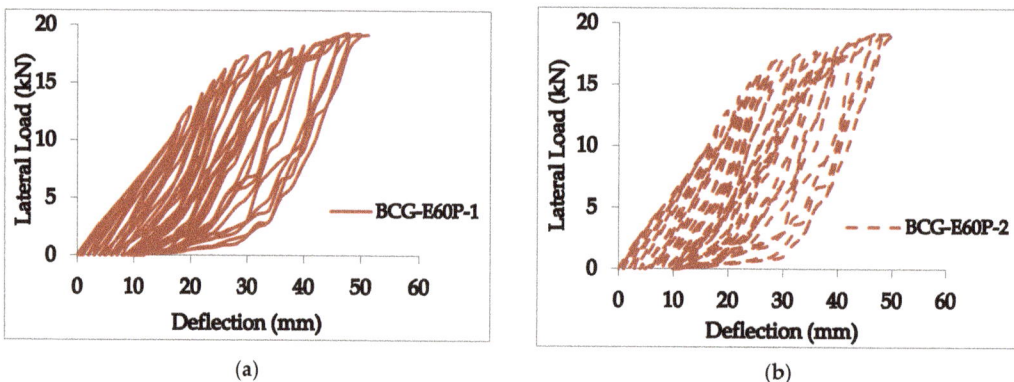

Figure 10. The hysteretic curve of the GFRP-ECC60P beam-column. (**a**) Specimen 1; (**b**) specimen 2.

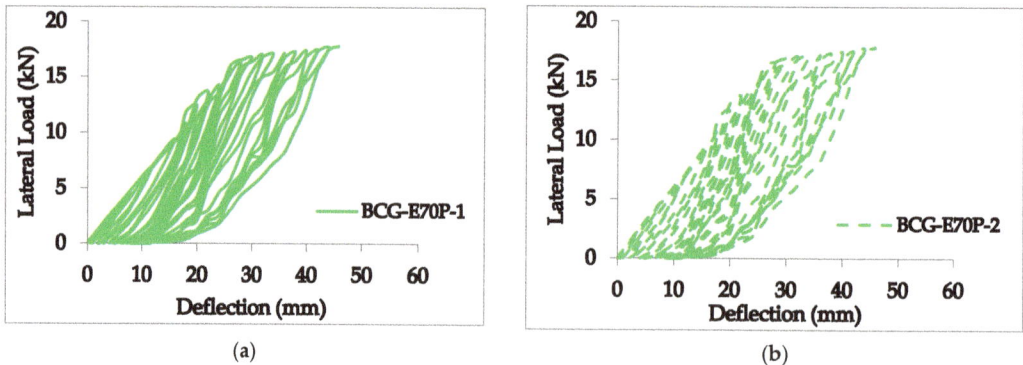

Figure 11. The hysteretic curve of the GFRP-ECC70P beam-column. (**a**) Specimen 1; (**b**) specimen 2.

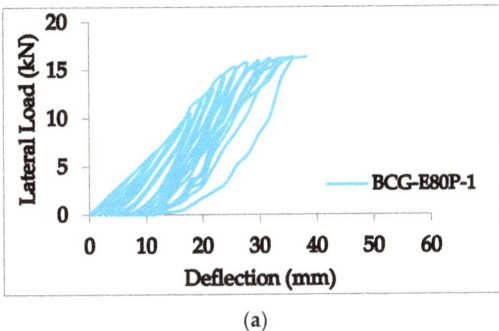
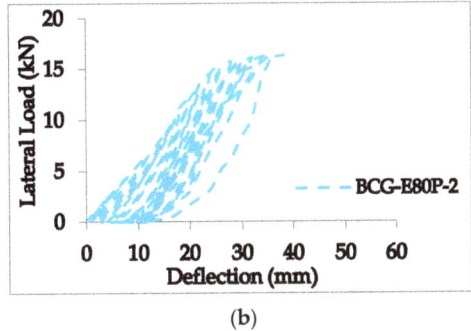

Figure 12. The hysteretic curve of the GFRP-ECC80P beam-column. (a) Specimen 1; (b) specimen 2.

3.2. Strength of the Beam-Column Specimens

Figure 13 depicts the cyclic envelope or P–Δ curves for the GFRP beam-column with or without HVFA-ECC. The lateral load capacity of the specimens is taken as the average of the load values when they were subjected to lateral loading in the forward direction. The BCG-H specimens attained an average ultimate lateral load of 13.6 kN with an average lateral displacement of 30 mm. The BCG-E60P specimens recorded an average lateral peak load of 19.5 kN with an average lateral displacement of 51.2 mm. The BCG-E70P specimens exhibited an average ultimate lateral strength of 17.82 kN with an average lateral displacement of 45.8 mm. The average lateral ultimate load of the BCG-E80P specimens was 16.35 kN with an average lateral displacement of 38.35 mm.

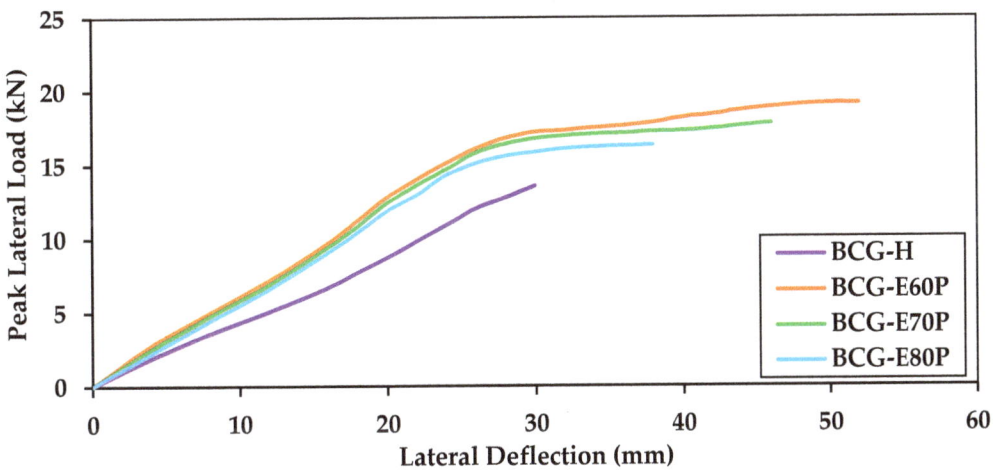

Figure 13. Cyclic envelope curve of GFRP with and without HVFA-ECC beam-columns.

3.3. Energy Dissipation Capacity

The dissipated energy in each cycle was calculated as the area bound by the hysteresis loop of that cycle from the load (P) versus displacement (Δ) curve, and the total dissipated energy is calculated as the summation of the energy dissipated in all the cycles up to the failure of the specimen. The energy dissipation curve of the beam-column is shown in Figure 14.

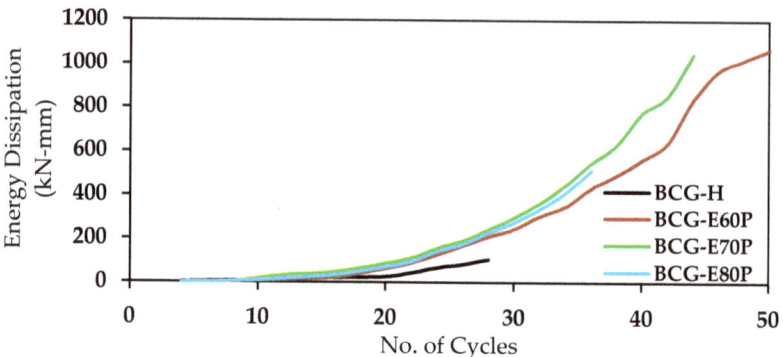

Figure 14. Energy dissipation curves.

The energy dissipation of BCG-H is 105 kN.mm. However, the energy dissipation of the BCG-E60P, BCG-E70P, and BCG-E80P was 1067 KN.mm, 1043 kN.mm, and 511 kN.mm, respectively. The energy dissipation of BCG-E60P observed was 10 times higher than that of the GFRP beam-column without infill.

3.4. "Pseudo-Ductile" Behavior

Ductility is one of the characteristics of a material that undergoes plastic deformations. However, non-plastic or non-ductile materials do not exhibit plasticity, and they could be characterized by a pseudo-ductility displacement index. The pseudo-ductility displacement index was calculated using the following Equation (1):

$$\mu = (d_u - d_y)/d_u \tag{1}$$

where μ—pseudo-ductility displacement index, d_u—failure displacement, d_y—displacement at yield.

However, the yield displacement is replaced by the displacement corresponding to the first peak load, while the failure displacement is assumed to be equal to the displacement corresponding to the last peak load of the load–displacement curve (just before the GFRP rupture). It should be pointed out that pseudo-ductility is not a measure of material plastic behavior, but rather an indicator of the post-peak load residual strength and concomitant deformation after significant damage in the material, component, or connection [37]. The values of μ of the beam-column tested are given in Table 8. BCG-E60P, BCG-E70P, and BCG-E80P respectively exhibited 67%, 48%, and 31% more pseudo-ductility than BCG-H.

Table 8. Pseudo-ductility index for all beam-column specimens.

Sl. No.	Beam-Column ID	d_u (mm)	d_y (mm)	μ (-)
1	BCG-H	30	30	0
2	BCG-E60P	52	28	0.46
3	BCG-E70P	46	28	0.39
4	BCG-E80P	38	24	0.36

4. Numerical Investigations

Numerical investigations were carried out using Autodesk robot structural analysis (RSA) software. RSA is a structural analysis software that verifies different code compliance and uses build information modeling (BIM) integrated workflows to exchange data with other software. The RSA has wind simulation, extensive analysis capabilities, finite element analysis (FEA) with auto meshing, country-specific design standards, and an open and flexible application programming interface. The integration option in RSA enables the import of structural members and connection profiles from software such as Auto-CADD,

Revit, and Advanced Steel. The connection profiles from other software can also be imported to RSA.

4.1. Modeling and Meshing

The material properties of GFRP and HVFA-ECC imported into RSA were based on the results obtained from the test. Figure 15 shows the material properties assigned in the RSA software.

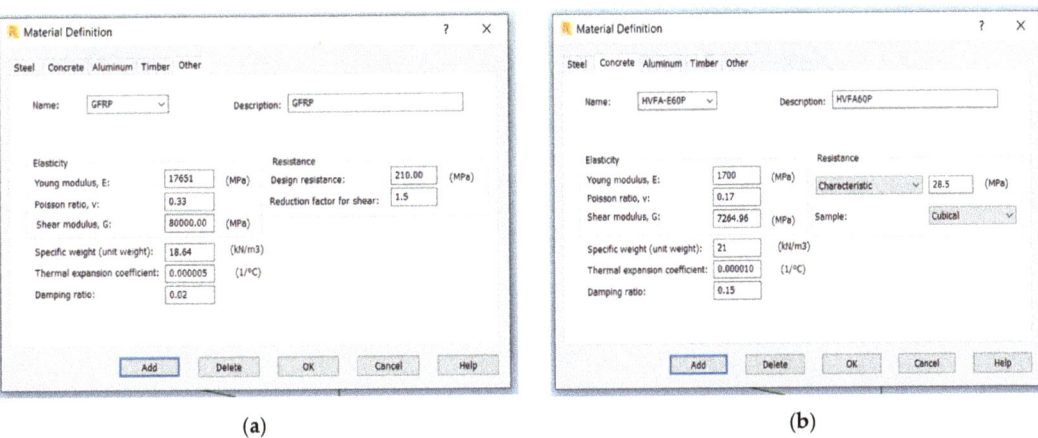

(a)   (b)

Figure 15. Material properties used in the FE model. (a) GFRP; (b) HVFA-ECC.

The Section definition tool option in RSA enables the creation of composite sections with different materials. The GFRP sections infilled with HVFA-ECC were created and the material properties were assigned. The contact behavior between the GFRP and HVFA-ECC was modeled as Coulomb friction. The beam model of the GFRP sections was created using RSA software. Figure 16 shows the creation of the GFRP composite section and modeling of the beam-column made of the GFRP section with and without infill with HVFA-ECC. However, the connection profile of the beam-column was modeled in Autodesk advanced steel software and imported to RSA. Figure 17 shows the modeling of the connection profile in Autodesk advanced steel software.

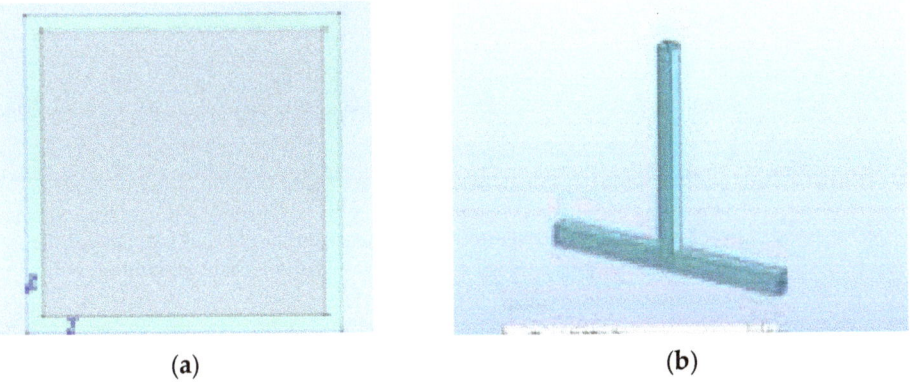

(a)   (b)

Figure 16. Creation of GFRP sections. (a) Composite section; (b) modeling of beam-column in RSA.

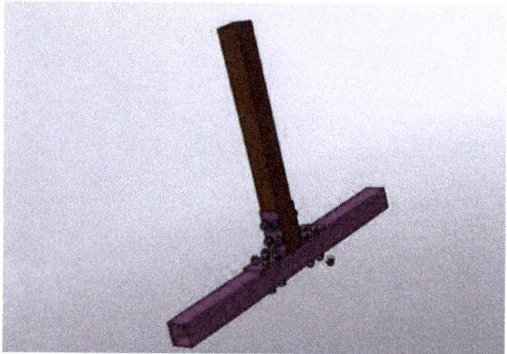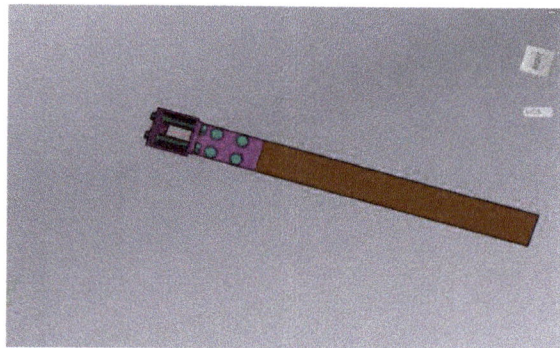

Figure 17. Creation of connection profile using Autodesk advanced steel software.

The coons meshing type with squares in rectangular contour meshing options were given to create meshing of the members. To obtain meshes with a fine size, four-noded quadrilaterals for surface and four-noded tetrahedrons for the volumetric type of meshing were given. Figure 18 shows the mesh type given and the meshing of the GFRP beam-column.

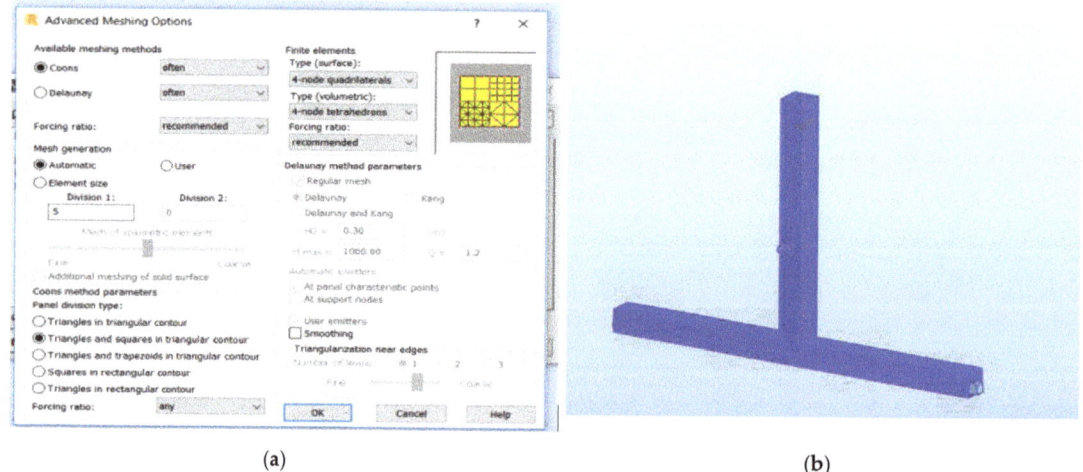

(a) (b)

Figure 18. Details of the FEA meshing. (**a**) Meshing data; (**b**) meshing of GFRP beam-column in RSA.

4.2. Support and Loading Condition

In the beam-column, both ends of the beam were assigned as fixed conditions, and the load was applied at the top of the column as applied in the experimental investigation. Nonlinear analysis was performed to understand the behavior of the GFRP beam-column. Figure 19 shows the support condition provided for the beam-column.

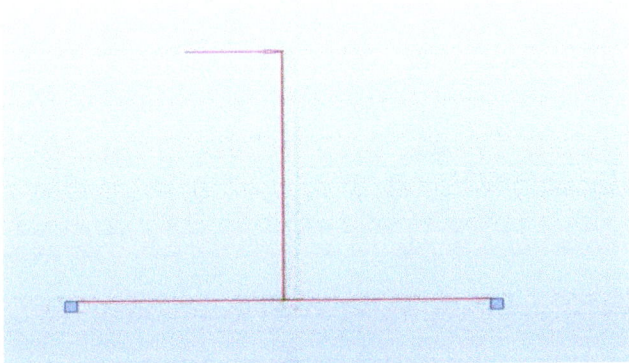

Figure 19. Support and loading condition of the beam-columns assigned in RSA.

4.3. Load–Deflection Behavior of Beam-Columns

The failure pattern of the BCG−H, BCG−E60P, BCG−E70P, and BCG−E80P beam-column is shown in Figure 20, obtained from RSA. The BCG−H exhibited an ultimate load of 12.98 kN, but BCG-E60P, BCG-E70P, and BCG-E80P showed a peak load of 17.70 kN, 19.09 kN, and 16.01 kN, respectively. Figure 21 shows the comparison of the load–deflection curves of the BCG-H, BCG-E60P, BCG-E70P, and BCG-E80P beam-columns obtained from experimental and analytical investigations.

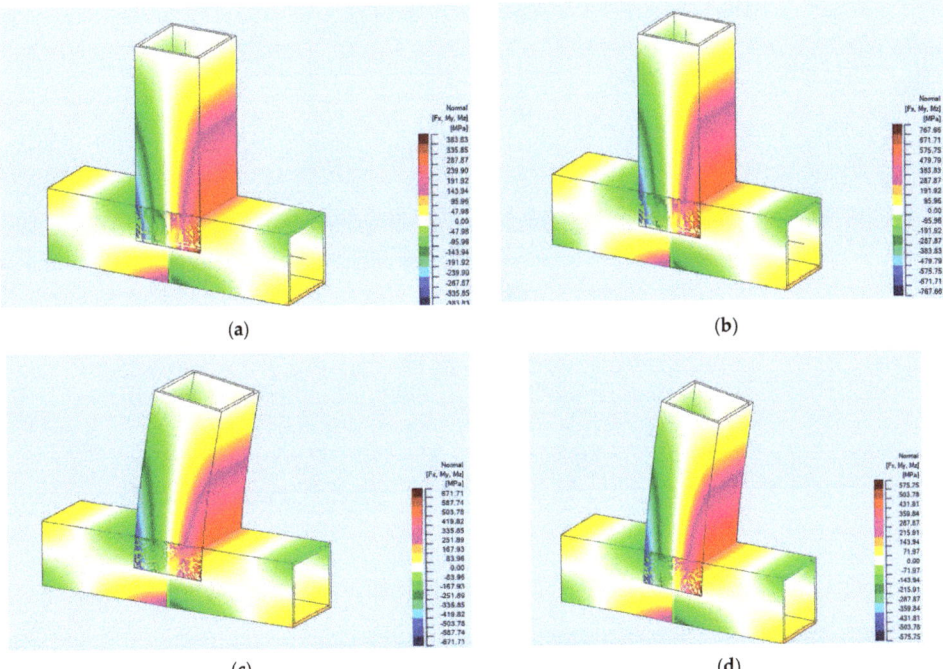

Figure 20. Failure modes of beam-columns obtained from the RSA. (**a**) BCG−H; (**b**) BCG−E60P; (**c**) BCG−E70P; (**d**) BCG−E80P.

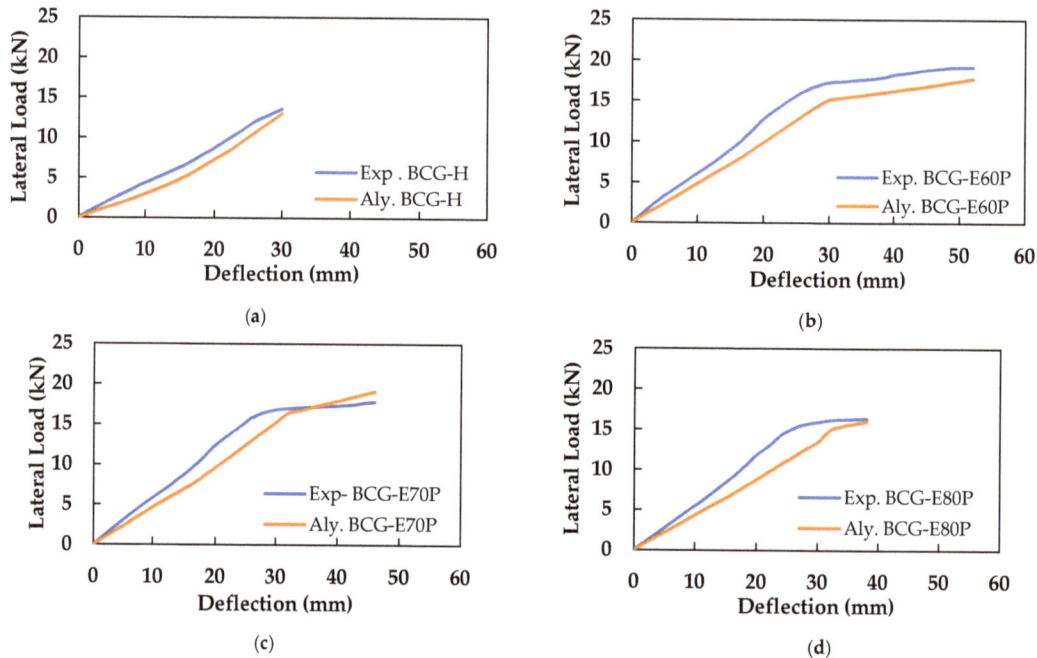

Figure 21. Comparison of load–displacement curves obtained from the experimental and numerical investigations. (**a**) BCG-H; (**b**) BCG-E60P; (**c**) BCG-E70P; (**d**) BCG-E80P.

5. Conclusions

Experimental and analytical investigations have been carried out on the performance of beam-columns made of a GFRP section with and without HVFA-ECC infill. Investigations into the mechanical properties of the GFRP section and HVFA-ECC were carried out. The GFRP beam-column load–displacement hysteretic behavior, capacity, pseudo-ductile behavior, and energy dissipation capacity are summarized below.

- The average ultimate tensile strength, compressive strength, elastic modulus, flexural strength, flexural modulus, shear strength, and shear modulus are 387.5 MPa, 150 MPa, 17.2 Mpa, 215 MPa, 1.1 GPa, 29 MPa, and 3 GPa, respectively.
- In the direct tensile strength test, the ultimate tensile strength of ECC-60P, ECC-70P, and ECC-80P was, respectively, 7%, 9%, and 11.5% less than ECC-0, and the ultimate tensile strains of ECC-60P, ECC-70P, and ECC-80P were 20% less than that of ECC-0.
- The average lateral load-carrying capacity of BCG-E60P, BCG-E70P, and BCG-E80P was found to be, respectively, 43%, 31%, and 20% higher than that of BCG-H.
- The energy dissipation of the BCG-E60P, BCG-E70P, and BCG-E80P beam-column specimens was, respectively, 100%, 39%, and 23% higher than that of the BCG-H specimen.
- Further, BCG-E60P, BCG-E70P, and BCG-E80P exhibited, respectively, 67%, 48%, and 31% more pseudo-ductility than BCG-H.
- ECC with fly ash up to 70% as a replacement for cement could be utilized in infilling the GFRP sections.
- The analytical results obtained from RSA show good agreement with the experimental results.

Thus, the GFRP beam-column infilled with high-volume ECC having cement replacement up to 70% with fly ash exhibited good lateral load-carrying capacity, energy dissipation capacity, and improved pseudo-ductility behavior compared with the hollow section. The use of high-volume fly ash, a byproduct of coal-burning power plants used in the engineered cementitious composite used in ECC, reduces CO_2 emissions, and manufactured sand was used in ECC due to the scarcity of river sand.

Author Contributions: Y.C. and P.S.J.: Conceptualized the model, established the empirical theorem, and conducted the experiments. P.S.J., B.G.A.G., and K.R.: Supervised the research as well as the analysis of results. P.S.J., and K.K.: Introduced the idea of cyclic loading in this project, designed the beam-column, wrote, reviewed, and submitted the paper, and collaborated in and coordinated the research. P.S.J., B.G.A.G., and K.R.: Suggested and chose the journal for submission. P.S.J., B.G.A.G., and K.R.: Participated in the manuscript revision phase. All authors have read and agreed to the published version of the manuscript.

Funding: This research received no external funding.

Institutional Review Board Statement: Not applicable.

Informed Consent Statement: Not applicable.

Data Availability Statement: The data presented in this study are available on request from the corresponding author.

Conflicts of Interest: This manuscript has not been submitted to, nor is it under review by, another journal or other publishing venue. The authors have no affiliation with any organization with a direct or indirect financial interest in the subject matter discussed in the manuscript. The authors declare no conflict of interest.

References

1. Hollaway, L.C. A Review of the Present and Future Utilisation of FRP Composites in the Civil Infrastructure with Reference to Their Important In-Service Properties. *Constr. Build. Mater.* **2010**, *24*, 2419–2445. [CrossRef]
2. Singh, S.B.; Chawla, H. An Investigation of Material Characterization of Pultruded FRP H- and I-Beams. *Mech. Adv. Mater. Struct.* **2016**, *25*, 124–142. [CrossRef]
3. Landesmann, A.; Seruti, C.A.; Batista, E.D.M. Mechanical Properties of Glass Fiber Reinforced Polymers Members for Structural Applications. *Mater. Res.* **2015**, *18*, 1372–1383. [CrossRef]
4. da S. Santos Neto, A.B.; Lebre La Rovere, H. Flexural Stiffness Characterization of Fiber Reinforced Plastic (FRP) Pultruded Beams. *Compos. Struct.* **2007**, *81*, 274–282. [CrossRef]
5. Zaghloul, M.M.Y.M. Mechanical Properties of Linear Low-Density Polyethylene Fire-Retarded with Melamine Polyphosphate. *J. Appl. Polym. Sci.* **2018**, *135*, 46770. [CrossRef]
6. Zaghloul, M.M.Y.; Zaghloul, M.Y.M.; Zaghloul, M.M.Y. Experimental and Modeling Analysis of Mechanical-Electrical Behaviors of Polypropylene Composites Filled with Graphite and MWCNT Fillers. *Polym. Test.* **2017**, *63*, 467–474. [CrossRef]
7. Correia, J.R.; Branco, F.A.; Silva, N.M.F.; Camotim, D.; Silvestre, N. First-Order, Buckling and Post-Buckling Behaviour of GFRP Pultruded Beams. Part 1: Experimental Study. *Comput. Struct.* **2011**, *89*, 2052–2064. [CrossRef]
8. Vieira, P.R.; Carvalho, E.M.L.; Vieira, J.D.; Toledo Filho, R.D. Experimental Fatigue Behavior of Pultruded Glass Fibre Reinforced Polymer Composite Materials. *Compos. Part B Eng.* **2018**, *146*, 69–75. [CrossRef]
9. Zhang, S.; Caprani, C.C.; Heidarpour, A. Strain Rate Studies of Pultruded Glass Fibre Reinforced Polymer Material Properties: A Literature Review. *Constr. Build. Mater.* **2018**, *171*, 984–1004. [CrossRef]
10. Al-saadi, A.U.; Aravinthan, T.; Lokuge, W. Effects of Fibre Orientation and Layup on the Mechanical Properties of the Pultruded Glass Fibre Reinforced Polymer Tubes. *Eng. Struct.* **2019**, *198*, 109448. [CrossRef]
11. Sirajudeen, R.S.; Sekar, R. Buckling Analysis of Pultruded Glass Fiber Reinforced Polymer (GFRP) Angle Sections. *Polymers* **2020**, *12*, 2532. [CrossRef] [PubMed]
12. Fam, A.; Cole, B.; Mandal, S. Composite Tubes as an Alternative to Steel Spirals for Concrete Members in Bending and Shear. *Constr. Build. Mater.* **2007**, *21*, 347–355. [CrossRef]
13. Rozylo, P. Stability and Failure of Compressed Thin-Walled Composite Columns Using Experimental Tests and Advanced Numerical Damage Models. *Int. J. Numer. Methods Eng.* **2021**, *122*, 5076–5099. [CrossRef]
14. Liu, X.; Karami, B.; Shahsavari, D.; Civalek, Ö. Elastic Wave Characteristics in Damped Laminated Composite Nano-Scaled Shells with Different Panel Shapes. *Compos. Struct.* **2021**, *267*, 113924. [CrossRef]
15. Mohamed, H.M.; Masmoudi, R. Flexural Strength and Behavior of Steel and FRP-Reinforced Concrete-Filled FRP Tube Beams. *Eng. Struct.* **2010**, *32*, 3789–3800. [CrossRef]
16. Aydın, F.; Sarıbıyık, M. Investigation of Flexural Behaviors of Hybrid Beams Formed with GFRP Box Section and Concrete. *Constr. Build. Mater.* **2013**, *41*, 563–569. [CrossRef]
17. Muttashar, M.; Manalo, A.; Karunasena, W.; Lokuge, W. Influence of Infill Concrete Strength on the Flexural Behaviour of Pultruded GFRP Square Beams. *Compos. Struct.* **2016**, *145*, 58–67. [CrossRef]
18. Li, V.C. On Engineered Cementitious Composites (ECC). *J. Adv. Concr. Technol.* **2003**, *1*, 215–230. [CrossRef]

19. Yang, E.-H.; Li, V.C. Strain-Hardening Fiber Cement Optimization and Component Tailoring by Means of a Micromechanical Model. *Constr. Build. Mater.* **2010**, *24*, 130–139. [CrossRef]
20. Kang, S.-B.; Tan, K.H.; Zhou, X.-H.; Yang, B. Experimental Investigation on Shear Strength of Engineered Cementitious Composites. *Eng. Struct.* **2017**, *143*, 141–151. [CrossRef]
21. Sivanantham, P.; Gurupatham, B.G.A.; Roy, K.; Rajendiran, K.; Pugazhlendi, D. Plastic Hinge Length Mechanism of Steel-Fiber-Reinforced Concrete Slab under Repeated Loading. *J. Compos. Sci.* **2022**, *6*, 164. [CrossRef]
22. Şahmaran, M.; Bilici, Z.; Ozbay, E.; Erdem, T.K.; Yucel, H.E.; Lachemi, M. Improving the Workability and Rheological Properties of Engineered Cementitious Composites Using Factorial Experimental Design. *Compos. Part B Eng.* **2013**, *45*, 356–368. [CrossRef]
23. Pan, Z.; Wu, C.; Liu, J.; Wang, W.; Liu, J. Study on Mechanical Properties of Cost-Effective Polyvinyl Alcohol Engineered Cementitious Composites (PVA-ECC). *Constr. Build. Mater.* **2015**, *78*, 397–404. [CrossRef]
24. Meng, D.; Lee, C.K.; Zhang, Y.X. Flexural and Shear Behaviours of Plain and Reinforced Polyvinyl Alcohol-Engineered Cementitious Composite Beams. *Eng. Struct.* **2017**, *151*, 261–272. [CrossRef]
25. Lin, C.; Kayali, O.; Morozov, E.V.; Sharp, D.J. Development of Self-Compacting Strain-Hardening Cementitious Composites by Varying Fly Ash Content. *Constr. Build. Mater.* **2017**, *149*, 103–110. [CrossRef]
26. Jia, Y.; Zhao, R.; Liao, P.; Li, F.; Yuan, Y.; Zhou, S. Experimental Study on Mix Proportion of Fiber Reinforced Cementitious Composites. *AIP Conf. Proc.* **2017**, *1890*, 020002. [CrossRef]
27. Rajamony Laila, L.; Gurupatham, B.G.A.; Roy, K.; Lim, J.B.P. Effect of Super Absorbent Polymer on Microstructural and Mechanical Properties of Concrete Blends Using Granite Pulver. *Struct. Concr.* **2020**, *22*, E898–E915. [CrossRef]
28. Yu, J.; Mishra, D.K.; Wu, C.; Leung, C.K. Very High Volume Fly Ash Green Concrete for Applications in India. *Waste Manag. Res. J. A Sustain. Circ. Econ.* **2018**, *36*, 520–526. [CrossRef]
29. Shanour, A.S.; Said, M.; Arafa, A.I.; Maher, A. Flexural Performance of Concrete Beams Containing Engineered Cementitious Composites. *Constr. Build. Mater.* **2018**, *180*, 23–34. [CrossRef]
30. Ismail, M.K.; Abdelaleem, B.H.; Hassan, A.A.A. Effect of Fiber Type on the Behavior of Cementitious Composite Beam-Column Joints under Reversed Cyclic Loading. *Constr. Build. Mater.* **2018**, *186*, 969–977. [CrossRef]
31. Pakravan, H.R.; Ozbakkaloglu, T. Synthetic Fibers for Cementitious Composites: A Critical and In-Depth Review of Recent Advances. *Constr. Build. Mater.* **2019**, *207*, 491–518. [CrossRef]
32. Wang, Q.; Lai, M.H.; Zhang, J.; Wang, Z.; Ho, J.C.M. Greener Engineered Cementitious Composite (ECC)—The Use of Pozzolanic Fillers and Unoiled PVA Fibers. *Constr. Build. Mater.* **2020**, *247*, 118211. [CrossRef]
33. Li, S.W.V.C.; Wu, C. Tensile Strain-Hardening Behavior of Polyvinyl Alcohol Engineered Cementitious Composite (PVA-ECC). *ACI Mater. J.* **2001**, *98*, 483–492. [CrossRef]
34. Rajamony Laila, L.; Gurupatham, B.G.A.; Roy, K.; Lim, J.B.P. Influence of Super Absorbent Polymer on Mechanical, Rheological, Durability, and Microstructural Properties of Self-Compacting Concrete Using Non-Biodegradable Granite Pulver. *Struct. Concr.* **2020**, *22*, E1093–E1116. [CrossRef]
35. Yoganantham, C.; Joanna, P.S. Effect of High Volume Fly Ash Concrete in Self-Curing Engineered Cementitious Composite (ECC). *Int. J. Adv. Res. Sci. Eng. Technol.* **2020**, *11*, 268–276. [CrossRef]
36. Lowe, D.; Roy, K.; Das, R.; Clifton, C.G.; Lim, J.B.P. Full Scale Experiments on Splitting Behaviour of Concrete Slabs in Steel Concrete Composite Beams with Shear Stud Connection. *Structures* **2020**, *23*, 126–138. [CrossRef]
37. Yang, E.-H.; Yang, Y.; Li, V.C. Use of High Volumes of Fly Ash to Improve ECC Mechanical Properties and Material Greenness. *ACI Mater. J.* **2007**, *104*, 620–628. [CrossRef]
38. Tosun-Felekoğlu, K.; Gödek, E.; Keskinateş, M.; Felekoğlu, B. Utilization and Selection of Proper Fly Ash in Cost Effective Green HTPP-ECC Design. *J. Clean. Prod.* **2017**, *149*, 557–568. [CrossRef]
39. Yoganantham, C.; Helen Santhi, M. Performance of Self-Compacting Self Curing Concrete with Fly Ash and M Sand. *Int. J. Earth Sci. Eng.* **2015**, *8*, 491–497.
40. Thiruchelve, S.R.; Sivakumar, S.; Raj, M.; Shanmugaraja, G.; Nallathambi, M. Effect of Polyethylene Glycol as Internal Curing Agent in Concrete. *Int. J. Innov. Res. Sci. Eng. Technol.* **2017**, *6*, 3521–3524.
41. Ascione, F.; Lamberti, M.; Razaqpur, A.G.; Spadea, S.; Malagic, M. Pseudo-Ductile Failure of Adhesively Joined GFRP Beam-Column Connections: An Experimental and Numerical Investigation. *Compos. Struct.* **2018**, *200*, 864–873. [CrossRef]
42. Madan, C.S.; Munuswamy, S.; Joanna, P.S.; Gurupatham, B.G.A.; Roy, K. Comparison of the Flexural Behavior of High-Volume Fly AshBased Concrete Slab Reinforced with GFRP Bars and Steel Bars. *J. Compos. Sci.* **2022**, *6*, 157. [CrossRef]
43. Madan, C.S.; Panchapakesan, K.; Anil Reddy, P.V.; Joanna, P.S.; Rooby, J.; Gurupatham, B.G.A.; Roy, K. Influence on the Flexural Behaviour of High-Volume Fly-Ash-Based Concrete Slab Reinforced with Sustainable Glass-Fibre-Reinforced Polymer Sheets. *J. Compos. Sci.* **2022**, *6*, 169. [CrossRef]
44. Yoganantham, C.; Joanna, P.S. Flexural Behaviour of Pultruded GFRP Beams Infilled with HVFA ECC. *Mater. Today Proc.* **2021**, *45*, 5978–5981. [CrossRef]
45. *ASTM D3039*; International Standard Test Method for Tensile Properties of Polymer Matrix Composite Materials. ASTM International: West Conshohocken, PA, USA, 2007.
46. *ASTM D3410*; Standard Test Method for Compressive Properties of Polymer Matrix Composite Materials with Unsupported Gage Section by Shear Loading. ASTM International: West Conshohocken, PA, USA, 2016.

47. *ASTM D790*; Standard Test Methods for Flexural Properties of Unreinforced and Reinforced Plastics and Electrical Insulating Materials. Astm International: West Conshohocken, PA, USA, 2010.
48. *ASTM D2344*; Standard Test Method for Short-Beam Strength of Polymer Matrix Composite Materials and Their Laminates. ASTM International: West Conshohocken, PA, USA, 2013.
49. *ASTM C1273*; Standard Test Method for Tensile Strength of Monolithic Advanced Ceramics at Ambient Temperatures. ASTM International: West Conshohocken, PA, USA, 2015.

Article

Influence on the Flexural Behaviour of High-Volume Fly-Ash-Based Concrete Slab Reinforced with Sustainable Glass-Fibre-Reinforced Polymer Sheets

Chinnasamy Samy Madan [1], Krithika Panchapakesan [1], Potlapalli Venkata Anil Reddy [1], Philip Saratha Joanna [1,*], Jessy Rooby [1], Beulah Gnana Ananthi Gurupatham [2] and Krishanu Roy [3,*]

1 Department of Civil Engineering, Hindustan Institute of Technology and Science, Chennai 603103, India; chinna_3_2001@yahoo.com (C.S.M.); keethu17111998@gmail.com (K.P.); p.v.anilreddy19181@gmail.com (P.V.A.R.); jessyr@hindustanuniv.ac.in (J.R.)
2 Department of Civil Engineering, Anna University, Chennai 600025, India; beulah28@annauniv.edu
3 School of Engineering, Civil Engineering, The University of Waikato, Hamilton 3216, New Zealand
* Correspondence: joanna@hindustanuniv.ac.in (P.S.J.); krishanu.roy@waikato.ac.nz (K.R.)

Abstract: Concrete structures provided with steel bars may undergo deterioration due to fatigue and corrosion, which leads to an increase in repair and maintenance costs. An innovative approach to eliminating these drawbacks lies in the utilisation of glass-fibre-reinforced polymer (GFRP) sheets as reinforcement in concrete structures instead of steel bars. This article relates to the investigation of the flexural behaviour of ordinary portland cement (OPC) concrete slabs and high-volume fly ash (HVFA) concrete slabs reinforced with bi-directional GFRP sheets. Slab specimens were cast with 60% fly ash as a replacement for cement and provided with a 1 mm-thick GFRP sheet in 2, 3 and 4 layers. The flexural behaviour of slabs reinforced with GFRP sheets was compared with that of the slabs reinforced with steel bars. Experiment results such as cracking behaviour, failure modes and load–deflection, load–strain and moment–curvature relationships of the slab specimens are presented. Subsequently, the nonlinear finite-element method (NLFEM) using ANSYS Workbench 2022-R1 was carried out and compared with the experimental results. The results obtained from the numerical investigation correlated with the experimental results. The experimental investigation showed that the HVFA concrete slabs reinforced with GFRP sheet provided a better alternative compared to the steel reinforcement, which led to sustainable construction.

Keywords: glass-fibre-reinforced polymer (GFRP) sheets; flexural behaviour; high-volume fly ash; cracking behaviour; load–deflection

1. Introduction

There has been a significant increase worldwide in the utilisation of fibre-reinforced polymer (FRP) [1]. FRP has been accepted as an alternate material to traditional steel reinforcement. The various types of FRP composites include aramid-fibre-reinforced polymers (AFRP), carbon-fibre-reinforced polymers (CFRP) and glass-fibre-reinforced polymers (GFRP). FRPs are available in various forms such as rods, sheets and plates. FRP offers several applications in concrete structures as they offer high resistance to corrosion, lightweight, ease of handling and high strength [2,3]. GFRP is most often used because of its lower cost than that of other FRP materials. GFRP sheet is used as an external reinforcement on the top surface of the concrete [4].

The structural elements reinforced with GFRP bars/sheets are usually over reinforced sections, which exhibit brittle failure. The use of nonmagnetic GFRP rebars as reinforcement is gaining importance in preventing deterioration in the structural integrity in concrete structures due to its corrosion resistance, lower maintenance costs and higher tensile strength than steel reinforcement [5–8]. Bidirectional binding of GFRP sheets with concrete attains superior mechanical performance, acting as a substitute for steel rods. Research carried out on slabs, beams and columns with the GFRP rebars as reinforcement has reported their structural performance as on par with the steel reinforcement [9–12]. The GFRP rods could be utilised as reinforcement in prestressed concrete members and reinforced concrete members, ground anchors and for strengthening the existing concrete structures [13–16]. The ultimate load-carrying capacity of the concrete slab reinforced with GFRP mesh is more than the engineered cementitious composite (ECC) slab made by polyvinyl alcohol fibres with 60% fly ash was used as a replacement for cement. Hence, reinforcing the concrete slab with GFRP mesh would be a better choice when compared to the ECC slab [17].

In the construction industry, concrete consists of Ordinary Portland Cement (OPC). It is the most commonly used construction material because of its raw material availability and low cost. However, OPC production requires argillaceous and calcareous materials and is energy-intensive. The main reasons for the emission of greenhouse gas during the production of OPC are calcination and fossil-fuel combustion [18]. The manufacturing of OPC contributes to around 8% of global carbon-dioxide emissions [19]. Waste materials from the industries act as an ingredient for conventional concrete, which helps in bringing down waste disposal problems. Many industrial waste materials such as ground granulated blast-furnace slag (GGBS), fly ash and micro silica have the potential to replace cement in concrete [20].

Fly ash, a by-product of the thermal power plant, is the widely accepted pozzolanic material for the replacement of OPC in concrete. The use of fly ash in concrete is increasing due to improvements in workability, strength and durability. Reinforced concrete beams with 50% fly ash show a 10% increase in moment capacity compared with conventional concrete [21]. Some drawbacks are seen when cement is replaced by fly ash as it attains poor strength at its earlier stage due to slow polymerization action [22–25]. Incorporation of micro silica (MS) in concrete enhances the mechanical properties related to uniformity, workability, strength, impermeability, durability, constructability, resistance to chemical attacks and reinforcement corrosion, and increases its compressive strength more than that of cementitious materials [26]. To enhance the workability, a chemical admixture known as superplasticiser (SP) was added in order to reduce the water content of the concrete mixtures [27].

Beams with 50% of fly ash as a replacement to cement attain a strength less than the conventional concrete at 28 days of curing [28,29]. A durable structure with less greenhouse-gas emission and with less energy could be obtained by the addition of fly ash to the concrete [30–32]. The electrical strain gauges were attached to measure the upward movement of the slabs on one corner. This arrangement of electrical gauges was kept constant throughout all the testing of the slab specimens [33]. The replacement of GFRP rods in place of steel as reinforcement in both OPC and HVFA slab specimens improves the flexural strength [34] Test results show lower split tensile and compressive strength for higher mix percentage influencing the minimum strength of the concrete. Structural elements with 50% fly ash have been found at later ages [35]. The wrapping of GFRP sheets drastically improves the stress-strain, strength and behaviour of fibres under various cooling regimes and heating temperatures [36]. With the application of GFRP sheets, a significant increase in the load-carrying capacity of the column was found. With the increase in the number of layers of GFRP, the load-carrying capacity was found to be increased [37].

An extensive literature review shows the potential of using fly ash in concrete. Despite the extensive use of GFRP sheets in the strengthening and repair of concrete structures, utilisation of GFRP sheets as reinforcement in structural elements is scanty. Hence, this paper investigates the possibility of using GFRP sheets as reinforcement in OPC/fly ash concrete slabs. Experimental investigations were carried out on 16 slabs, in which 12 slabs were reinforced with GFRP sheets 1 mm thick in 2, 3 and 4 layers, and 4 slabs were reinforced with steel bars. Parameters such as load-deflection behaviour, crack pattern, failure modes, moment-curvature behaviour and load–strain relationship were used for examination of all the slabs. This study also implements a nonlinear finite-element method

(FEM) using ANSYS Workbench 2022-R1 [38] software to numerically investigate the overall structural performance of the slab specimens with reference to the ultimate load and deflection of slab specimens reinforced with steel/GFRP sheets.

2. Materials and Methods

2.1. Ingredients of OPC/HVFA Concrete

The slabs cast with M25 grade concrete consisted of 53-grade ordinary portland cement (OPC) having a specific gravity of 3.1, crushed granite coarse aggregate of 20 mm nominal size conforming to IS:383, manufactured sand (M-sand) as fine aggregate and 10% micro silica by weight of cementitious material. In the fly-ash concrete, 60% of cement was replaced by Class F-type fly ash. The mix design of the concrete is shown in Table 1. To increase workability, 0.3% of Master Glenium sky 8233 superplasticizers were added with concrete as per IS 9103. Details of the chemical composition of Class F fly ash are listed in Table 2. The mix design of concrete arrived as per the Indian standard IS: 10262 and IS: 456.

Table 1. Mix proportion of concrete.

Materials/Type of Concrete	Cement	Fly Ash	Microsilica	M Sand	Aggregate	Water	Super Plasticiser (%)
OPC concrete	1	-	-	2.16	3.42	0.5	0.3
60% HVFA	0.4	0.6	0.1	2.1	3.32	0.5	0.3

Table 2. Chemical properties of fly ash.

Chemical Composition	Content (% by Mass)
SiO_2	52.52
Al_2O_3	32.63
Fe_2O_2	6.16
SO_3	4.95
LOI	1.08
MnO	0.03
NAI-20	0.02
Cao	Nil

2.2. Reinforcing System

The glass-fibre-reinforced polymer sheets were used as the reinforcing members in the OPC/fly ash-based concrete slabs. The GFRP sheets (E-glass fibre type) are of woven-type bidirectional mat, having a thickness of 1 mm with a fibre density of 2.6 g/cm^3. The first layer of GFRP sheets was laid on the fresh concrete at a depth of 20 mm from the bottom, and then it was folded to form the second layer, and then the concreting was completed. Similarly, it was laid for slabs with 3 layers and 4 layers of GFRP sheets. Figure 1 shows the GFRP sheets used as reinforcement in the slabs and the schematic view of placing the sheets in the slab. The test methods were conducted by manufacturers concerning the ASTM D3774/D3801 and ISO 10119/10618 standards. The specifications of the GFRP sheets are shown in Table 3. For comparison, the OPC/fly-ash-based slabs reinforced with conventional steel bars were also cast and tested. The steel rods of grade Fe 550D and having a diameter of 10 mm were reinforced with the centre-to-centre spacing of 130 mm along the longer direction and 240 mm along the shorter direction. Steel rods were placed at 20 mm depth from the bottom.

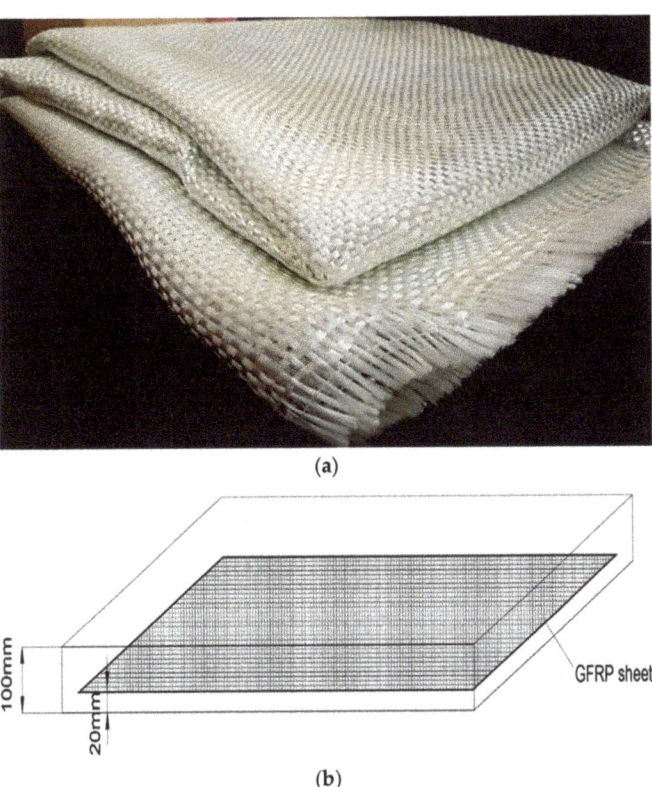

Figure 1. (a) Glass-fibre-reinforced polymer (GFRP) sheet (b) Schematic view of placing the sheet.

Table 3. Specification of the GFRP sheets.

Particulars	Specification
Aerial weight (GSM)	400
Tensile strength (N/mm^2)	2700
Modulus of elasticity (kN/mm^2)	73
Poisson's ratio	0.3
The thickness of GFRP sheet (mm)	1
Elongation at break (%)	5
Fibre density (g/cm^3)	2.6

3. Experimental Investigation

3.1. Specimen Geometry and Detailing

In this experimental work, a total of 16 slabs 1000 mm long with a cross-section of 450 mm × 100 mm were cast and tested at the end of 56 days of curing. They are categorised into two groups, of which Group I consists of eight slabs of OPC concrete reinforced with steel bars/GFRP sheets in layers 2, 3 and 4. Group II consists of eight slabs made of HVFA concrete reinforced with steel bars/GFRP sheets in layers 2, 3 and 4. Two slabs were cast in each series. A five-lettered designation was allotted to the slab specimens, where the first two letters indicate the reinforcement type as steel-reinforced (SR)/glass-fibre-reinforced polymer sheets (GS). The third letter indicates the type of concrete, i.e., OPC concrete as (C)/HVFA-based concrete as (F). The fourth identity denotes the number of layers of GFRP sheets as 2, 3 and 4, and the fifth identity denotes the trial numbers.

3.2. Experimental Set-Up

The one-way slabs were subjected to two-point flexural loading. The slab specimens were tested with roller support at one end and hinge support at the other end. A spreader beam was placed for the application of two-point loading to the slab specimens. The specimens were subjected to static load through a loading frame of 400 kN capacity. Electrical strain gauges were placed at the bottom of the GFRP sheet, and they were protected using coating tape to avoid any accidental damage while pouring concrete; and also on the top concrete surface of the slab for measurement of the compressive strain at midspan. The slabs were instrumented with linear voltage displacement transducers (LVDT), which were placed at the mid-span to monitor the deflection. A load of 2 kN/min was applied incrementally through a hydraulic jack via load cell up to the failure of the slabs. Electrical signals captured from the strain gauges and the LVDT were transmitted to the computer via a data logger. The schematic view of the experimental set-up for testing the slab specimens is shown in Figure 2 and the testing of the slab specimens is shown in Figure 3.

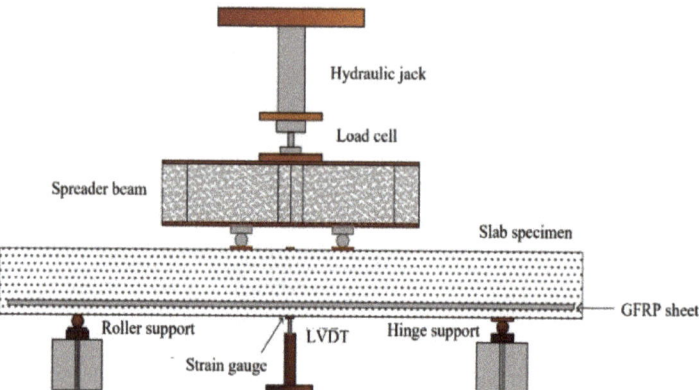

Figure 2. Schematic view of experimental set-up.

Figure 3. Testing of the slab specimens.

4. Results and Discussion

4.1. Cracking Behaviour

Propagation of cracks and the failure mode of the slab specimens are shown in Figures 4–7. The summary of the test results of the slab specimens is shown in Table 4. The average initial crack load of Group-I slab specimens GSC-2, GSC-3, GSC-4 and SRC were 8.5 kN, 14.5 kN, 6.5 kN and 16 kN, respectively. The average first crack load of Group-II slab specimens GSF-2, GSF-3, GSF-4 and SRF were 5.8 kN, 14.1 kN, 5.4 kN and 18.8 kN, respectively. Four modes of failure were observed in the slab specimens. The slab specimens SRC and SRF exhibited flexural cracks under both the loading points (Mode-I). Slab specimens GSC-2 and GSF-2 exhibited flexural crack with concrete crushing under the mid-span of the slab specimens (Mode-II). At the bottom of GSC-3 and GSF-3 slabs, fine vertical cracks began at an average load of 14.5 kN and 14.1 KN, respectively. With an increase in load, the crack propagated towards the top of the slab with crack widening. At an average load of 21.5 KN, which is 90% of the ultimate average load under the loading point, a flexural crack with the initiation of horizontal cracks was formed at the junction of GFRP sheet and concrete, which may be due to the debonding of GFRP sheets and concrete surface (Mode-III). In the case of GSC-4 and GSF-4, flexural cracks at an average load of 6.5 kN and 5.4 kN, respectively, with the subsequent formation of horizontal cracks were noticed over the entire span of the slab (Mode-IV). In the slabs reinforced with GFRP sheets, the cracks propagated from the bottom of the slab to the top of the slab exhibited brittle failure at the ultimate load level.

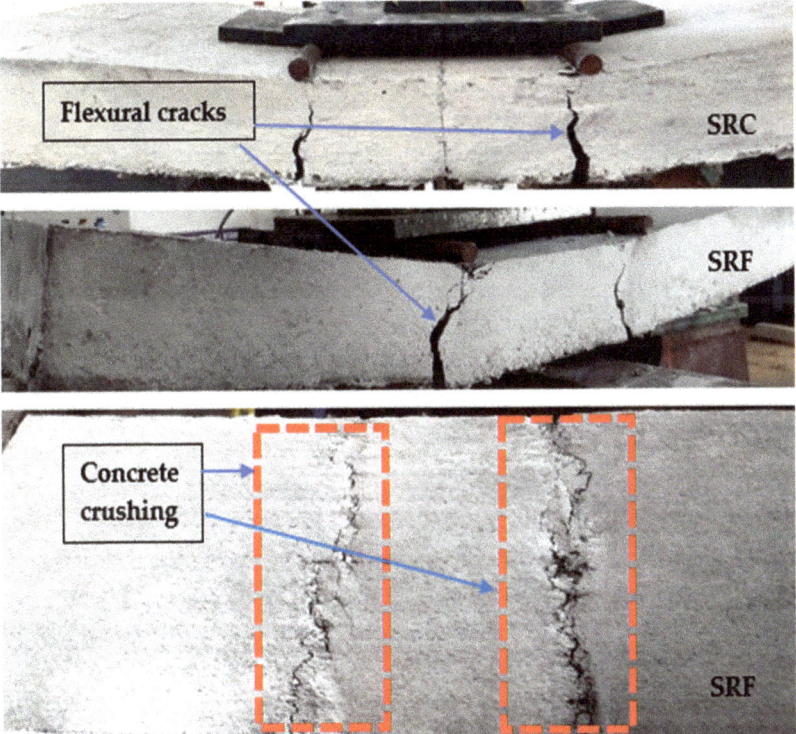

Figure 4. Crack propagation and failure mode of Mode-I slab specimens.

Figure 5. Crack propagation and failure mode of Mode-II slab specimens.

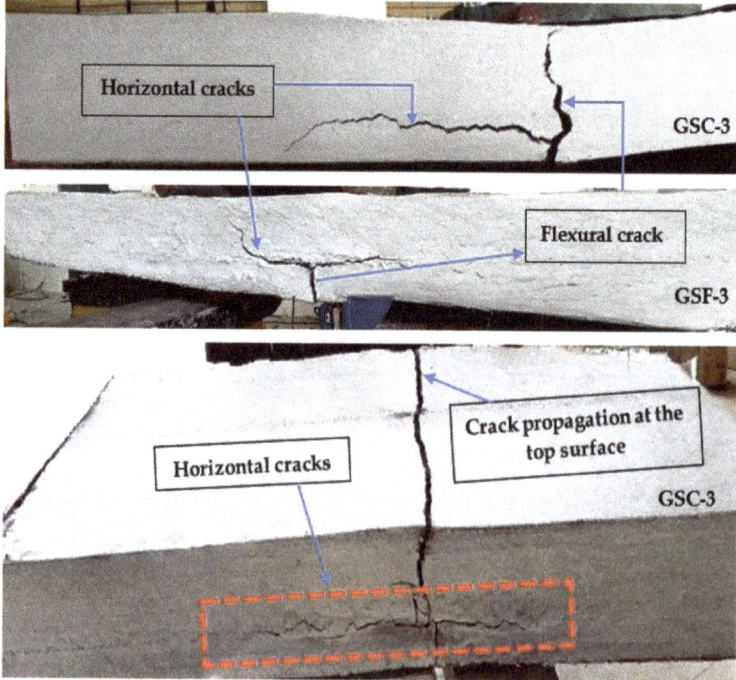

Figure 6. Crack propagation and failure mode of Mode-III slab specimen.

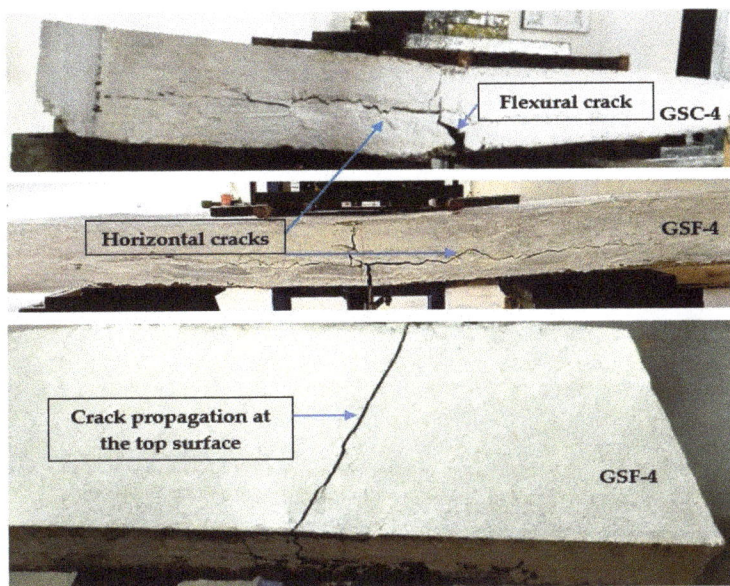

Figure 7. Crack propagation and the failure modes Mode-IV slab specimens.

Table 4. Summary of test results of the slab specimens.

Category	Slab Designation	Trial Numbers	Initial Crack Load (KN)	Ultimate Load (KN)	Modes of Failure
GROUP-I	GSC-2	Trial 1	8.3	16.8	Mode-II
		Trial 2	8.7	17.5	Mode-II
	GSC-3	Trial 1	14.2	23.7	Mode-III
		Trial 2	14.7	24	Mode-III
	GSC-4	Trial 1	6.7	16.4	Mode-IV
		Trial 2	6.3	15.5	Mode-IV
	SRC	Trial 1	16.3	23.8	Mode-I
		Trial 2	15.7	24	Mode-I
GROUP-II	GSF-2	Trial 1	5.9	17.1	Mode-II
		Trial 2	5.7	16.7	Mode-II
	GSF-3	Trial 1	14.1	23.9	Mode-II
		Trial 2	14	23.7	Mode-II
	GSF-4	Trial 1	5.2	15.3	Mode-IV
		Trial 2	5.6	15.9	Mode-IV
	SRF	Trial 1	18.5	27.3	Mode-I
		Trial 2	19.1	28.5	Mode-I

4.2. Load–Deflection Behaviour

Details relating to load–deflection of the Group-I and Group-II slab specimens tested after 56 days of curing were plotted. All the slab specimens showed linear elastic behaviour up to the initial crack, and beyond that, the behaviour was nonlinear. Details of the load-deflection behaviour of slabs GSC-2 and GSF-2; GSC-3 and GSF-3; GSC-4 and GSF-4; and SRC and SRF are shown in Figures 8–11, respectively. The average ultimate load-carrying capacity of the Group-I specimens (OPC slab) GSC-2, GSC-3, GSC-4 and SRC was 17.15 kN, 23.85 kN, 15.95 kN and 23.9 kN, respectively. The average ultimate load-carrying capacity of the Group-II specimens (fly ash slab) GSF-2, GSF-3, GSF-4 and SRF was 16.9 kN, 23.8 kN, 15.6 kN and 27.9 kN, respectively. The SRF slab specimens reinforced with steel showed a 17% increase in their average ultimate load-carrying capacity compared with SRC slabs.

The average ultimate load-carrying slab specimens GSC-2, GSC-4, GSF-2 and GSF-4 showed 28%, 33%, 29% and 34%, respectively, less than the SRC slab specimens. However, the average ultimate load-carrying capacity of GSC-3 and GSF-3 was the same as that of SRC. In the case of GSC-4 and GSF-4, flexural cracks at an average load of 6.5 kN and 5.4 kN, respectively, with the subsequent formation of horizontal cracks over the entire span of the slab, were observed. As the horizontal cracks formed at the earlier stage due to the debonding of the sheets, the reduction in ultimate load was therefore observed. The deflection in the slab specimens reinforced with GFRP sheets was less than the deflection in the slab specimens reinforced with steel bars.

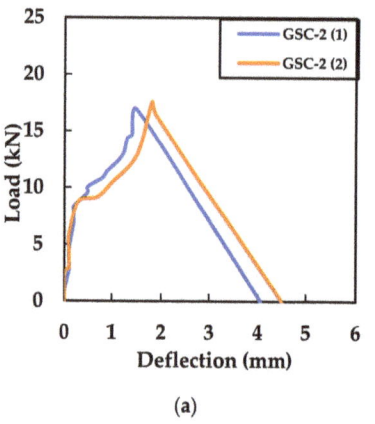
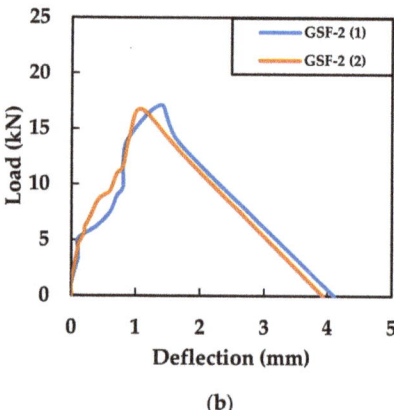

Figure 8. Load–deflection behaviour of (**a**) GSC-2 and (**b**) GSF-2.

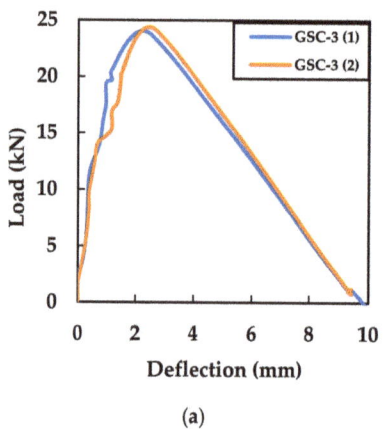
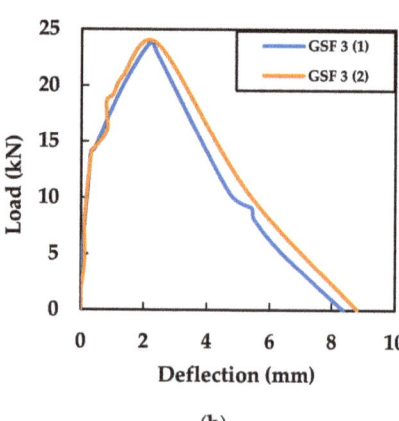

Figure 9. Load–deflection behaviour of (**a**) GSC-3 and (**b**) GSF-3.

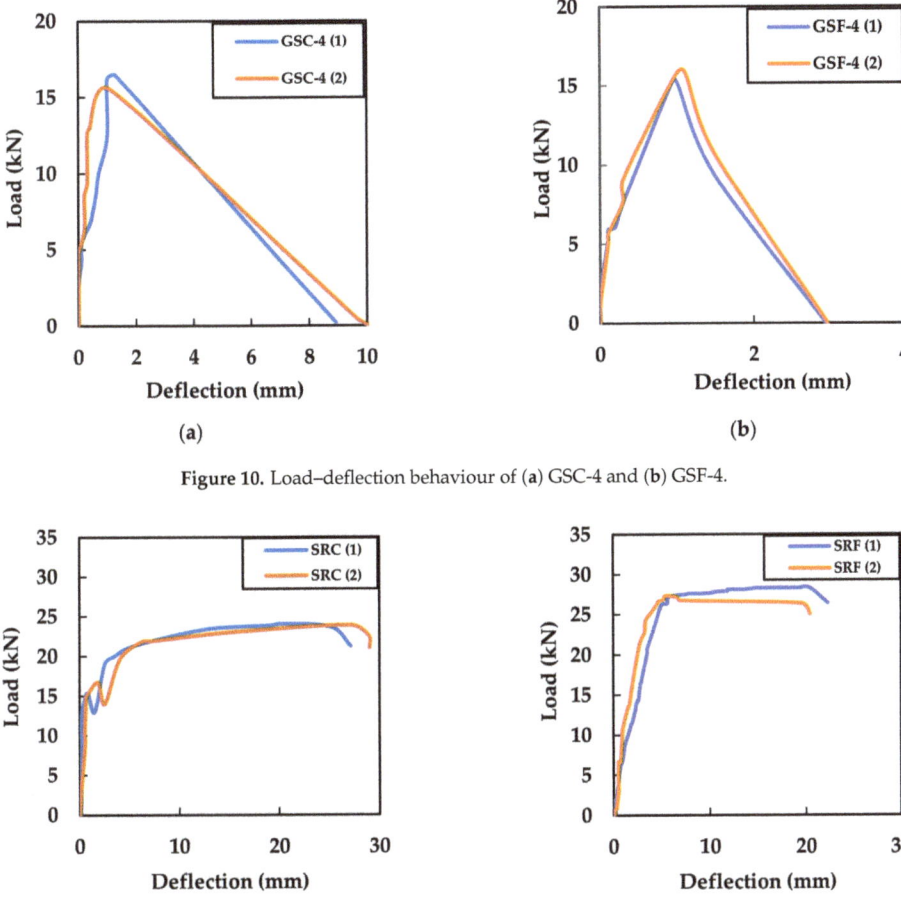

Figure 10. Load–deflection behaviour of (**a**) GSC-4 and (**b**) GSF-4.

Figure 11. Load–deflection behaviour of (**a**) SRC and (**b**) SRF.

4.3. Strain Distribution

The top positive strain indicates the compressive strain in concrete, while the bottom negative strain indicates the tensile strain in the GFRP sheets/steel bars. For each load increment, the strain values experienced by both strain gauges at the mid-span region of the slab were plotted. The load vs. strain variations in the slab specimens were GSC-2 and GSF-2; GSC-3 and GSF-3; GSC-4 and GSF-4; and SRC and SRF are shown in Figures 12–15, respectively. The top strain development in the Group-I specimens (OPC slab) and Group-II specimens (HVFA slab) reinforced with steel bars/GFRP sheets range from 2985 µ to 3099 µ. The bottom strain development in the Group-I specimens (OPC slab) and Group-II specimens (HVFA slab) reinforced with GFRP sheets ranges from 3317 µ to 4315 µ. However, the strain measured at the bottom of the steel bars in the SRC and SRF at failure ranges from 19,080 µ to 21,900 µ, which was almost 21% of the failure strain of the GFRP sheets. The result shows that the top and the bottom strain of GSC-3 and GSF-3 is higher when compared with the slab reinforced with GFRP sheets of two layers and four layers.

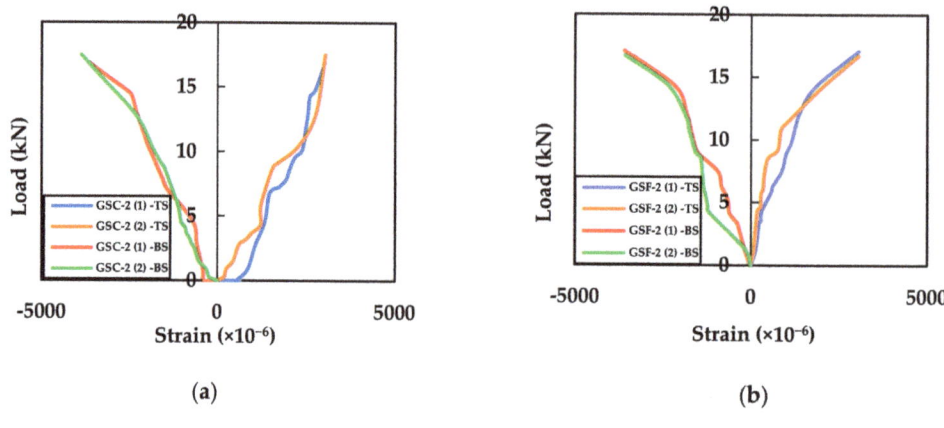

Figure 12. Load–strain behaviour of (**a**) GSC-2 and (**b**) GSF-2. TS—top strain, BS—bottom strain.

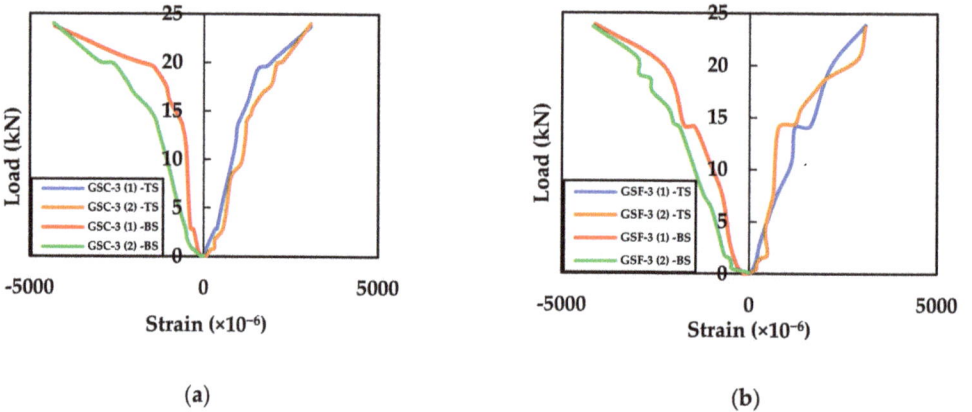

Figure 13. Load–strain behaviour of (**a**) GSC-3 and (**b**) GSF-3. TS—top strain, BS—bottom strain.

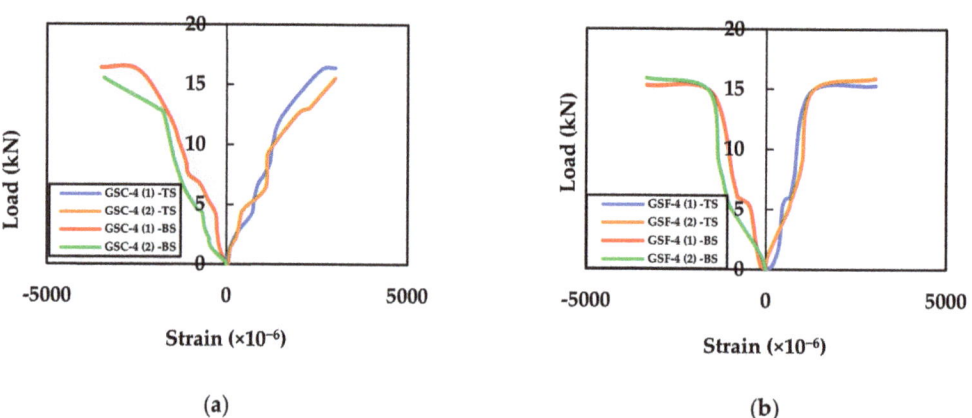

Figure 14. Load–strain behaviour of (**a**) GSC-4 and (**b**) GSF-4. TS—top strain, BS—bottom strain.

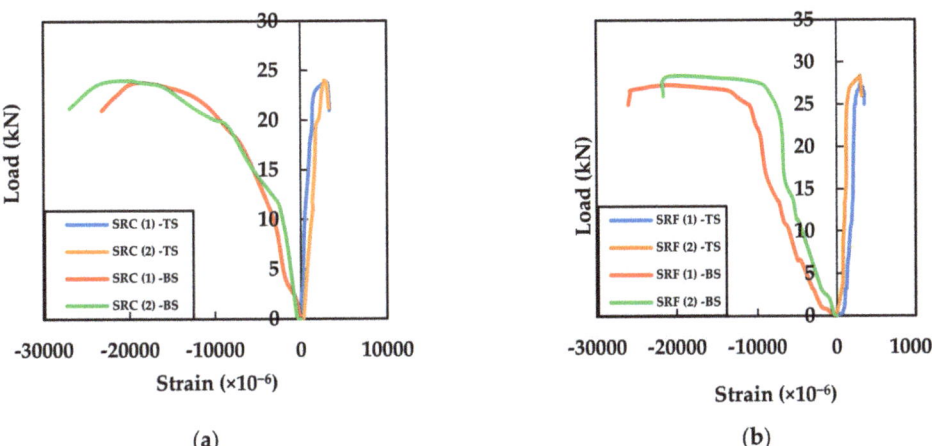

Figure 15. Load–strain behaviour of (a) SRC and (b) SRF. TS—top strain, BS—bottom strain.

4.4. Moment–Curvature

The moment–curvature diagram defines the ultimate capacity of the slab elements and is also used to access the energy absorption capacity of the slab elements. The moment–curvature relationship was calculated for all the slab specimens based upon the top strain (OPC/HVFA concrete) and the bottom strain (steel/GFRP sheets). The moment–curvature relationship of GSC-2 and GSF-2; GSC-3 and GSF-3; GSC-4 and GSF-4 and SRC and SRF are shown in Figures 16–19, respectively.

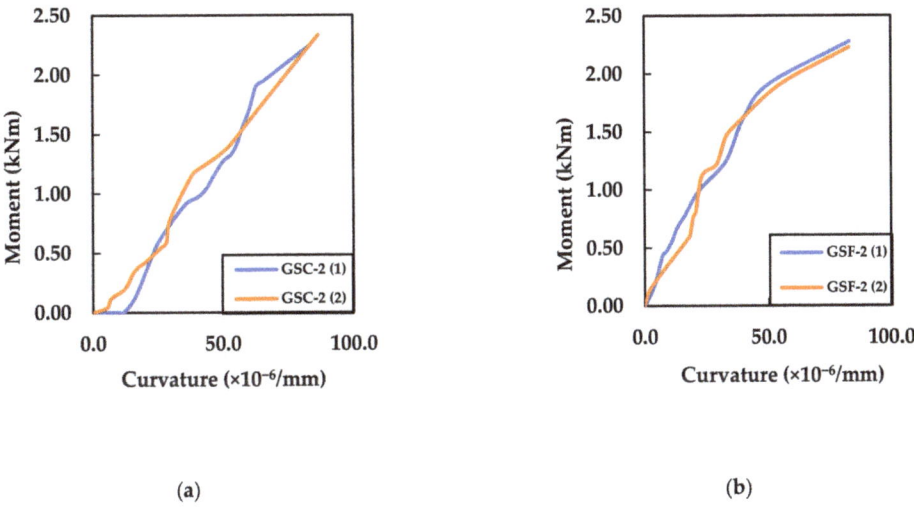

Figure 16. Moment–curvature behaviour of (a) GSC 2 and (b) GSF 2.

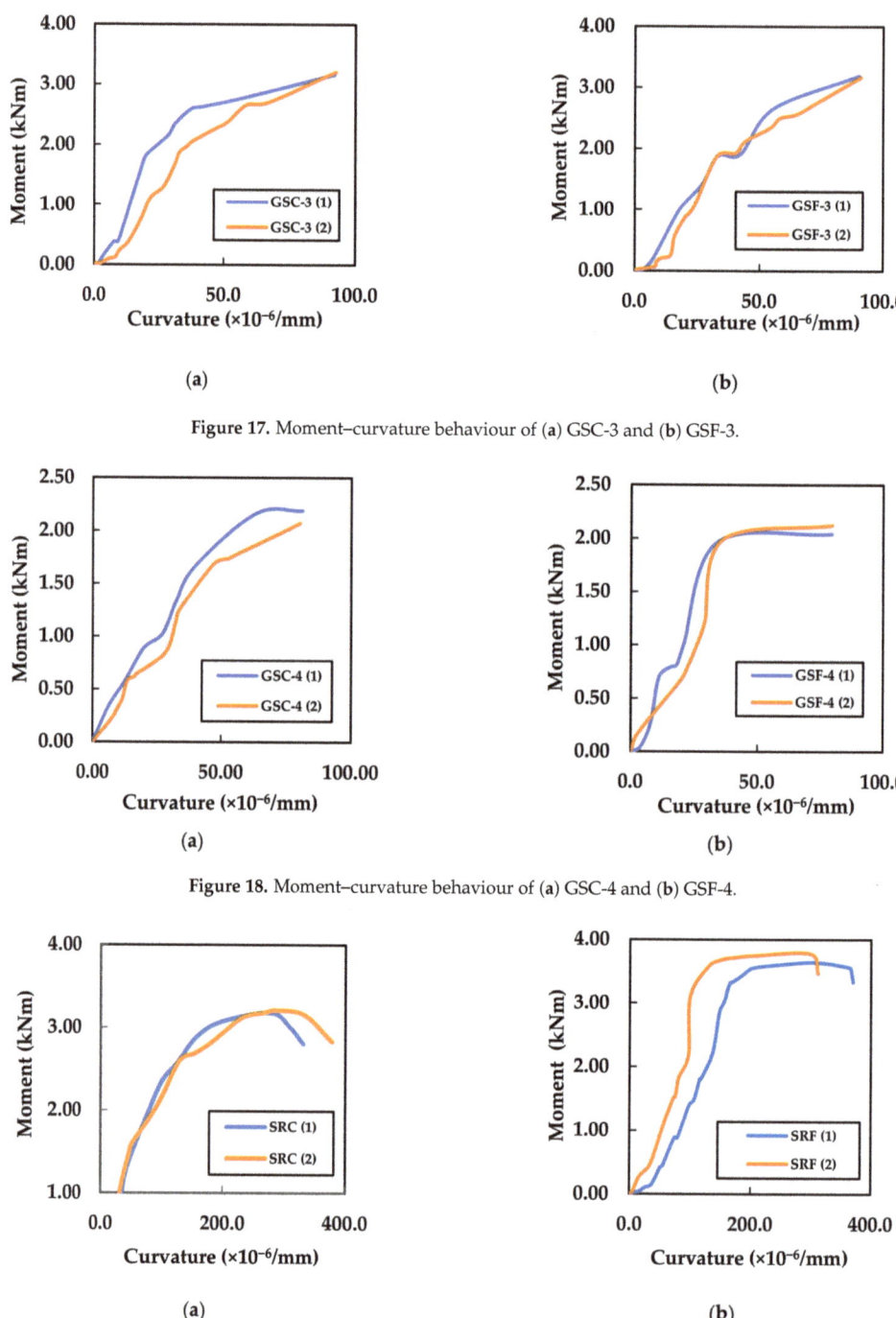

Figure 17. Moment–curvature behaviour of (a) GSC-3 and (b) GSF-3.

Figure 18. Moment–curvature behaviour of (a) GSC-4 and (b) GSF-4.

Figure 19. Moment–curvature behaviour of (a) SRC and (b) SRF.

The following equation was used for the calculation of curvature (∅),

$$\varnothing = \frac{\varepsilon_c + \varepsilon_r}{d} \quad (1)$$

where,
    d—Effective depth of the slab
    ε_r—Tensile strain in the reinforcement (steel/GFRP sheets)
    ε_c—Compressive strain in concrete

Moment vs. curvature relationships showed the average ultimate moment-carrying capacity of the slabs of Group-I: GSC-2, GSC-3, GSC-4 and SRC as 2.28 kNm, 3.18 kNm, 2.13 kNm and 3.18 kNm, respectively. The average ultimate moment-carrying capacity of the slabs of Group-II: GSF-2, GSF-3, GSF-4 and SRF were 2.25 kNm, 3.17 kNm, 2.08 kNm and 3.25 kNm, respectively. The moment-carrying capacity of the slabs reinforced with GFRP sheets (GSC-3 and GSF-3) was the same as that of the slab specimen reinforced with the steel bars (SRC). Table 5 shows the details of the overall performance details of the OPC/HVFA concrete slab reinforced with steel rod/GFRP sheets.

Table 5. Overall performance details of the concrete slab reinforced with steel rod/GFRP sheets.

Category	Slab Designation	Max. Load (P_u) (kN)	Ultimate Moment (M_{Exp}) (kNm)	Ultimate Strain in Concrete at Max Load (ε_{cu}) %	Ultimate Strain in Reinforcement at Max Load (ε_f) %
GROUP-I	GSC-2 (1)	16.8	2.24	0.31	0.36
	GSC-2 (2)	17.5	2.33	0.31	0.39
	GSC-3 (1)	23.7	3.16	0.31	0.43
	GSC-3 (2)	24	3.20	0.31	0.43
	GSC-4 (1)	16.4	2.19	0.30	0.35
	GSC-4 (2)	15.5	2.07	0.30	0.34
	SRC (1)	23.8	3.17	0.29	2.01
	SRC (2)	24.0	3.20	0.31	1.91
GROUP-II	GSF-2 (1)	17.1	2.28	0.31	0.36
	GSF-2 (2)	16.7	2.22	0.30	0.36
	GSF-3 (1)	23.9	3.19	0.31	0.41
	GSF-3 (2)	23.7	3.16	0.31	0.42
	GSF-4 (1)	15.3	2.04	0.30	0.33
	GSF-4 (2)	15.9	2.12	0.30	0.33
	SRF (1)	27.3	3.70	0.30	2.03
	SRF (2)	28.5	3.64	0.31	2.19

5. Numerical Analysis and Consecutive Models

The nonlinear finite-element analysis (NLFEA) comprises modelling of the slab specimens, introducing the element type, material properties, boundary conditions, meshing and loading. To obtain accurate results from the numerical simulations, all the necessary components such as OPC/HVFA concrete, steel rods and GFRP sheets were modelled properly with the aid of nonlinear stress–strain graphs and the material properties. Numerical analysis using ANSYS Workbench 2022-R1 was carried out to simulate the OPC/HVFA concrete slabs reinforced with steel bars/GFRP sheets.

From the experimental investigation, it was observed that the OPC/HVFA concrete slabs reinforced with three layers of GFRP sheets had the highest ultimate load-carrying capacity when compared with the slabs reinforced with two and four layers of GFRP sheets. Hence, the slab specimens SRC, SRF, GSC-3 and GSF-3 were analysed using nonlinear finite-element analysis.

5.1. Considerations for Element Types

In ANSYS, M25 grade concrete was modelled using SOLID 65, which is an eight-noded element consisting of 3 degrees of freedom in x, y and z directions and capable of cracking in three orthogonal directions. The GFRP sheet was modelled with four-noded SHELL 181 elements. Two-noded LINK 180 element was used to model the steel reinforcement [39]. The bonded contact was used between the GFRP sheet and concrete to prevent separation between them.

5.2. Modelling and Numerical Solution

The modelling of the slab specimens was performed via a geometry design modeller in the ANSYS Workbench 2022-R1. The effect of crack pattern, stress, strain, ultimate load, ultimate deflection and displacement of concrete with different types of end conditions could be analysed in the ANSYS [40–42]. Geometry with the support conditions and load points of application and 3D meshed modelling of the slab specimen are shown in Figure 20. Steel bars/GFRP sheet reinforcement with a sheet thickness of 3 mm was provided at 20 mm from the bottom.

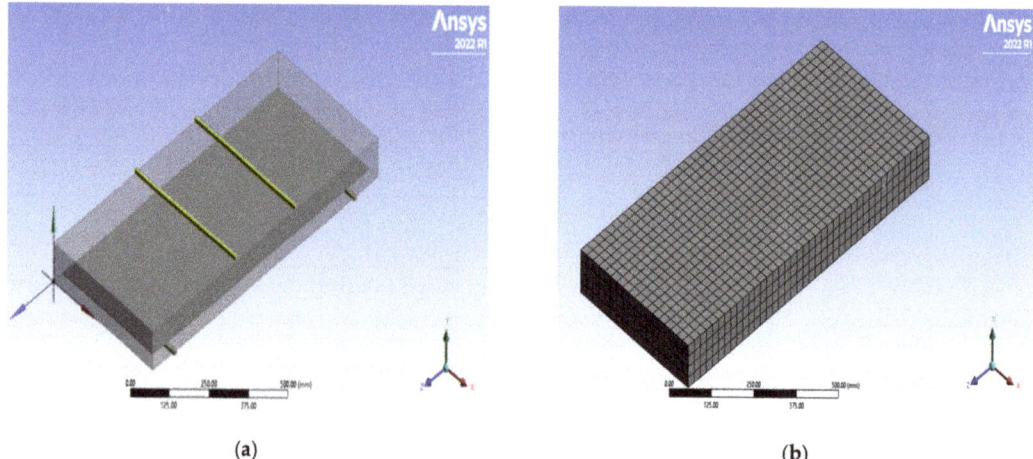

(a)   (b)

Figure 20. Slab specimen (**a**) Geometry and (**b**) 3D meshed model.

In NLFEM, an incremental loading that was the same as the sequence of loading used for the experiment was applied until the failure of the specimens. The load–deflection parameters were recorded during the loading step. The ultimate deflection of the SRC, SRF, GSC-3 and GSF-3 slabs obtained from the numerical analysis is shown in Figure 21. The ultimate load of the slab specimen SRC, SRF, GSC-3 and GSF-3 are 23 kN, 27 kN, 23.5 kN and 23 kN, respectively, with the ultimate deflection of 15.9 mm, 16.4 mm, 2.5 mm and 2.2 mm, respectively.

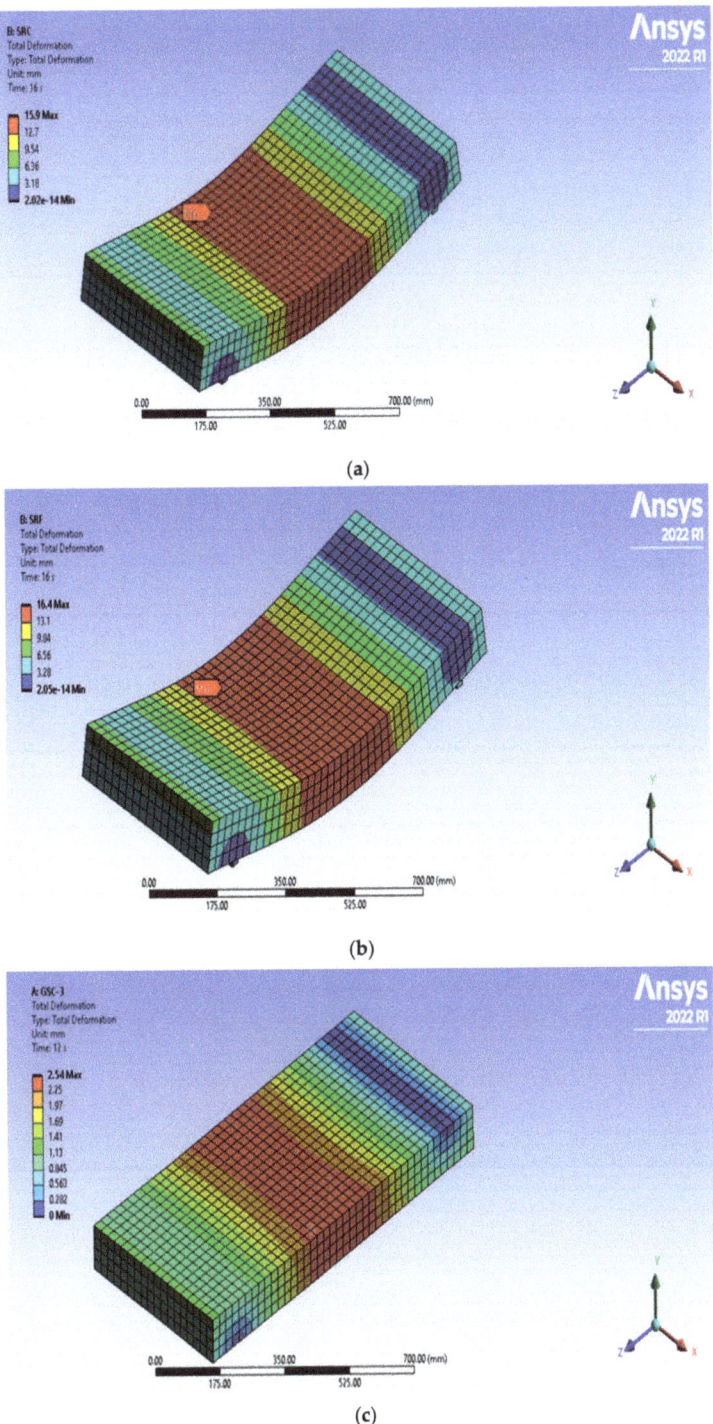

Figure 21. *Cont.*

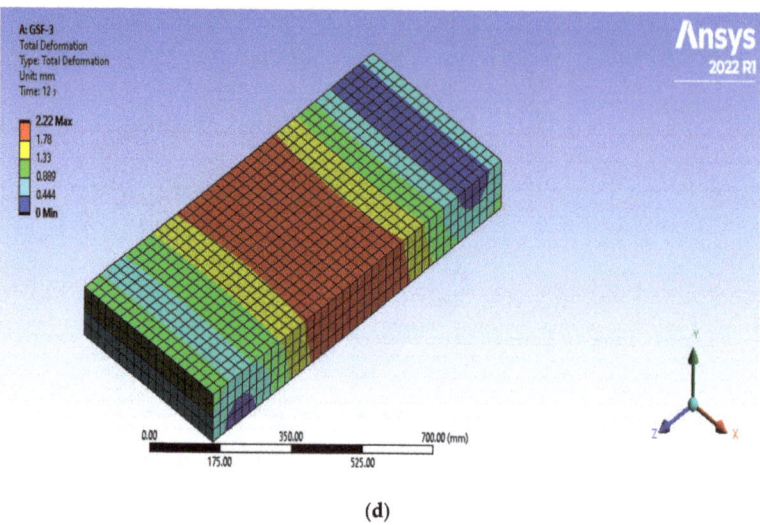

(d)

Figure 21. Ultimate deflection of slab (**a**) SRC, (**b**) SRF, (**c**) GSC-3, (**d**) GSF-3.

5.3. Comparison of Experimental Results with NLFEA Results

The comparison between experimental and NLFEA results is shown in Table 6. Less than 10% difference in the ultimate load and ultimate deflection of the SRC, SRF, GSC-3 and GSF-3 was noticed between the experimental and NLFEA results. Figure 22 shows the comparison of the load–deflection relationship of the SRC and SRF, and Figure 23 shows the comparison of the load–deflection relationship of the GSC-3 and GSF-3 obtained from the NLFEA analysis and the experimental investigation. From the results, it is observed that both the experimental and numerical results are in good correlation. Hence, ANSYS 2022-R1 software can be used for the analysis of fly-ash concrete slabs reinforced with a GFRP sheet.

Table 6. Comparison between experimental and numerical results.

Specimen	Ultimate Load (kN)		Deflection at Mid-Span (mm)	
	Experimental	NLFEA (ANSYS)	Experimental	NLFEA (ANSYS)
SRC	24	23	16.2	15.9
SRF	28.5	27	17.9	16.4
GSC-3	24	23.5	2.7	2.5
GSF-3	23.9	23	2.4	2.2

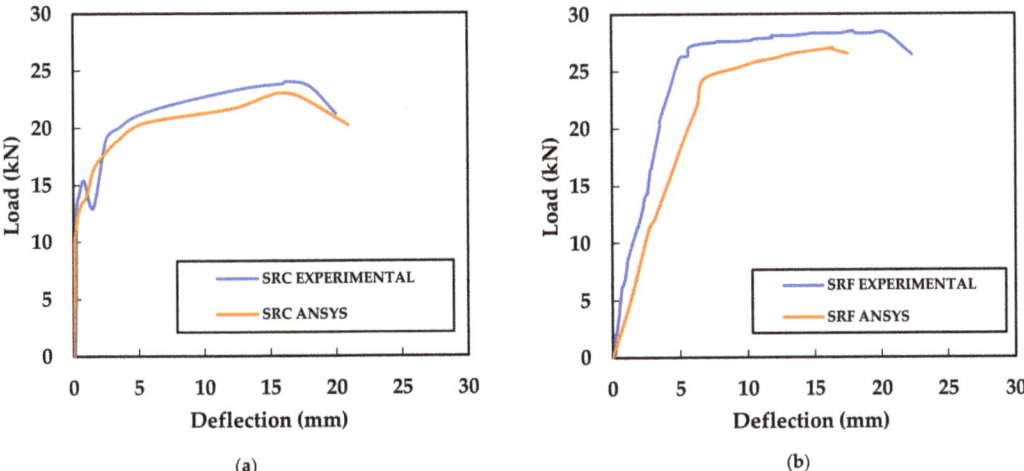

Figure 22. Comparison between experimental and numerical load–deflection relationship of (**a**) SRC and (**b**) SRF.

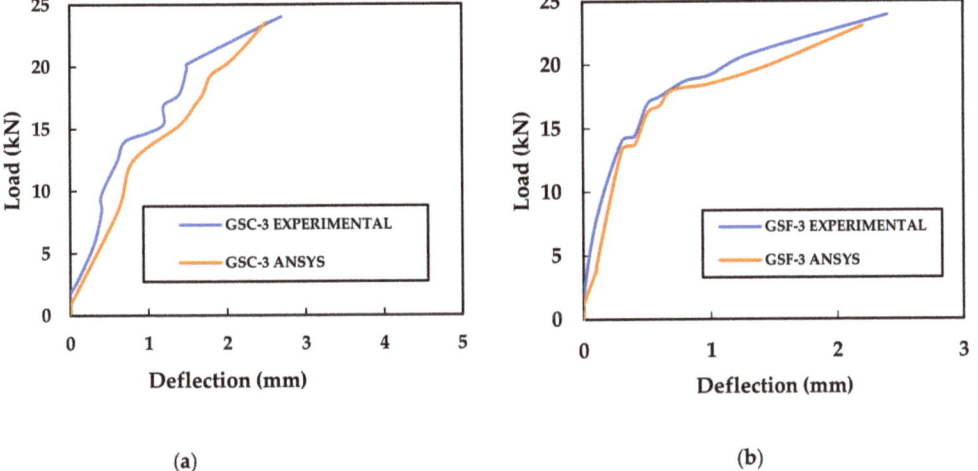

Figure 23. Comparison between experimental and numerical load–deflection behaviour of (**a**) GSC-3 and (**b**) GSF-3.

6. Conclusions

This study presents the results of an experimental investigation involving sixteen simply supported slab specimens made of OPC/HVFA concrete reinforced with steel bars/GFRP sheets.

1. HVFA slabs reinforced with the steel bars (SRF) recorded a 17% increase in their ultimate load-carrying capacity compared with the OPC slabs reinforced with the steel bars (SRC).
2. All the specimens failed due to the formation of flexural cracks that propagate to the top surface at failure with concrete crushing. Slabs reinforced with two layers of GFRP sheets failed in the formation of flexural cracks under the two-loading point. However,

the slab reinforced with three layers and four layers of GFRP sheets showed flexural cracks as well as horizontal cracks.
3. The average ultimate load-carrying capacity of OPC/HVFA concrete slabs reinforced with three layers of GFRP sheets (GSC-3/GSF-3) has the same strength as that of slabs reinforced with the steel bars (SRC).
4. The ultimate average load-carrying capacity of a slab reinforced with three layers of GFRP sheets (GSC-3 and GSF-3) is more than that of the slabs reinforced with two and four layers (GSC-2, GSC-4, GSF-2 and GSF-4) by 39%, 49%, 41% and 53%, respectively.
5. Less than 10% difference in the ultimate load and ultimate deflection of SRC, SRF, GSC-3 and GSF-3 was observed between the experimental and NLFEM results. Hence, ANSYS Workbench 2022-R1 software could be used for the numerical analysis of fly-ash concrete slabs reinforced with a GFRP sheet.

From this study, it is evident that a one-way slab cast with OPC concrete/high-volume fly-ash concrete could be reinforced with GFRP sheets instead of steel bars. This study also reinstates the potential use of high-volume fly ash as a replacement of cement in concrete slab. Thus, a reduction in OPC content in concrete could be an effective way of mitigating the effect of greenhouse-gas emissions, leading to sustainable construction.

Author Contributions: C.S.M., K.P. and P.V.A.R.: Conceptualised the model and conducted the experiments. P.S.J., J.R., B.G.A.G. and K.R.: Supervised the research as well as the analysis of results. P.S.J., J.R., C.S.M. and K.P.: Introduced the idea of static loading in this project; designed the slab; wrote, reviewed and submitted the paper; and collaborated in and coordinated the research. P.S.J., J.R., B.G.A.G. and K.R.: Suggested and chose the journal for submission. P.S.J., B.G.A.G. and K.R.: Participated in the manuscript revision phase. All authors have read and agreed to the published version of the manuscript.

Funding: This research received no external funding.

Institutional Review Board Statement: Not applicable.

Informed Consent Statement: Not applicable.

Data Availability Statement: The data presented in this study are available on request from the corresponding author.

Conflicts of Interest: This manuscript has not been submitted to, nor is it under review by, another journal or other publishing venue. The authors have no affiliation with any organisation with a direct or indirect financial interest in the subject matter discussed in the manuscript. The authors declare no conflict of interest.

References

1. Ji, H.; Son, B.; Ma, Z. Evaluation of Composite Sandwich Bridge Decks with Hybrid FRP-Steel Core. *J. Bridg. Eng.* **2009**, *14*, 36–44. [CrossRef]
2. Shin, Y.S.; Lee, C. Flexural behaviour of reinforced concrete beams strengthened with carbon fibre-reinforced polymer laminates at different levels of sustaining load. *ACI Struct. J.* **2003**, *100*, 231–239.
3. Teng, J.G.; Chen, J.F.; Smith, S.T.; Lam, L. Behaviour and strength of FRP-strengthened RC structures: A state-of-the-art review. *Proc. Inst. Civ. Eng.-Struct. Build.* **2003**, *156*, 51–62. [CrossRef]
4. Djamaluddin, R.; Irmawaty, R.; Tata, A. Flexural Capacity of Reinforced Concrete Beams Strengthened Using GFRP Sheet after Fatigue Loading for Sustainable Construction. *Key Eng. Mater.* **2016**, *692*, 66–73. [CrossRef]
5. Sethi, A.K.; Kinjawadekar, T.A.; Nagarajan, P.; Shashikala, A.P. Design of Flexural Members Reinforced with GFRP Bars. *IOP Conf. Ser. Mater. Sci. Eng.* **2020**, *936*, 012036. [CrossRef]
6. Abdalla, H.A. Evaluation of deflection in concrete members reinforced with fibre reinforced polymer (FRP) bars. *Compos. Struct.* **2002**, *56*, 63–71. [CrossRef]
7. Ferdous, W.; Manalo, A.; Aravinthan, T. Effect of beam orientation on the static behaviour of phenolic core sandwich composites with different shear span-to-depth ratios. *Compos. Struct.* **2017**, *168*, 292–304. [CrossRef]
8. Manalo, A. Behaviour of fibre composite sandwich structures under short and asymmetrical beam shear tests. *Compos. Struct.* **2013**, *99*, 339–349. [CrossRef]

9. Maranan, G.; Manalo, A.; Benmokrane, B.; Karunasena, W.; Mendis, P. Evaluation of the flexural strength and serviceability of geopolymer concrete beams reinforced with glass-fibre-reinforced polymer (GFRP) bars. *Eng. Struct.* **2015**, *101*, 529–541. [CrossRef]
10. Bouguerra, K.; Ahmed, E.; El-Gamal, S.; Benmokrane, B. Testing of full-scale concrete bridge deck slabs reinforced with fibre-reinforced polymer (FRP) bars. *Constr. Build. Mater.* **2011**, *25*, 3956–3965. [CrossRef]
11. ACI. *ACI Guide for the Design and Construction of Concrete Reinforced with FRP Bars*; Report 440R-96; ACI: Detroit, MI, USA, 2001; pp. 1023–1034.
12. ACI. *State-of-the-Art Report on Fibre Reinforced Plastic (FRP) Reinforcement for Concrete Structures*; ACI: Detroit, MI, USA, 2004.
13. Grace, N.F.; Abdel-Sayed, G.; Ragheb, W.F. Strengthening of Concrete Beams Using Innovative Fibre-Reinforced Polymer Fabric. *ACI Struct. J.* **2002**, *99*, 692–700.
14. Li, V.C.; Wang, S. Flexural behaviours of glass fibre reinforced polymer (GFRP) reinforced engineered cementitious composite beams. *ACI Mater. J.* **2002**, *99*, 11–20.
15. Razaqpur, A.G.; Sevecova, D.; Cheung, M.S. Rational method for calculating deflection of fibre-reinforced polymer reinforced beams. *ACI Struct. J.* **2000**, *97*, 175–184.
16. Sen, R.; Mullins, G.; Salem, T. Durability of E-glass/vinyl ester reinforcement in alkaline solution. *ACI Struct. J.* **2002**, *99*, 369–375.
17. Chinnasamy, M.; Ajithkumar, R.; Singh, A.; Yangzom, D.; Parvati, T.; Joanna, P. Comparative study on the behaviour of textile reinforced concrete slab with engineered cementitious composite slab. *Mater. Today Proc.* **2020**, *33*, 1175–1180. [CrossRef]
18. Laila, L.R.; Gurupatham, B.G.A.; Roy, K.; Lim, J.B.P. Effect of super absorbent polymer on microstructural and mechanical properties of concrete blends using granite pulver. *Struct. Concr.* **2020**, *22*, E898–E915. [CrossRef]
19. He, Z.; Zhu, X.; Wang, J.; Mu, M.; Wang, Y. Comparison of CO_2 emissions from OPC and recycled cement production. *Constr. Build. Mater.* **2019**, *211*, 965–973. [CrossRef]
20. Sivaramakrishnan, R.; Anbarasu, E. Experimental Study on High-Performance Concrete by 40% Partial Replacement of Cementitious Material with Micro Silica, GGBS & Fly-Ash. *IJESC* **2020**, *10*, 25227–25231.
21. Joanna, P.S.; Rooby, J.; Prabhavathy, A.; Preetha, R.; Pillai, C.S. Behaviour of reinforced concrete beams with 50 per cent fly ash. *Int. J. Civ. Eng. Technol.* **2013**, *4*, 36–48.
22. Partha, S.D.; Pradip, N.; Prabir, K.S. Strength and Permeation Properties of Slag Blended Fly Ash Based Geopolymer Concrete. *Adv. Mater. Res.* **2013**, *651*, 168–173. [CrossRef]
23. Nazari, A.; Riahi, S. Improvement compressive strength of concrete in different curing media by Al_2O_3 nanoparticles. *Mater. Sci. Eng. A* **2011**, *528*, 1183–1191. [CrossRef]
24. Hosseini, P.; Hosseinpourpia, R.; Pajum, A.; Khodavirdi, M.M.; Izadi, H.; Vaezi, A. Effect of nano-particles and aminosilane interaction on the performances of cement-based composites: An experimental study. *Constr. Build. Mater.* **2014**, *66*, 113–124. [CrossRef]
25. Maravelaki-Kalaitzaki, P.; Agioutantis, Z.; Lionakis, E.; Stavroulaki, M.; Perdikatsis, V. Physico-chemical and mechanical characterization of hydraulic mortars containing nano-titania for restoration applications. *Cem. Concr. Compos.* **2013**, *36*, 33–41. [CrossRef]
26. Laila, L.R.; Gurupatham, B.G.A.; Roy, K.; Lim, J.B.P. Influence of super absorbent polymer on mechanical, rheological, durability, and microstructural properties of self-compacting concrete using non-biodegradable granite pulver. *Struct. Concr.* **2020**, *22*, E1093–E1116. [CrossRef]
27. Rana, A.K.; Rana, S.; Kumari, A.; Kiran, V. Significance of nanotechnology in construction engineering. *IJRTE* **2009**, *1*, 46.
28. Arezoumandi, M.; Volz, J.S.; Myers, J.J. Shear Behavior of High-Volume Fly Ash Concrete versus Conventional Concrete. *J. Mater. Civ. Eng.* **2013**, *25*, 1506–1513. [CrossRef]
29. Rao, R.M.; Mohan, S.; Sekar, S.K. Shear Resistance of High Volume Fly ash Reinforced Concrete Beams without Web Reinforcement. *Int. J. Civ. Struct. Eng.* **2001**, *1*, 986–993.
30. Agarwal, V.; Gupta, S.M.; Sachdeva, S.N. High volume fly ash concrete—A green concrete. *J. Environ. Res. Dev.* **2012**, *6*, 884–887.
31. Lowe, D.; Roy, K.; Das, R.; Clifton, C.; Lim, J. Full-scale experiments on splitting behaviour of concrete slabs in steel-concrete composite beams with shear stud connection. *Structures* **2020**, *23*, 126–138. [CrossRef]
32. Madan, C.S.; Munuswamy, S.; Joanna, P.S.; Gurupatham, B.G.A.; Roy, K. Comparison of the Flexural Behavior of High-Volume Fly AshBased Concrete Slab Reinforced with GFRP Bars and Steel Bars. *J. Compos. Sci.* **2022**, *6*, 157. [CrossRef]
33. Kim, H.-K.; Lee, H. Use of power plant bottom ash as fine and coarse aggregates in high-strength concrete. *Constr. Build. Mater.* **2011**, *25*, 1115–1122. [CrossRef]
34. Balakrishnan, B.; Awal, A.A. Mechanical Properties and Thermal Resistance of High Volume Fly Ash Concrete for Energy Efficiency in Building Construction. *Key Eng. Mater.* **2016**, *678*, 99–108. [CrossRef]
35. Aravind Raj, P.S.; Divahar, R.; Sangeetha, S.P.; Naveen Kumar, K.; Ganesh, D.; Sabitha, S. Sustainable Development of Structural Joint made using High Volume Fly-Ash concrete. *Int. J. Adv. Sci. Technol.* **2020**, *29*, 6850–6857.
36. Abadel, A.; Abbas, H.; Albidah, A.; Almusallam, T.; Al-Salloum, Y. Effectiveness of GFRP strengthening of normal and high strength fibre reinforced concrete after exposure to heating and cooling. *Eng. Sci. Technol. Int. J.* **2022**, *36*, 101147. [CrossRef]
37. Shukla, S.; Waghmare, M.V. Strengthening of RC Column Using GFRP. *Int. J. Res. Appl. Sci. Eng. Technol.* **2022**, *10*, 1217–1224. [CrossRef]
38. *ANSYS Mechanical APDL Verification Set*; ANSYS Inc.: Canonsburg, PA, USA, 2014.

39. Sandrasekaran, S.; Praveen Kumar, A. Numerical Modeling of Square Steel Members Wrapped by CFRP Composites. *Int. J. Innov. Technol. Explor. Eng.* **2019**, *8*, 3082–3087. [CrossRef]
40. Adam, M.A.; Erfan, A.M.; Habib, F.A.; El-Sayed, T.A. Structural Behavior of High-Strength Concrete Slabs Reinforced with GFRP Bars. *Polymers* **2021**, *13*, 2997. [CrossRef]
41. Jayajothi, P.; Kumutha, R.; Vijai, K. Finite element analysis of FRP strengthened RC beams using Ansys. *Asian J. Civ. Eng.* **2013**, *14*, 631–642.
42. Gherbi, A.; Dahmani, L.; Boudjemia, A. Study on two way reinforced concrete slab using Ansys with different boundary conditions and loading. *World Acad. Sci. Eng. Technol. Int. J. Civ. Environ. Eng.* **2018**, *12*, 1151–1156.

Article

Plastic Hinge Length Mechanism of Steel-Fiber-Reinforced Concrete Slab under Repeated Loading

Pradeep Sivanantham [1], Beulah Gnana Ananthi Gurupatham [2], Krishanu Roy [3,*], Karthikeyan Rajendiran [1] and Deepak Pugazhlendi [1]

[1] Department of Civil Engineering SRMIST, Chennai 603203, India; pradeeps@srmist.edu.in (P.S.); karthikeyan9000@gmail.com (K.R.); deepakpugazhlendi@gmail.com (D.P.)
[2] Division of Structural Engineering, College of Engineering Guindy Campus, Anna University, Chennai 600025, India; beulah28@annauniv.edu
[3] School of Engineering, The University of Waikato, Hamilton 3216, New Zealand
* Correspondence: krishanu.roy@waikato.ac.nz

Abstract: The plastic hinge is the most critical damaging part of a structural element, where the highest inelastic rotation would occur. In particular, flexural members develop maximum bending abilities at that point. The current paper experimentally investigates the influence of steel fiber reinforcement at the plastic hinge length of the concrete slab under repeated loading, something which has not been reported by any researcher. Mechanical properties such as compressive strength and tensile strength of M20-grade concrete that are used for casting specimens are tested through the compressive strength test and the split tensile strength test. Six different parameters are considered in the slab while carrying out this study. First, the conventional concrete slab and then the steel-fiber-reinforced slab were cast. The plastic hinge length of the slab was calculated through different empirical expressions taken from methods by Baker, Sawyer, Corley, Mattock, Paulay, Priestley and Park. Finally, the steel fiber was added as per methods detailed by Paulay, Priestley and Park in the plastic hinge length mechanism in the concrete slab at 70 mm and 150 mm separately. The results arrived through experimental investigation by applying repeated loads to the slab, indicating that steel fibers used at critical sections of plastic hinge length provide similar strength, displacement, and performance as that of the conventional RCC slab and fully steel-fiber-reinforced concrete slabs. Steel fiber at a plastic hinge length of slab has a better advantage over a conventional slab.

Keywords: SFRC (steel-fiber-reinforced concrete); conventional RCC slab; plastic hinge length; repeated loading and composite material

Citation: Sivanantham, P.; Gurupatham, B.G.A.; Roy, K.; Rajendiran, K.; Pugazhlendi, D. Plastic Hinge Length Mechanism of Steel-Fiber-Reinforced Concrete Slab under Repeated Loading. *J. Compos. Sci.* **2022**, *6*, 164. https://doi.org/10.3390/jcs6060164

Academic Editor: Francesco Tornabene

Received: 28 April 2022
Accepted: 31 May 2022
Published: 2 June 2022

Publisher's Note: MDPI stays neutral with regard to jurisdictional claims in published maps and institutional affiliations.

Copyright: © 2022 by the authors. Licensee MDPI, Basel, Switzerland. This article is an open access article distributed under the terms and conditions of the Creative Commons Attribution (CC BY) license (https://creativecommons.org/licenses/by/4.0/).

1. Introduction

The maximum bending moment occurs at the plastic hinge length, which is considered to be the most critically damaged location of the RCC member and will experience more inelastic deflection in the member. In recent years, researchers have become more interested in the length of plastic hinges. The experimental study of plastic hinge length examines the load-carrying capacity and its deformation capability. When the RCC member is subjected to load, the deformation is inelastic, and these inelastic zones result in places where the bending moment is greater. As the applied load is raised further, the zones rotate till the final hinge is formed, and this results in the collapse of the structural member [1].

The plastic hinge length can be obtained by various formulas devised by scientists such as Sawyer, Baker, Priestley, Park, etc. These formulas can be used to find the required plastic hinge for the slab, which is a factor of the concrete grade and reinforcement detailing. It is also a factor of support distance, contra flexure distance, and the geometry of the member [2,3]. Nazaripoor et al. [4] have studied the acoustic emission damage detection by performing three-point bend tests and demonstrating the accumulation of flexural damage for composite panels of different sizes and fiber volume content. Paulay et al. [5]

studied over one thousand specimens representing various types of reinforced concrete (RC) members (beams, columns, and walls). This was then used to develop expressions for the deformations of RC members at yielding or failure under cyclic loading in terms of member geometric and mechanical properties. The yield and ultimate curvature expressions based on the plane-section assumption agree on averages within the test results, but with a large scatter. The same was found to be true for models based on curvatures and the concept of ultimate drift or chord-rotation capacity [5].

The previous study by Paulay et al. [4] was created for analytical purposes, and the analysis was created to study the behavior of RCC members subjected to various loadings and plastic hinge lengths. The properties of steel-fiber-reinforced concrete (SFRCs) are determined by the amount and percentage of fibers introduced into the concrete. The length of the plastic hinge zone is an important design parameter that should be provided with intense confinement to increase the ductility of the member to survive extreme events such as earthquakes. The behavior of plastic hinges is extremely complicated due to the high nonlinearity of the materials, interaction, relative movement between the constituent materials, and strain localization. As a result, the majority of researchers [6,7] used experimental testing to investigate the problem. The plastic hinge zone's performance is crucial for flexural members since it regulates the load-bearing and deformation capabilities of the member. A computational model is developed and validated using existing experimental data, such as load-deflection response, rotational capacity, and reinforcement stress and strain distributions detailed by Qin et al. [8].

The length of the rebar yielding zone is thought to represent the upper bound of the three physical inelastic deformation zones, with its value never exceeding more than twice the effective depth of the cross-section in any of the conditions investigated by Zhao et al. [9]. The diameter of the rebar under tension has a reduced impact on the plastic hinge length and flexural capacity of the member due to its impact on bond strength. The true plastic hinge zone is much smaller than the yielding zone of the member, which comprises the majority of the plastic rotation. None of the existing empirical models for forecasting plastic hinge length have taken all of these crucial aspects into account [9]. A constitutive model has been developed for material non-linear analysis of steel-fiber-reinforced concrete slabs supported on the soil. The energy absorption capacity provided by fiber reinforcement is taken into account in the material constitutive relationship. The plasticity theory is used to explain the elastoplastic behavior of concrete [10].

The RCC member would fail immediately without providing a prior warning when this plastic hinge is formed. For these reasons, understanding the behavior of plastic hinge length formations is critical in RCC construction. Steel fibers have been used as replacements for conventional RCC structures as they have good tensile strength and delay the brittle failure of the concrete. The goal is to contribute to the development of design guidelines that can accurately predict the punching resistance of SFRC flat slabs. Past investigation data show that SFRC has better performance than conventional concrete. To have better performance [11–14], the steel fibers are used at 1.5% of the volume of the concrete.

Holschemacher et al. [15] tested 28 steel-fiber-reinforced concrete (SFRC) slabs under flexure to see the effect length of steel fiber percentage on the energy absorption capacity of concrete slabs with varying concrete strengths. According to the findings from the tests, longer fibers with a greater fiber content were found to absorb more energy. The findings are contrasted with a theoretical prediction based on fiber distribution randomness. The theoretical technique yielded a larger energy absorption than the experimental method. Within the range of fiber volumetric percentages employed in the study, a design technique based on permissible deflection is provided for SFRC slabs [15].

Fiber concrete is a technology that has been studied for several decades, mostly to prevent cracking in specific reinforced concrete constructions. Fiber concrete has recently been examined for use in various applications, such as the total replacement of steel-reinforcing rebar. The behavior of a fiber concrete slab was investigated using a validated numerical model [16–18]. The findings demonstrated that fiber concrete slabs without rebar reinforcement met the limit states required by the different dwellings (spans ranging from 4–5 m). The effect mechanisms of SFRC's strength parameters and slab's structural parameters are clarified using both finite element and theoretical analysis and a technique are devised [16–18]. Service stress limits in SFRC slabs were achieved at higher demands than in RC slabs. When service stress limitations in SFRC slabs were achieved, crack widths were substantially less than conventional crack width limits, indicating that designing for crack widths may be an effective way of addressing serviceability in SFRC slabs [16–18].

Paramasivam et al. [19] conducted 50 tests on reinforced concrete slab specimens; both repaired and unrepaired slabs were examined under static and cyclic stress conditions. The specimens had a rectangular cross-section of 300 × 80 mm and were tested in flexure on a span of 600 mm with a line load at mid-span. Four different kinds of specimen were tested. The performance of statically loaded to failure specimens was tracked, whereas the deterioration of flexural rigidity of cyclically loaded specimens was examined after a preset number of cycles at various stress levels [19]. Steel-fiber-reinforced concrete (SFRC) is used to improve the flexural and cyclic responses of reinforced concrete bridge deck slabs. First, cyclic loads were applied to one plain concrete slab and two SFRC slabs reinforced with mill-cut steel fibers and corrugated steel fibers, respectively. Load-deflection responses, cyclic deformation behaviors, strain and fracture development are all examined. The results show that employing SFRC in deck slabs improves cyclic deformation performance, reduces residual strain in the slab section, and improves crack behavior by decreasing residual fracture width and improving cracking stiffness [20,21].

Experimental and numerical analyses were used to investigate the impact of fiber clustering on the fatigue behavior of steel-fiber-reinforced concrete (SFRC) beams with reinforcements. Furthermore, to better understand the underlying mechanism, the fiber distribution in the cross-section of concrete was experimentally investigated. The results showed that when the fiber volume percentage grew, the fatigue life of the beam increased [22]. To forecast the shear strength of steel-fiber-reinforced concrete beams, a mechanics-based mathematical model that considers the effect to resist shear was suggested (without transverse reinforcement). The suggested model's efficacy was tested using a wide number of datasets, and it was discovered that it had good correlations with experimental results, with mean, standard deviation, and coefficient of variation of 0.94%, 0.22%, and 22.99%, respectively. In addition, each effect to resist shear contribution was recommended [23].

Although the importance of fiber distribution to the characteristics of SFRC members was highlighted in the literature listed above, the underlying mechanism of fiber distribution to the fatigue performance of SFRC, as well as the behavior of the beam and slab under static and repetitive loading, was not. Even though there have been some studies on identifying plastic hinges, the contribution of steel fibers to plastic hinge length has had little attention. There has been no research on the impacts of fiber distribution at the slab's plastic hinge length under repeated stress.

In this paper, the concrete was tested experimentally with various SFRC parameters at various plastic hinge lengths. Using five different slabs of varying parameters, the behavior of steel fiber at the plastic hinge length under repetitive loads was investigated. The mechanical properties of the RCC slab were determined for standard concrete and SFRC. The properties of the steel fiber slab at plastic hinge length were determined under repeated loading becomes the novelty of this work. In the future, the plastic hinge length concept along with different strengthening techniques may be applied to beams and slabs under different types of loading.

The plastic hinge will form at the yielding zone of any structural member, which is the location of the highest bending moment. The most crucial damaged region of the structural element is the plastic hinge zone, where the more inelastic rotation would occur in the structural element. The shape of the bending moment diagram (BMD), the support distance, the contra flexure distance, and the geometry of the member all influence the plastic hinge.

The BMD considers the concrete's durability and the amount of steel required. The plastic hinge length of the RCC member has been calculated using a variety of empirical formulae. Table 1 contains a collection of formulas for determining the length of a plastic hinge.

Table 1. Empirical formula to calculate plastic hinge length.

Description	Empirical Formula
Baker (1956)	$k(z/d)^{1/4} d$
Herbert and Sawyer (1964)	$0.25 d + 0.075 z$
Corley (1966)	$0.5 d + 0.2 \sqrt{d} (z/d)$
Priestley and Park (1987)	$0.08 z + 6 d_b$
Paulay and Priestley (1992)	$0.08 z + 0.022 d_b f_y$

The empirical formulas of Corley, Herbert and Baker provide the same results. The empirical formula provides a high value for plastic hinge length and a factor of the diameter of the rebar. Figure 1 shows the plastic hinge length formed due to the application of load 'w'.

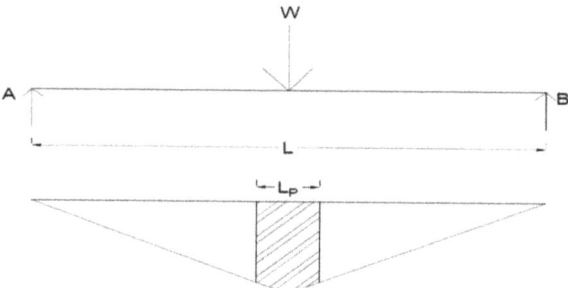

Figure 1. Plastic hinge length (L_p).

2. Materials

Table 2 shows the M20 grade mix design used in the experimental studies. The steel fibers until utilized in the casting had a hook end of 30 mm and a diameter of 0.5, and they were cast in M20 concrete. Steel fiber accounts for 1.5% of the weight of concrete in the specimen. The steel fiber parameters used in this study are listed in Table 3.

Table 2. M20 grade mix design as per IS 10262-2009.

Component Quantity	Quantity (kg/m^3)
Cement	438.7
F.A	757.28
C.A	1071.11
Steel Fiber	36.96
W/C	197

Table 3. Properties of steel fibers.

Description	Properties
Shape	Straight
L(length)	30 mm
Dia	0.5 mm
L/Dia	1/60

Figure 2 depicts the picture of the steel fibers, of which the weight is about 1.5% of the concrete.

Figure 2. Steel fibers (30 mm length and 0.5 mm diameter).

Slab Specifications

The slabs cast for the experimentation have the dimensions 850 mm × 300 mm × 80 mm as per the Indian standard IS456-2000. Five distinct slabs are cast, each with their own sets of criteria, such as minimum reinforcement and plastic hinge length. Mattock's (0.5 d + 0.05 z) and Priestly and Park's (0.08 z + 6 db) empirical formulas are used to compute the plastic hinge length of slabs. The minimum reinforcement considered for the slab is a 10 mm diameter bar of three numbers as longitudinal reinforcement at 100 mm c/c anan d 8 mm diameter bar of 6 numbers as transverse reinforcement with 150 mm c/c based on the requirement and laboratory condition as per the design. Figure 3a,b represent the conventional RCC concrete slab with minimum reinforcement and Figure 3c shows the steel-fiber-reinforced concrete slab. Figure 3d shows the combination of RCC concrete slab and steel-fiber-reinforced concrete slab of 1.5% of the weight of the concrete. Consecutively, the steel-fiber-reinforced concrete with 1.5% weight of the concrete at 150 mm and 70 mm plastic hinge length are shown in Figure 3e,f, respectively. The complete specifications and details of slabs are mentioned in Table 4, along with plastic hinge length.

Table 4. Dimensions and specifications of slabs.

Sl.No	Description	Dimension (mm)	Plastic Hinge Length (mm)	Reinforcement Details
1	RCC Slab	850 × 300 × 80	-	Longitudinal-3 no's 10 mm diameter bar at 100 mm c/c Transverse-6 no's 8 mm diameter bar at 150 mm c/c
2	SFRC Slab	850 × 300 × 80	-	-
3	SFRC Slab + Min Reinforcement	850 × 300 × 80	-	Longitudinal-3 no's 10 mm diameter bar at 100 mm c/c Transverse-6 no's 8 mm diameter bar at 150 mm c/c
4	SFRC + L_p @150 mm	850 × 300 × 80	150	-
5	SFRC + Lp @70 mm	850 × 300 × 80	70	-

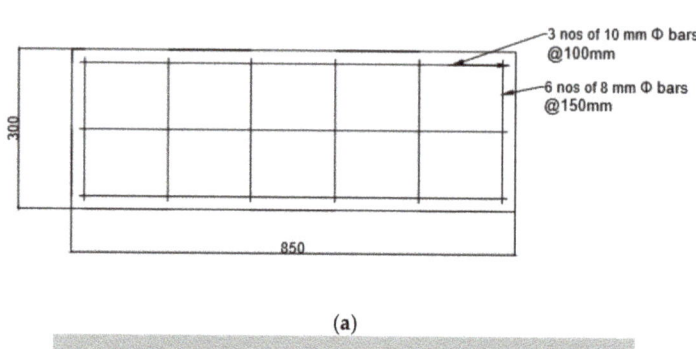

(a)

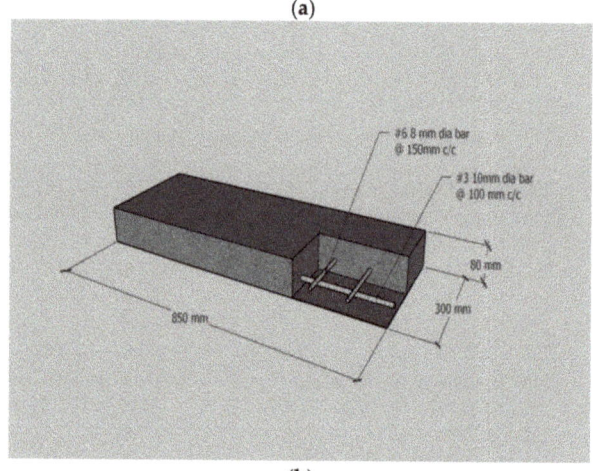

(b)

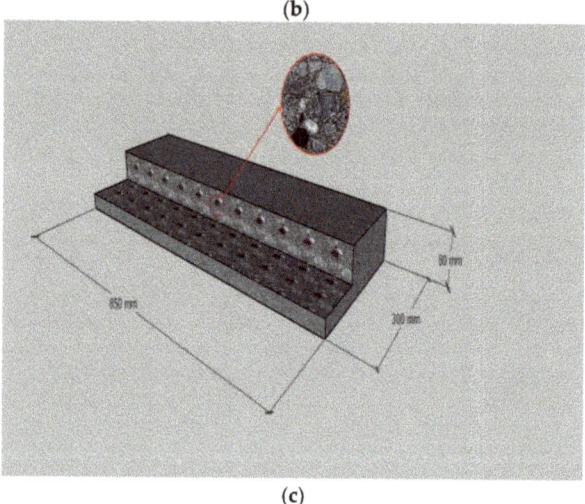

(c)

Figure 3. *Cont.*

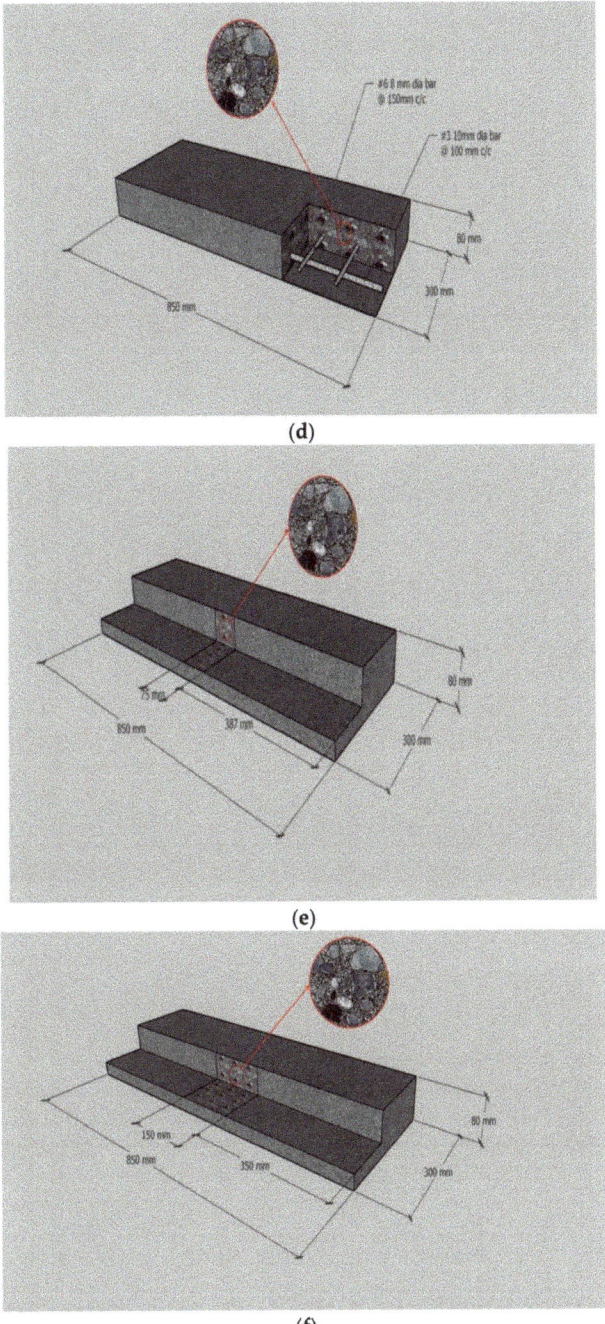

Figure 3. Schematic representation of different types of slabs. (a) Reinforcement details of slab; (b) RCC Slab; (c) SFRC Slab; (d) SFRC Slab + Min Reinforcement; (e) SFRC + L_p @150 mm; (f) SFRC + Lp @70 mm.

3. Experimental Test Procedure and Results

3.1. Test Specimens

Before casting the slabs, the compressive strength of concrete, split tensile strength and flexure strength of concrete are investigated as per codal standards to finalize the concrete used for casting the specimens. Ten cubes of size 150 mm × 150 mm × 150 mm were cast, out of which 5 cubes are made of conventional concrete of grade M20. Meanwhile, the 5 remaining same-sized cubes were cast with steel-fiber-reinforced concrete of 1.5% by weight of concrete. Cylinders of size 100 mm × 200 mm of 10 numbers are cast to determine the tensile strength of concrete through a split tensile strength test. In total, 5 cylinders are made of conventional concrete and the remaining 5 are made of steel-fiber-reinforced concrete. In total, 6 flexure beams of size 500 mm × 150 mm × 150 mm were cast, out of which 3 beams are made of conventional concrete and 3 beams are made of SFRC to determine the flexural strength of concrete.

Testing slabs of different configurations based on the parameters of the study are cast. The size of all the 6 specimens of slabs remain the same as 850 mm × 300 mm × 80 mm, but the change in parameters is carried out through reinforcement detailing. Minimum reinforcement is provided in the first conventional concrete slab. Meanwhile, the second specimen was completely replaced with steel-fiber-reinforced concrete, and the third specimen was a combination of both the cases mentioned above. The last 2 specimens are dosed with steel fibers placed at the plastic hinge length of 70 mm and 150 mm, and the remaining portion of the slab is completely made of plain cement concrete. Figure 4a,b show the cube specimens and cylinder specimens to be tested after 28 days of curing. Figure 4c, shows the flexure beam specimen (RCC beam and SFRC beam) and Figure 4d–h, shows the 5 slab specimens (RCC slab, SFRC slab, SFRC + min rein, SFRC + 150 mm, SFRC + 70 mm) to be tested.

3.2. Experimental Setup

A hydraulic jack is connected with a load cell of 200 kN capacity fixed to the self-straining testing frame, and the slab is fixed on all the sides. The repeated loading is applied to the top of the slab at the center, as shown in Figure 5a,b. A mechanical dial gauge is fixed below the loading area of the slab to measure the deflection of the slab under repeated loading.

3.3. Repeated Loading

Figure 6 shows the repeated loading processes for slab specimens. Plastic hinge lengths were calculated to be 70 mm and 150 mm, respectively, based on the theoretical study. The slab loads applied are always kept between load levels of 0 kN to 100 kN. For all cyclic loading stages, the force-controlled mode with rates of 10 kN was used. Upon loading 10 cycles at the maximum of 10 kN per cycle were applied on the slab.

3.4. Result and Discussion

The compressive strength, split tensile strength and flexural strength of both RCC and SFRC are determined by a compression testing machine. The average strength of the specimen for both concrete types tested is presented in Figure 7a–c. The concrete strength is high in all aspects when the steel fiber reinforcement was used as mentioned in Table 5.

After carrying out the compressive strength test in the testing machine for cast cubes, which are cured in a concrete tank for 28 days, the strength of concrete after crushing was found to be 22 N/mm^2, 21 N/mm^2, and 22 N/mm^2 for all three cubes, respectively. Therefore, the target strength of the M20 grade is achieved and the same is used for casting the concrete beams and slabs. Cylinders with 1.5% steel fiber amount of concrete are cast and cured for 7, 14, and 28 days. The same has been tested for split tensile strength. For all the three curing periods, SFRC specimens outperformed plain cement concrete specimens in terms of strength. 28 days cured, SFRC concrete cylinders achieved around 4 N/mm^2 when compared with plain cement concrete cylinder's strength of 2.7 N/mm^2, as mentioned in Figure 7a. A flexure beam with 1.5% steel fiber amount by weight of concrete was cast and cured for 7, 14, and 28 days.

The same has been tested for flexural strength, as mentioned in Figure 7c. For all the three curing periods, SFRC specimens outperformed plain cement concrete specimens in terms of strength. After curing for 28 days, SFRC concrete beams achieved around 3.92 N/mm^2 when compared with plain cement concrete flexural strength of 2.86 N/mm^2.

Figure 4. Test specimens (**a,b**) the cube specimens and cylinder specimens; (**c**) flexure beam specimen; (**d**) RCC slab specimen; (**e**) SFRC slab specimen; (**f**) SFRC + min rein specimen; (**g**) SFRC + 150 mm specimen; (**h**) SFRC + 70 mm specimen.

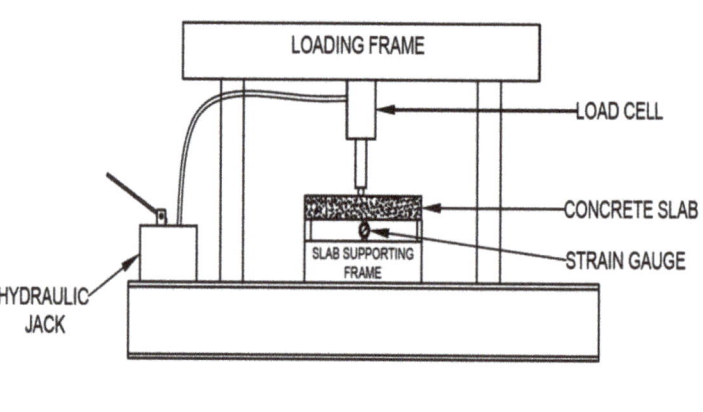

Figure 5. (a) Experimental setup for repeated loading; (b) schematic representation of loading Frame.

The modulus of elasticity was calculated by applying uniaxial compression to the cylinder specimen and measuring the deformations with a dial gauge set between the 200 mm gauge length, as illustrated in Figure 7d. The tests were carried out using a compressometer in accordance with the IS 516-1959. The cylinder specimens were put on a compression testing machine, and a uniform load was applied until the cylinder reached its failure. The target load and deflection were taken into consideration.

The deflection values are calculated as a strain based on length change. The strain is calculated by dividing the dial gauge readings by the gauge length, and the stress is calculated by dividing the load applied by the area of the cylinder's cross-section. The deformation of various loads was recorded and the findings were plotted graphically against the tension to determine the Young's modulus of concrete of both the conventional and SFRC specimen as shown in Figure 7e,f. Later, the modulus of elasticity of conventional concrete and of steel-fiber-reinforced concrete (SFRC) was derived as 25.47 N/mm^2 and 29.025 N/mm^2, as mentioned in Table 6.

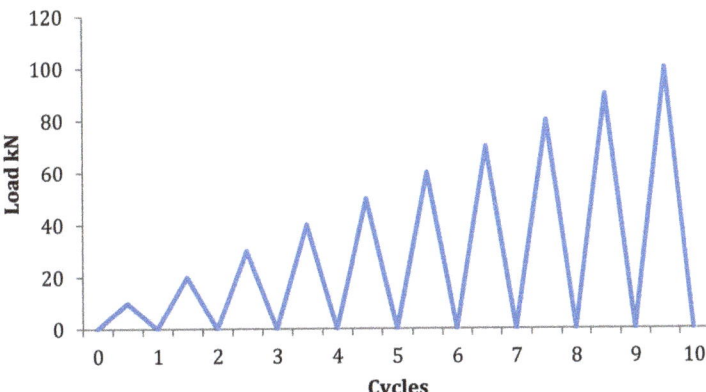

Figure 6. Repeated loading procedures.

The five slabs were subjected to repeated loading, and the experimental setup is presented in Figure 5b. The deformation of the slabs was measured by a digital dial gauge. The repeated load was applied on all the types of slabs and the obtained results are tabulated in Table 5. A maximum load of 96 kN was achieved by a steel-fiber-reinforced slab with a minimum reinforcement and the deflection observed in the dial was 5.78 mm. Meanwhile, an SFRC slab with a steel fiber amount of 1.5% at 150 mm plastic hinge length achieves an 80 kN maximum load with a deflection of 7.29 mm. A minimum of 40 kN was achieved at a 70 mm plastic hinge length as given in Table 7.

Table 5 shows the test results of the five different slab specimens subjected to repeated loading. The RCC slab has less deflection compared with the other slab; however, the maximum load at the failure for RCC is less compared with the SFRC slab and SFRC Slab + Min Reinforcement and SFRC + L_p @150 mm. Hence, the load-carrying capacity for the SFRC slab, SFRC Slab + Min Reinforcement, and SFRC + L_p @150 mm is greater compared with the RCC slab. This shows that the performance of the SFRC slab, SFRC Slab + Min Reinforcement and SFRC + L_p @150 mm is better than the RCC slab.

Figure 8a depicts the load versus deflection for a repeatedly loaded RCC slab. Figure 8b shows the load versus deflection for the SFRC slab under repeated loading (half cycle), which has less deflection than the RCC slab because steel fibers increase the strength of the concrete slab. The load versus deflection curve for the SFRC + minimum reinforced slab subjected to repeated loading (half cycle) is shown in Figure 8c, and it outperforms the RCC slab. Figure 8d depicts the load versus deflection curve for an SFRC + Lp 150 mm slab subjected to repeated loading (half cycle), demonstrating improved performance as steel fibers are added to the plastic hinge length. Figure 8e depicts the load versus deflection curve for an SFRC + Lp 70 mm slab subjected to repeated loading (half cycle), which performs satisfactorily but has more deflection than an SFRC + Lp 150 mm slab.

From the load–displacement curve, it is very clear that up to 60 kN, a conventional RCC slab behaves well, and the curves show the even distribution of load. After that, there is an uneven scattering of load due to the strength-losing character of the slab. Once the steel fibers are added to the slab, the curves are evenly scattered due to the improvement in the ductility of the slab. Steel fibers at 150 mm plastic hinge length and steel fibers amounting for overall slab behavior are almost identical, which comes to around 5.78 and 7.29 mm with a maximum load of 96 kN and 80 kN.

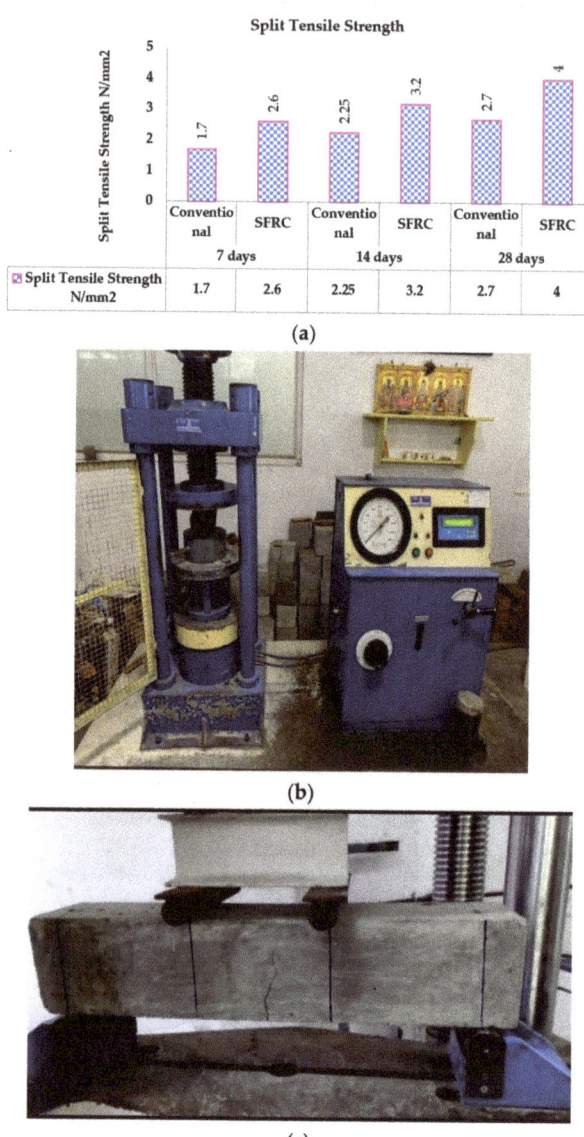

Figure 7. Cont.

(d)

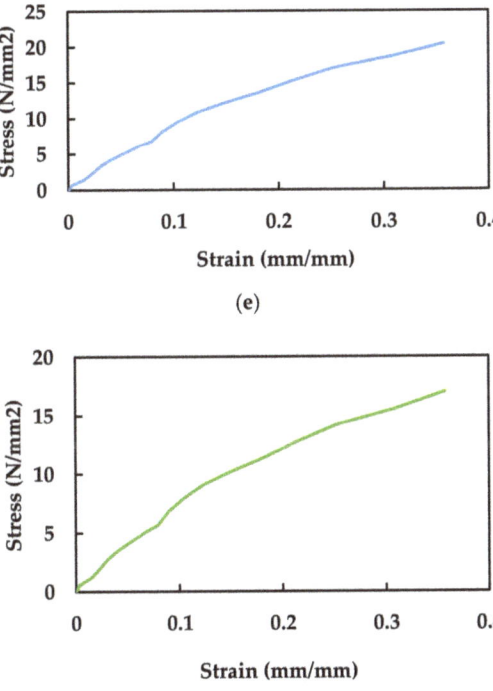

(e)

(f)

Figure 7. (a) Split tensile strength; (b) split tensile strength testing machine; (c) flexural strength test; (d) compression test of cylinder specimens with compressometer; (e) stress–strain behavior of conventional concrete; (f) stress–strain behavior of SFRC.

Table 5. Mechanical properties of concrete.

Sl.No	Type	Compressive Strength	Split Tensile Strength	Flexural Strength
			28 Days Curing	
1	Conventional Concrete	22 N/mm^2	2.7 N/mm^2	2.86 N/mm^2
2	Steel-Fiber-Reinforced concrete (SFRC)	26 N/mm^2	4 N/mm^2	3.92 N/mm^2

Table 6. Modulus of elasticity of specimens.

Sl.No	Specimen	Modulus of Elasticity
1	Conventional Concrete	25.47 N/mm^2
2	Steel-Fiber-Reinforced concrete (SFRC)	29.025 N/mm^2

Table 7. Test load deflection of the slab under cyclic loading.

Slab Specimen	Description	Max Load (kN)	Deflection of the Slab (mm)
1	RCC Slab	68	2.31
2	SFRC Slab	72	4.58
3	SFRC Slab + Min Reinforcement	96	5.78
4	SFRC + L_p @150 mm	80	7.29
5	SFRC + L_p @70 mm	40	6.75

Deflection of the specimen SFRC with minimum reinforcement was decreased and the corresponding ultimate load was increased than RC Slab and SFRC slab. The addition of steel fibers increased the stiffness and ultimate load. At the initial stage of loading, the stiffness of the slabs was high. As the load increased, the stiffness of the slab was reduced, and cracks were formed. SFRC at plastic hinge length with minimum reinforcement has a higher initial crack load than SFRC at plastic hinge length without minimum reinforcement. Additionally, the deflection of SFRC at plastic hinge length with minimum reinforcement was lesser than SFRC at plastic hinge length without minimum reinforcement. Comparing SFRC at plastic hinge length with minimum reinforcement and SFRC with minimum reinforcement, deflection of SFRC at plastic hinge length with minimum reinforcement was thus equal to SFRC with minimum reinforcement. Specimens with bar reinforcement had lower deflection and higher ultimate load than specimens without reinforcement. As a result, steel-fiber-reinforced concrete increased the flexure strength of the slab under loading.

Figure 9a depicts the crack pattern on the RCC slab due to repeated loading (half cycle). Figure 9b depicts the crack pattern on the SFRC slab as a result of repeated loading (half cycle), demonstrating that cracks are minimal in comparison to the RCC slab. Figure 9c depicts the crack pattern on the SFRC + min reinforced slab as a result of repeated loading (half cycle), demonstrating that the cracks are very small in comparison to the RCC slab. Figure 9d depicts the crack pattern on the SFRC + Lp 150 mm slab as a result of repeated loading (half cycle), demonstrating that cracks are kept to the minimum because steel fibers are included in the plastic hinge length. Figure 9e depicts the crack pattern on the SFRC + Lp 70 mm slab as a result of repeated loading (half cycle), demonstrating that cracks are more prevalent than on the SFRC + Lp 150 mm slab.

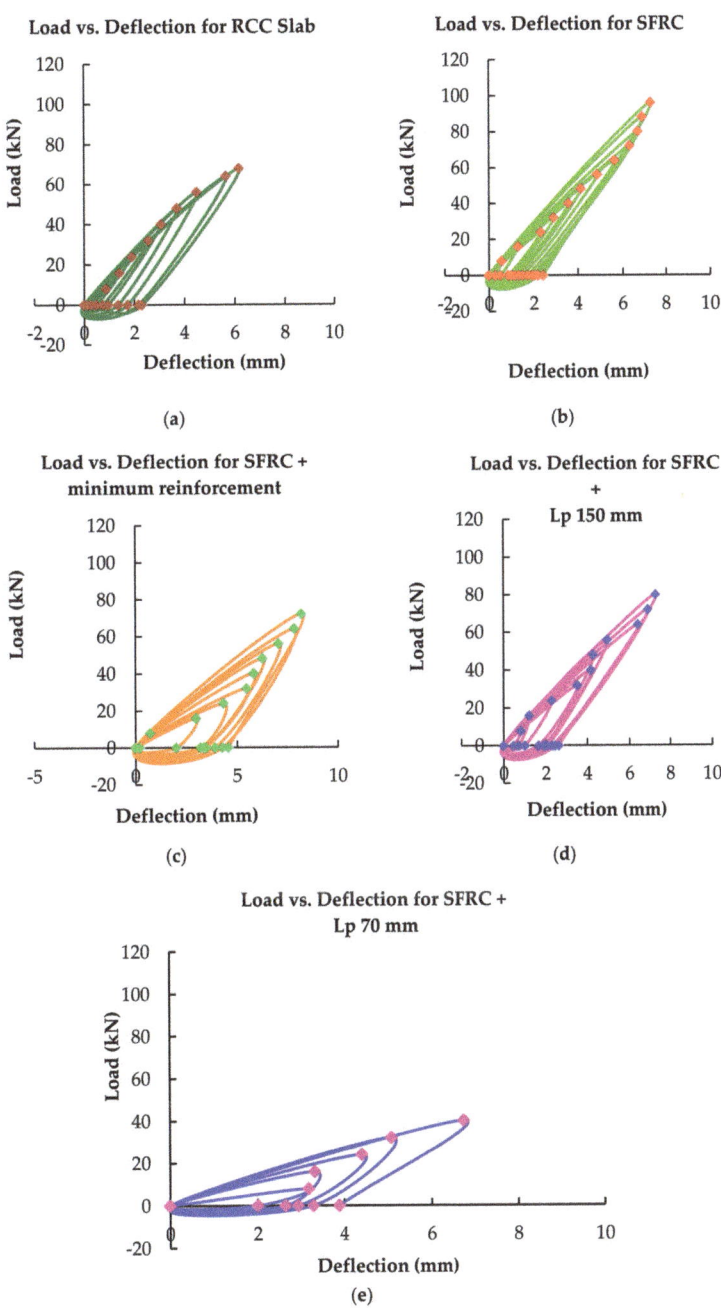

Figure 8. (**a**) Load versus deflection curve for RCC Slab; (**b**) load versus deflection curve for SFRC Slab; (**c**) load versus deflection curve for SFRC + Minimum reinforcement Slab; (**d**) load versus deflection curve for SFRC + Lp 150 mm; (**e**) load vs. deflection curve for SFRC + Lp 70 mm.

The conventional RCC slab developed a 2 mm diagonal crack running almost the whole length of the specimen, and the SFRC Slab at 70 mm plastic hinge length shows the wider brittle cracks of 4 mm at the middle of the slab where the maximum deflection took place. Meanwhile, the SFRC Slab with minimum reinforcement and the SFRC Slab with 150 mm plastic hinge length shows a 1 mm crack width and that crack moved away from the maximum deflection zone to another area.

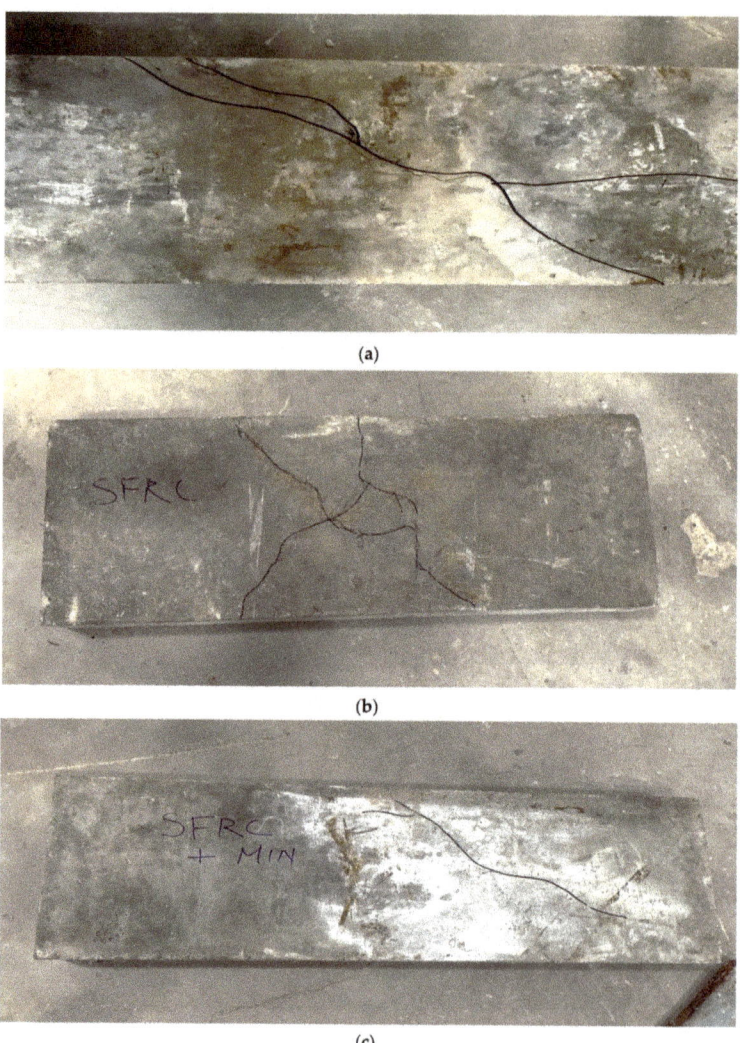

Figure 9. *Cont.*

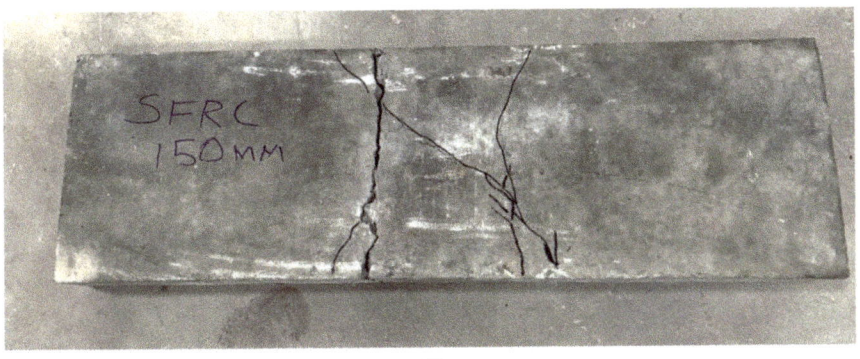

(d)

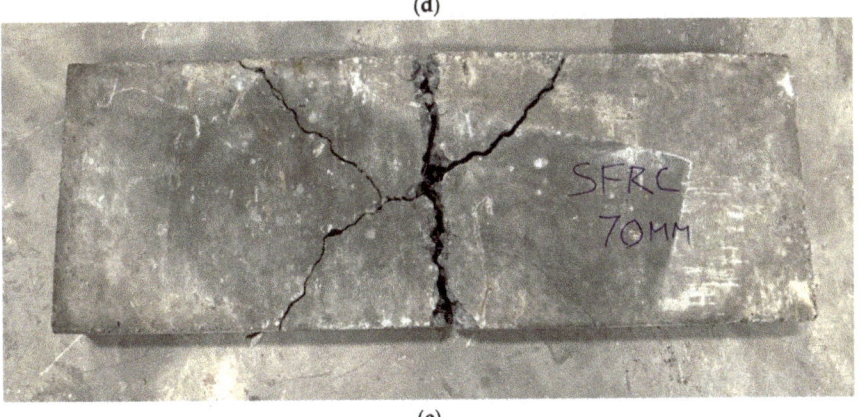

(e)

Figure 9. (**a**) Depiction of the crack pattern on the RCC slab due to repeated loading (Half cycle); (**b**) depiction of the crack pattern on the SFRC slab due to repeated loading; (**c**) depiction of the crack pattern on the SFRC + minimum reinforcement slab due to repeated loading; (**d**) depiction of the crack pattern on the SFRC + Lp 150 mm due to repeated loading; (**e**) depiction of the crack pattern on the SFRC + Lp 70 mm due to repeated loading.

4. Conclusions

The examination of steel fiber reinforcement at the plastic hinge length of slab is presented in this experimental investigation. The mechanical properties of five distinct slabs are studied after they are cast and subjected to repeated loading (half cycle). The conclusions obtained from this following are as follows:

1. Steel-fiber-reinforced concrete with the amount of 1.5% shows better performance in compressive as well as split-tensile strength compared with that of conventional concrete.
2. Split tensile strength of steel-fiber-reinforced concrete seems to be 1.5 times higher than that of conventional concrete due to the distribution of steel fibers in concrete, which influence the bonding and improves the ductility. The modulus of elasticity was calculated by applying uniaxial compression to the cylinder specimen showing that SFRC specimen with 1.5% steel fiber performs 1.14 times better than the conventional concrete specimen. Hence, the same was adopted for slab specimen casting. The behavior under bending is evident from the flexural strength test, where the flexure beam with steel fibers shows 1.39 times performance improvement than that of a conventional concrete beam.

3. The steel fiber reinforcement in the concrete provides higher ductility and can withstand intensive loads compared with conventional concrete when subjected to repeated loading (Half cycle). Steel-fiber-reinforced concrete at 150 mm plastic hinge length and steel fiber dosed throughout the slab span have similar types of crack pattern, and the failure occurred at the same time due to improved ductility; cracks are forming away from the maximum deflection zone due to the steel fiber at the hinge length.
4. The SFRC slab shows relatively fewer deflections compared with the RCC slab. A similar observation was noticed in the SFRC slab with the addition of minimum reinforcement because the steel fibers increase the strength of the slab. Steel fiber reinforcement at 150 mm plastic hinge length provides equal ductility, and resistance against the load is equal to that of SFRC and minimum reinforcement slab.
5. Crack formation in the steel-fiber-reinforced slab at a plastic hinge length of 150 mm shows 1 mm crack width, which moved away from the maximum stress zone. The crack pattern was similar to that of a fully steel-fiber-reinforced concrete slab.
6. Rather than using the steel fibers throughout the member, the steel fibers can be incorporated at the plastic hinge length alone. This provides a similar performance to that of a full SFRC slab. This would decrease the quantity of steel fiber and is more economical.
7. Based on the experimental results, it is evident that incorporating steel fibers into the 150 mm plastic hinge length of the slab alone will result in a more economical and efficient method of slab construction.
8. Further studies can be carried out for other structural elements with different fibers under different types of loading conditions.

Author Contributions: P.S. and K.R. (Karthikeyan Rajendiran): Conceptualized the model and established the empirical theorem, experimented. B.G.A.G. and K.R. (Krishanu Roy): Supervised the research as well as the analysis of results. K.R. (Karthikeyan Rajendiran) and D.P.: Introduced the idea of cyclic loading in this project, designed the slab, and wrote the paper. Review and submission, Collaborated in and Co-ordinated the research. B.G.A.G. and K.R. (Krishanu Roy): suggested and chose the journal for submission. B.G.A.G. and K.R. (Krishanu Roy): Participated in the manuscript revision phase. All authors have read and agreed to the published version of the manuscript.

Funding: This research received no external funding.

Institutional Review Board Statement: Not applicable.

Informed Consent Statement: Not applicable.

Data Availability Statement: The data presented in this study are available on request from the corresponding author.

Conflicts of Interest: This manuscript has not been submitted to, nor is it under review by, another journal or other publishing venue. The authors have no affiliation with any organization with a direct or indirect financial interest in the subject matter discussed in the manuscript. The authors declare no conflict of interest.

References

1. Gopinath, A.; Nambiyanna, B.; Nakul, R.; Prabhakara, R. Parametric study on rotation and plastic hinge formation in RC beams. *J. Civ. Eng. Technol. Res.* **2014**, *2*, 393–401.
2. Ko, M.-Y.; Kim, S.-W.; Kim, J.-K. Experimental study on the plastic rotation capacity of reinforced high strength concrete beams. *Mater. Struct.* **2001**, *34*, 302–311. [CrossRef]
3. Mendis, P. Plastic hinge lengths of normal and high-strength concrete in flexure. *Adv. Struct. Eng.* **2001**, *4*, 189–195. [CrossRef]
4. Nazaripoor, H.; Ashrafizadeh, H.; Schultz, R.; Runka, J.; Mertiny, P. Acoustic Emission Damage Detection during Three-Point Bend Testing of Short Glass Fiber Reinforced Composite Panels: Integrity Assessment. *J. Compos. Sci.* **2022**, *6*, 48. [CrossRef]
5. Paulay, T.; Priestley, M.J.N. *Seismic Design of Reinforced Concrete and Masonry Buildings*; John Wiley & Sons. Inc.: Hoboken, NJ, USA, 1992; pp. 135–146.
6. Mattock, A.H. Rotational Capacity of Hinging Regions in Reinforced Concrete Beams, Flexural Mechanics of Reinforced Concrete. In Proceedings of the ASCE-ACI International Symposium, Miami, FL, USA, 10–12 November 1964; pp. 143–180.

7. Pradeep, S.; Vengai, V.U.E.; Florence, D. Experimental Investigation on the Usage of Steel Fibres and Carbon Fibre Mesh at Plastic Hinge Length of Slab. *Mater. Today Proc.* **2019**, *14*, 248–256. [CrossRef]
8. Qin, S.; Gao, D.; Wang, Z.; Zhu, H. Research on the fracture behavior of steel-fiber-reinforced high-strength concrete. *Materials* **2022**, *15*, 135. [CrossRef] [PubMed]
9. Zhao, X.; Wu, Y.F.; Leung, A.Y.; Lam, H.F. Plastic hinge length in reinforced concrete flexural members. *Procedia Eng.* **2011**, *14*, 1266–1274. [CrossRef]
10. Zhao, X.M.; Wu, Y.F.; Leung, A.Y.T. Analyses of plastic hinge regions in reinforced concrete beams under monotonic loading. *Eng. Struct.* **2012**, *34*, 466–482. [CrossRef]
11. Barros, J.A.O.; Figueiras, J.A. Model for the Analysis of Steel Fiber Reinforced Concrete Slabs on Grade. *Comput. Struct.* **2001**, *79*, 97–106. Available online: www.elsevier.com/locate/compstruc (accessed on 25 April 2022).
12. Neto, B.N.M.; Barros, J.A.O.; Melo, G.S.S.A. A model for the prediction of the punching resistance of steel fibre reinforced concrete slabs centrically loaded. *Constr. Build. Mater.* **2013**, *46*, 211–223. [CrossRef]
13. Eligehausen, R.; Langer, P. *The Rotation Capacity of Plastic Hinges in Reinforced Concrete Beams and Slabs*; Institute for Building Materials University of Stuttgart: Stuttgart, Germany, 2012.
14. Michels, J.; Waldmann, D.; Maas, S.; Zürbes, A. Steel fibers as only reinforcement for flat slab construction—Experimental investigation and design. *Constr. Build. Mater.* **2012**, *26*, 145–155. [CrossRef]
15. Holschemacher, K.; Mueller, T.; Ribakov, Y. Effect of steel fibres on mechanical properties of high-strength concrete. *Mater. Des.* **2010**, *31*, 2604–2615. [CrossRef]
16. Khaloo, A.R.; Afshari, M. Flexural behaviour of small steel fibre reinforced concrete slabs. *Cem. Concr. Compos.* **2005**, *27*, 141–149. [CrossRef]
17. Nguyen, N.T.; Bui, T.T.; Bui, Q.B. Fiber reinforced concrete for slabs without steel rebar reinforcement: Assessing the feasibility for 3D-printed individual houses. *Case Stud. Constr. Mater.* **2022**, *16*, e00950. [CrossRef]
18. Xiang, D.; Liu, Y.; Gu, M.; Zou, X.; Xu, X. Flexural fatigue mechanism of steel-SFRC composite deck slabs subjected to hogging moments. *Eng. Struct.* **2022**, *256*, 114008. [CrossRef]
19. McMahon, J.A.; Birely, A.C. Service performance of steel fiber reinforced concrete (SFRC) slabs. *Eng. Struct.* **2018**, *168*, 58–68. [CrossRef]
20. Paramasivam, P.; Ong, K.C.G.; Ong, B.G.; Lee, S.L. Performance of Repaired Reinforced Concrete Slabs Under Static and Cyclic Loadings. *Cement Concr.* **1995**, *17*, 37–45. [CrossRef]
21. Xiang, D.; Liu, S.; Li, Y.; Liu, Y. Improvement of flexural and cyclic performance of bridge deck slabs by utilizing steel fiber reinforced concrete (SFRC). *Constr. Build. Mater.* **2022**, *329*, 127184. [CrossRef]
22. Gao, D.; Gu, Z.; Wei, C.; Wu, C.; Pang, Y. Effects of fiber clustering on fatigue behavior of steel fiber reinforced concrete beams. *Constr. Build. Mater.* **2021**, *301*, 124070. [CrossRef]
23. Negi, B.S.; Jain, K. Shear resistant mechanisms in steel fiber reinforced concrete beams: An analytical investigation. *Structures* **2021**, *39*, 607–619. [CrossRef]

Article

Comparison of the Flexural Behavior of High-Volume Fly AshBased Concrete Slab Reinforced with GFRP Bars and Steel Bars

Chinnasamy Samy Madan [1], Swetha Munuswamy [1], Philip Saratha Joanna [1,*], Beulah Gnana Ananthi Gurupatham [2] and Krishanu Roy [3,*]

1. Department of Civil Engineering, Hindustan Institute of Technology and Science, Padur, Chennai 603103, India; chinna_3_2001@yahoo.com (C.S.M.); swethamunus@gmail.com (S.M.)
2. Department of Civil Engineering, Anna University, Chennai 600025, India; beulah28@annauniv.edu
3. School of Engineering, Civil Engineering, The University of Waikato, Hamilton 3216, New Zealand
* Correspondence: joanna@hindustanuniv.ac.in (P.S.J.); krishanu.roy@waikato.ac.nz (K.R.)

Abstract: Fiber-reinforced polymer (FRP) rods are advanced composite materials with high strength, light weight, non-corrosive properties, and superior durability properties. Under severe environmental conditions, for concrete structures, the use of glass-fiber-reinforced polymer (GFRP) rods is a cost-effective alternative to traditional steel reinforcement. This study compared the flexural behavior of an OPC concrete slab with a high-volume fly ash (HVFA) concrete slab reinforced with GFRP rods/steel rods. In the fly ash concrete slabs, 60% of the cement used for casting the slab elements was replaced with class F fly ash, which is emerging as an eco-friendly and inexpensive replacement for ordinary Portland cement (OPC). The data presented include the crack pattern, load–deflection behavior, load–strain behavior, moment–curvature behavior, and ductility of the slab specimens. Additionally, good agreement was obtained between the experimental and nonlinear finite element analysis results using ANSYS 2022-R1. The study also compared the experimental moment capacity with the most commonly used design standard ACI 440.1R-15. This investigation reveals that there is a huge potential for the utilization of GFRP rods as reinforcement in fly ash concrete slabs.

Keywords: glass-fiber-reinforced polymer (GFRP) rods; concrete slab; high-volume fly ash; moment–curvature; ductility

Citation: Madan, C.S.; Munuswamy, S.; Joanna, P.S.; Gurupatham, B.G.A.; Roy, K. Comparison of the Flexural Behavior of High-Volume Fly AshBased Concrete Slab Reinforced with GFRP Bars and Steel Bars. *J. Compos. Sci.* **2022**, *6*, 157. https://doi.org/10.3390/jcs6060157

Academic Editor: Francesco Tornabene

Received: 2 May 2022
Accepted: 24 May 2022
Published: 26 May 2022

Publisher's Note: MDPI stays neutral with regard to jurisdictional claims in published maps and institutional affiliations.

Copyright: © 2022 by the authors. Licensee MDPI, Basel, Switzerland. This article is an open access article distributed under the terms and conditions of the Creative Commons Attribution (CC BY) license (https://creativecommons.org/licenses/by/4.0/).

1. Introduction

Sustainability is an important factor for the well-being and continuous growth of society. The reduction in CO_2, the main ingredient of greenhouse gases, has become critical for ensuring a sustainable ecosystem. Cement is the prime ingredient in concrete, and its production leads to 7 to 10% of global carbon emissions. Because of its pozzolanic properties, fly ash can be utilized as a cement substitute in concrete slab elements. The pozzolanic properties of fly ash suggest that it could be utilized as a cement substitute in slab elements. Green concrete is becoming more popular in the construction business because of the disadvantages of traditional concrete. The various green concrete types available are geopolymer concrete, ultrahigh performance concrete, high-volume fly ash concrete, and lightweight concrete [1,2].

Fly ash has become an important ingredient in concrete, as the spherical shape of fly ash helps in improving the workability of fresh concrete, and its smaller particle size helps it fill voids in the concrete, leading to the production of dense and durable concrete. Class F fly ash is widely used by the construction industry [3–5]. Reinforced concrete (RC) beams can be prepared, which contain 50% fly ash instead of cement [6]. High-volume fly ash (HVFA) concrete, in which 60% of cement was replaced with fly ash, showed excellent mechanical properties with enhanced durability performance. Replacing 60% of cement with fly ash can produce adequate strength in self-compacting concrete [7]. Concrete containing up to 60% fly ash plus a chemical superplasticizer can be adjusted to meet structural-grade concrete strength and workability criteria. The early-age compressive strength of fly ash concrete was less than that of the control concrete, but the strength of the fly ash concrete gradually improved over a long period due to the pozzolanic reaction. However, the strong growth of control samples stopped after 56 days of curing [8–10].

In concrete structures, the use of non-metallic FRP reinforcement as an alternative to steel reinforcement is gaining acceptance due to its resistance to corrosion and its better mechanical properties. FRP composites have the advantage of high strength, light weight, and corrosion resistance and are considered for rehabilitation and strengthening purposes in concrete structural members [11].

The FRP rod assures better long-term performance than steel reinforcement [12]. Shave [13] suggested that to prevent corrosion, FRP rods can be used instead of steel reinforcement in precast concrete structures. Nowadays, the use of FRP is still not widespread despite its high potential for application in the field of civil engineering and the advantages it provides as an alternative to steel bars. One of the reasons for this may be the lack of design codes for the design of FRP structures.

GFRP rods can be used to improve the structural response of existing flat slabs and also to withstand higher longitudinal strains. They also provide remediation for punching shear failure [14]. The HVFA concrete element with GFRP composites could provide better fire protection up to 1100 °C [15]. In addition, the use of a flame retardant and nanoparticles on the fiber composite can improve the mechanical properties of the GFRP composite, which can also significantly improve the lifespan [16–21]. The fiber composite also showed promising electrical conductivity, a positive effect on the post-fracture residual stiffness, redundancy, resistance, load-bearing capacity, and flexural strength of structural beams [22–24].

The crack width limitations for reinforced concrete (RC) elements reinforced with FRP rods are more relaxed than for RC elements reinforced with steel rods. Less severe crack width limits could be adopted, as FRP bars are corrosion-resistant [25]. When the whole-life cost of GFRP bars is considered, they provide an economical solution, reduce maintenance costs, and increase the useful life of the structure. The installation of GFRP rods in structural elements is similar to that of steel rods, along with fewer handling, transporting, and storage problems [26–28]. As GFRP bars are anisotropic composite materials, they have higher tensile strength. The yield tensile strength of GFRP bars is 13% higher than that of steel rebar. The yield strain of the GFRP bar is 58% higher than that of steel rebar [29]. Research carried out on concrete structural elements reinforced with GFRP rods showed no reaction or degradation process in the presence of an alkaline and corrosive environment [30]. The mechanical properties of GFRP rods were better than those of the conventional steel reinforcement, and the GFRP rods provided good resistance against environmental effects such as a chemical attack, freeze–thaw cycles, etc. [31,32]. The flexural behavior of a beam reinforced with glass-fiber-reinforced polymer as an alternative to traditional steel reinforcement was studied. The results showed that GFRP rods had flexural behavior similar to that of steel rods. Besides acceptable shear properties, concrete elements reinforced with GFRP rods led to high bending properties [33,34].

A nonlinear finite element (NLFE) model using ANSYS software was used to investigate the structural performance of concrete elements reinforced with GFRP bars. The effects of the ultimate load, ultimate deflection, crack pattern, stress, strain, and displacement of concrete elements with different end conditions can be analyzed in ANSYS [35–37].

Only a few research works have been carried out with green concrete reinforced with GFRP rods. The present study compares the flexural behaviors of HVFA-based concrete slab and OPC concrete slab reinforced with GFRP rods and steel rods. Comparisons were made in terms of crack pattern, load–deflection behavior, load–strain behavior, ultimate moment capacity, and ductility. ANSYS 2022-R1 [38] software was used to investigate the structural performance in terms of the ultimate load and deflection of concrete slabs reinforced with steel/GFRP bars. The study also compared the experimental moment capacity with the most commonly used design standard ACI 440.1R-15.

2. Materials and Methods

2.1. Concrete

The materials used in the concrete mix were ordinary Portland cement of 53 grade, class F fly ash, manufactured sand as fine aggregate, gravel as coarse aggregate, 10% micro silica by weight of the binder, and 0.3% superplasticizer (Master Glenium sky 8233). The water–cement ratio adopted was 0.5. M25 grade concrete was used to cast the slabs. The properties of class F fly ash are given in Table 1. The compressive strength of the cube and the split tensile strength of the cylinder with OPC concrete and HVFA concrete after 28 days and 56 days of curing are shown in Table 2.

Table 1. Properties of fly ash (class F).

Chemical Composition	Content (% by Mass)
SiO_2	52.52
Al_2O_3	32.63
Fe_2O_3	6.16
CaO	Nil
NA1-20	0.02
SO_3	4.95
MnO	0.03
LOI	1.08

Table 2. Strength of Concrete.

Specimen Type	Compressive Strength f'_c (MPa)		Split Tensile Strength (MPa)	
	28 Days	56 Days	28 Days	56 Days
Control concrete	34.23	36.27	3.94	5.01
60% Fly ash concrete	26.39	37.74	3.16	5.79

Reinforcing Bars

The GFRP rods used in the slabs were manufactured as per ACI 440.6M-08 and ACI 440.3R-04. The steel rods used as reinforcement in the slab specimens were of Fe550D grade. The tests were performed by the manufacturers following the ASTM D7205/D7205M-06 standards, and the mechanical properties of the GFRP rod and steel rod were determined. The mechanical properties of steel and GFRP rods are shown in Table 3. The GFRP rod used in the experiment is shown in Figure 1.

Table 3. Mechanical properties of steel and GFRP rods.

Reinforcement Material	Diameter (mm)	Tensile Strength f_{fu} (MPa)	Modulus of Elasticity E_f (MPa)	Density (Kg/m^3)
STEEL rod	10	650	200,000	7800
GFRP rod	10	1100	55,000	1900

2.2. Experimental Program

2.2.1. Specimen Details

The experimental program consisted of four groups: OPC concrete slab with steel bars, OPC concrete slab with GFRP bars, HVFA concrete slab with steel bars, and HVFA concrete slab with GFRP bars. A total of eight slabs were tested, with two slabs in each group. All of the one-way slab specimens had a span of 1000 mm, a width of 450 mm, and a depth of 100 mm. The slabs were reinforced with GFRP/steel rods with a center–center spacing of 130 mm along the longer span and 240 mm along the shorter span. In the high-volume fly ash concrete slab, 60% of cement was replaced with fly ash (class F). All specimens were tested on the 56th day from the date of casting. A four-letter designation was given to the slab specimens. The first two letters refer to reinforcement type, which may take the value SR or GR: SR for steel reinforcement and GR for GFRP rods. The third letter indicates the type of concrete (C, F): C for OPC concrete and F for high volume fly ash concrete. The fourth identifier represents the trial number of the specimens in a particular series, as two specimens were tested in each series. The tested specimen details are given in Table 4.

Figure 1. Photograph of GFRP rods.

Table 4. Details of Test Specimens.

Specimen	Concrete Material	Reinforcement Material	Diameter (mm)
SRC	OPC (C)	Steel rod (SR)	10
SRF	Fly ash (F)	Steel rod (SR)	10
GRC	OPC (C)	GFRP rod (GR)	10
GRF	Fly ash (F)	GFRP rod (GR)	10

2.2.2. Experimental Setup

The RC slabs were tested in a loading frame with a capacity of 40 T. One end of the slab rested on the hinge support, and the other end rested on the roller support with an effective span of 800 mm. A spreader beam was used to apply two-point loading on the slab element. Static loads were applied to the slab specimen through the load cell. The deflections of the slabs were measured by linear voltage displacement transducers (LVDTs). An internal strain gauge was pasted on the surface of the steel/GFRP reinforcements with precaution during the casting of slabs to measure the tensile strain. For measuring the compressive strain, the strain gauge was pasted externally on the top surface of the slabs at the time of testing. The signals obtained from the LVDT and the electrical strain gauges were captured by a data acquisition system, which in turn was connected to a computer. The load was gradually applied with an increment of 2 kN/min until the failure of the slabs. The experimental setup for testing the slab is shown in Figure 2.

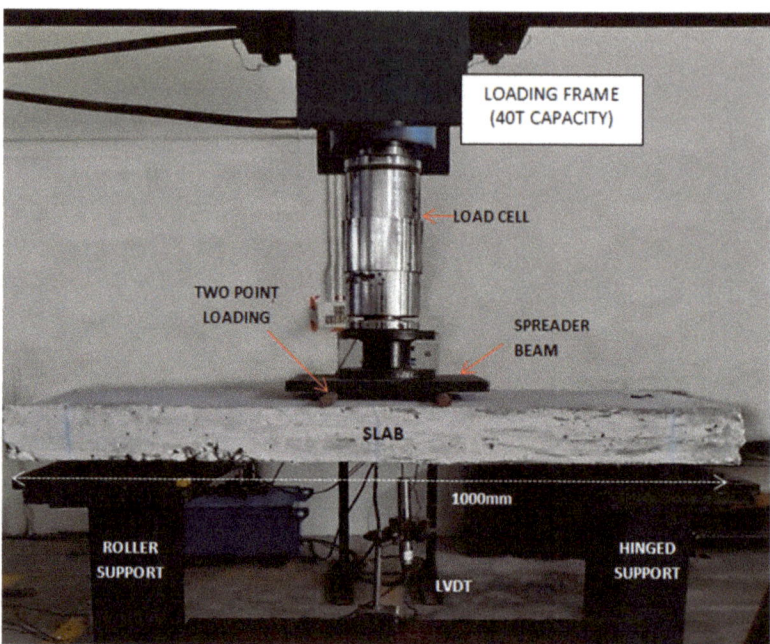

Figure 2. Experimental setup.

3. Results and Discussion

3.1. Crack Pattern

The typical crack patterns of the SRC and SRF slabs are shown in Figures 3 and 4, respectively, and the crack patterns of the GRC and GRF slabs are shown in Figures 5 and 6, respectively. The crack patterns and the modes of failure are similar in slabs with steel bars and GFRP bars [39]. All tested slabs were initially uncracked before loading, and the first cracking of slabs occurred in the constant moment region. The average first crack loads in SRC, SRF, GRC, and GRF slabs were 16 kN, 18.8 kN, 15.75 kN, and 18.05 kN, respectively. After the first cracking, new cracks were formed in the slabs, and the width of the existing cracks continued to enlarge with the load increment in all of the slabs. At the ultimate load, cracks from the bottom of the slabs propagated to the top surface of the slabs. In all slabs, the cracks were mainly vertical flexural cracks that were perpendicular to the longitudinal axis of the slabs. All slabs failed due to the crushing of the concrete at the top surface of the slabs, and no failure was noticed in the steel bars and GFRP bars. Table 5 shows the first crack load and the ultimate load of all the slabs.

Table 5. First crack load and ultimate load of slab reinforced with steel and GFRP rods.

Specimen ID	First Crack Load P_{cr} (kN)	Ultimate Load P_u (kN)
SRC 1	15.7	24
SRC 2	16.3	23.8
SRF 1	19.1	28.5
SRF 2	18.5	27.3
GRC1	15.2	29
GRC2	16.3	28.1
GRF 1	17.6	31.8
GRF 2	18.5	30.3

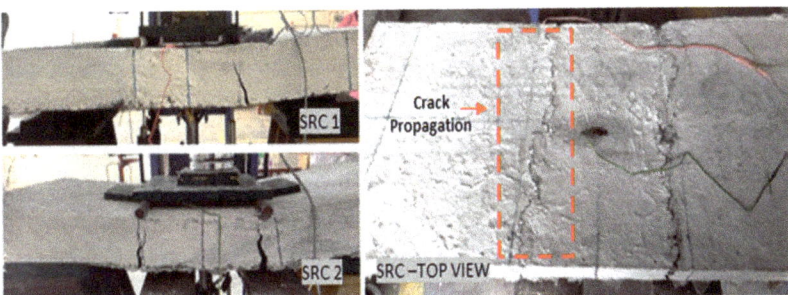

Figure 3. Crack pattern and failure mode of SRC slab.

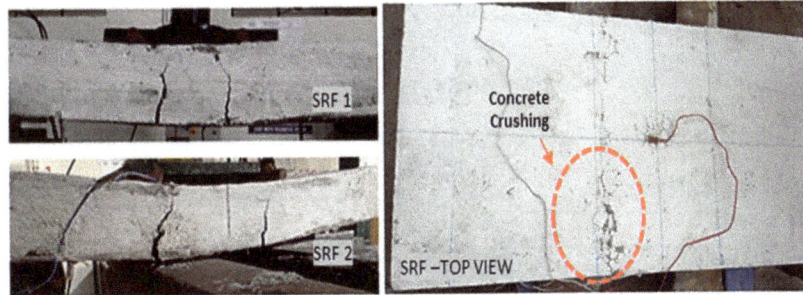

Figure 4. Crack pattern and failure mode of SRF slab.

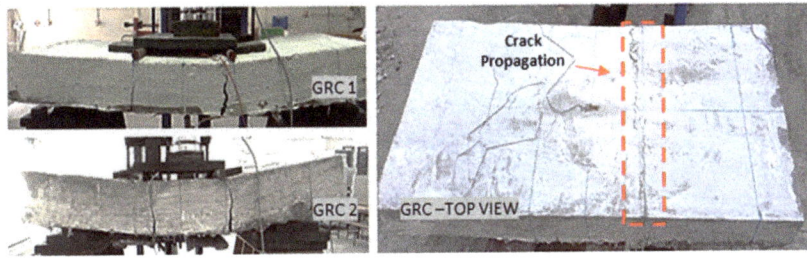

Figure 5. Crack pattern and failure mode of GRC slab.

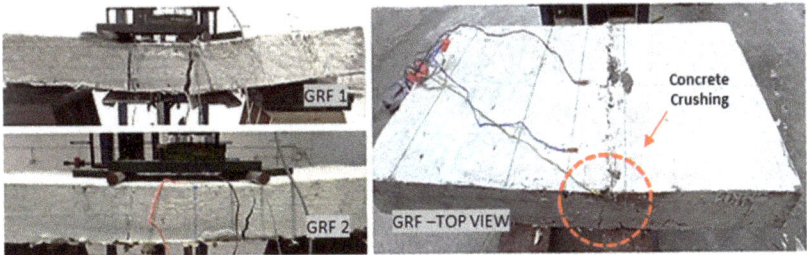

Figure 6. Crack pattern and failure mode of GRF slab.

3.2. Load–Deflection Behavior

The load–deflection behaviors of OPC and 60% fly ash concrete slabs reinforced with steel rods are shown in Figure 7, and those of OPC and 60% fly ash concrete slabs reinforced with GFRP rods are shown in Figure 8. The average ultimate load-carrying capacities of the concrete slab specimens SRC, SRF, GRC, and GRF were 23.9 kN, 27.9 kN, 28.55 kN, and 31.05 kN, respectively. The slab specimens SRF, GRC, and GRF showed 16.7%, 19.4%, and 29.9% increases in the ultimate load-carrying capacity when compared with the control specimen SRC [40]. The comparison of the ultimate load of slab specimens is shown in Figure 9. In the OPC concrete slab and fly ash concrete slab reinforced with steel rods, the initial loading phase corresponds to elastic behavior, in which no cracking appeared. A drop in the slope of the curve was observed after the cracking load due to progressive crack formation in the slab, and an almost linear segment was observed until failure. Steel-reinforced slabs and GFRP-reinforced slabs failed due to the crushing of concrete. In the concrete slab reinforced with GFRP rods, the deflection continued to increase with the increase in load and exhibited some ductility even though GFRP has a brittle nature [41].

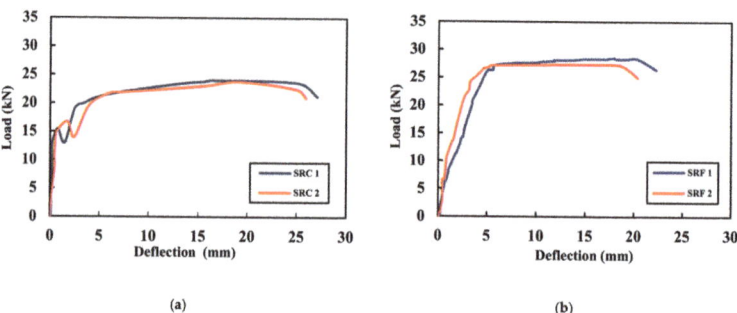

Figure 7. Load–deflection behavior of (**a**) SRC and (**b**) SRF.

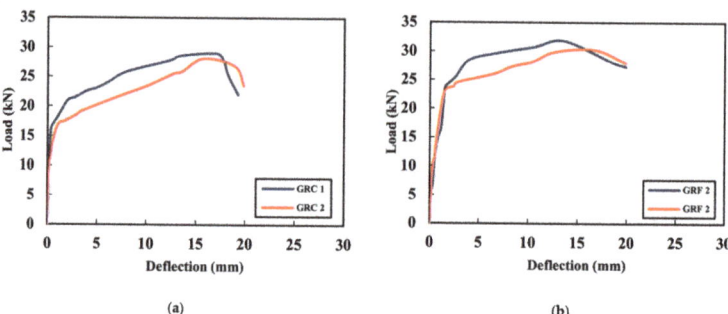

Figure 8. Load–deflection behavior of (**a**) GRC and (**b**) GRF.

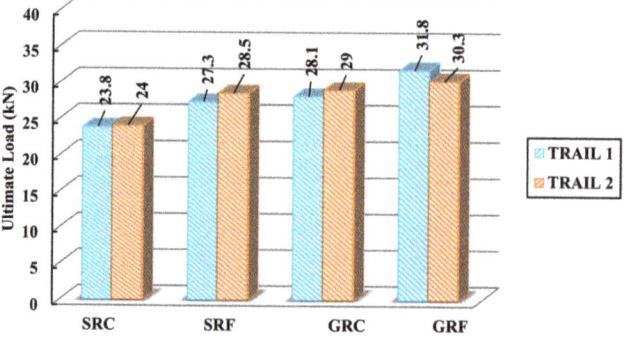

Figure 9. Comparison of ultimate load of the slab specimens.

3.3. Strain Distribution

The strain in OPC and fly ash concrete slabs at the top and bottom reinforcement was measured at the mid-span of the specimens. The positive strain represents the strain on the top surface of the OPC/fly ash concrete, and the negative strain represents the strain values of the steel rods/GFRP rods. The load–strain curves for SRC and SRF are shown in Figure 10, and the load–strain curves for GRC and GRF are shown in Figure 11. The strain in the steel rod and GFRP rod of the OPC concrete slab was $19{,}577 \times 10^{-6}$ mm/mm and $23{,}378 \times 10^{-6}$ mm/mm, respectively. The strain in the steel rod and GFRP rod of the fly ash concrete slab was $21{,}075 \times 10^{-6}$ mm/mm and $23{,}845 \times 10^{-6}$ mm/mm, respectively. The ultimate strain at the top of the OPC/fly ash concrete slabs lay in the range of 0.28% to 0.31% for the slab with steel reinforcement, and it ranged from 0.30% to 0.33% for the slab with GFRP reinforcement. The strain in the GFRP rods was higher than in steel rods in the OPC and fly ash concrete slab. From the graph, it can be observed that the concrete slab reinforced with GFRP rods has higher flexural strength than the steel-reinforced concrete slab [29,42].

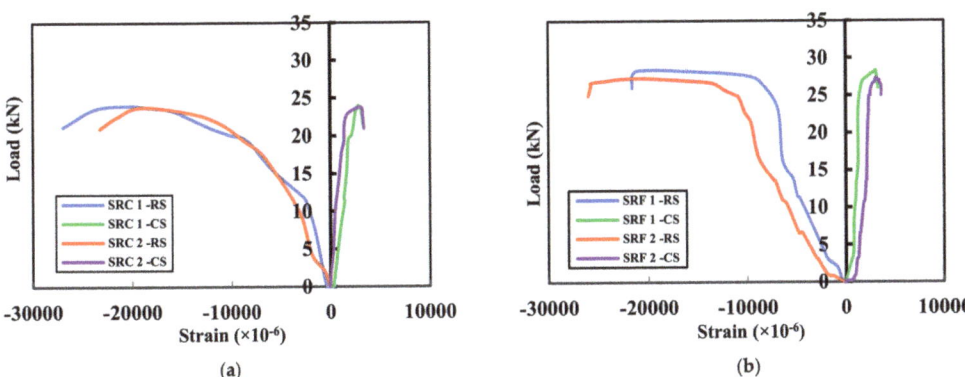

Figure 10. Load–strain behavior of (**a**) SRC and (**b**) SRF. RS—reinforcement strain; CS—concrete strain.

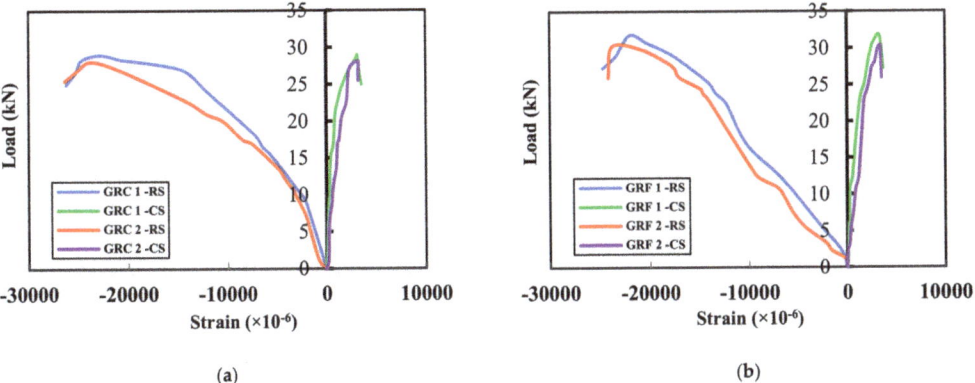

Figure 11. Load–strain behavior of (**a**) GRC and (**b**) GRF. RS—reinforcement strain; CS—concrete strain.

3.4. Moment–Curvature

The moment–curvature behaviors of the OPC and fly ash concrete slabs reinforced with steel rods are shown in Figure 12, and those of the OPC and fly ash concrete slabs reinforced with GFRP rods are shown in Figure 13. The moment vs. curvature relationship reflects ductility as well as provides the macroscopic mechanical properties of the slab elements.

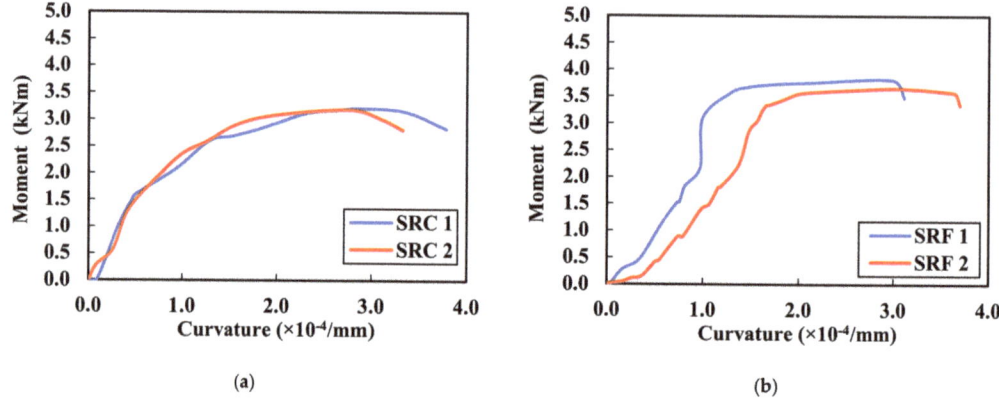

Figure 12. Moment–curvature behavior of (**a**) SRC and (**b**) SRF.

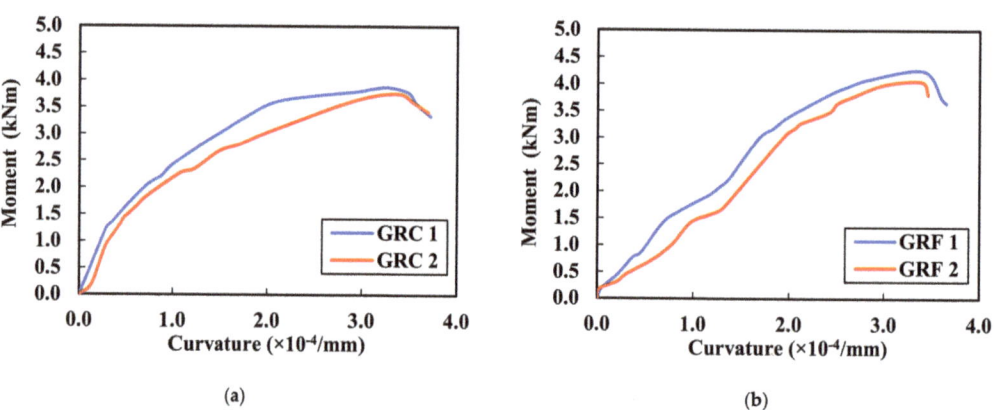

Figure 13. Moment–curvature behavior of (**a**) GRC and (**b**) GRF.

The curvature ($\varnothing$) was calculated using the following Equation (1):

$$\varnothing = \frac{\varepsilon_c + \varepsilon_r}{d} \quad (1)$$

where ε_c is the compressive strain in concrete, ε_r is the tensile strain in the reinforcement (steel/GFRP rod), and d is the effective depth of the slab.

The average ultimate moment capacities of the SRC, SRF, GRC, and GRF slabs were 3.18 kNm, 3.67 kNm, 3.81 kNm, and 4.12 kNm, respectively. The ultimate moment capacity of the GRF slab was 12% higher than that of the SRF slab, and the ultimate moment capacity of the GRC slab was 20% higher than that of the SRC slab. The provision of GFRP rods increased the slab curvature capacity more than the steel rods at the same moment. In a concrete slab reinforced with steel rods, the moment capacity was sustained with an increase in curvature before failure, but the slab reinforced with GFRP rods attained the peak moment capacity just before failure. However, it can be noted that slabs reinforced with GFRP rods are capable of exhibiting more curvature before failure when compared to slabs reinforced with steel rods [43]. The ultimate moment capacity of the tested slabs is provided in Table 6.

Table 6. Performance details of the slabs reinforced with steel rods and GFRP rods.

Slab Designation	Max. Load (P_u) (kN)	Deflection at Max Load (Δ) (mm)	Ultimate Moment (M_{Exp}) (kNm)	Ultimate Strain in Concrete (ε_{cu}) %	Ultimate Strain In Reinforcement (ε_f) %	Ductility Ratio
SRC 1	24.0	16.2	3.20	0.28	2.00	11.77
SRC 2	23.8	19.2	3.17	0.31	1.90	11.87
SRF 1	28.5	17.9	3.70	0.30	2.19	12.27
SRF 2	27.3	18.2	3.64	0.30	2.02	11.33
GRC 1	29.0	17.1	3.87	0.31	2.30	7.33
GRC 2	28.1	15.8	3.75	0.32	2.36	7.72
GRF 1	31.8	13.6	4.24	0.33	2.39	8.35
GRF 2	30.3	16.7	4.04	0.33	2.37	7.88

3.5. Displacement Ductility

Displacement ductility is defined as the ratio of ultimate deflection to first yield deflection [44]. A high ductility ratio implies that a structural member is capable of undergoing a large deflection before failure. The ductility calculation for both the steel-reinforced concrete slab and GFRP-reinforced concrete slab is shown in Figure 14. Despite the brittle nature of the GFRP composites, these GFRP rods showed ductility in the concrete slab [45]. The SRF and GRF slabs had an average ductility of 11.8 and 8.1, respectively, and the average ductility of SRC and GRC was 11.5 and 7.52, respectively. Since the ductility of GFRP rods is above 4, they can be used in structural members [6]. Both the slabs reinforced with steel and GFRP rods failed due to flexure with maximum strength and ductility, whereas the SRC and SRF slabs showed greater ductility than the GRC and GRF slabs. The ductility of the tested slabs is provided in Table 6.

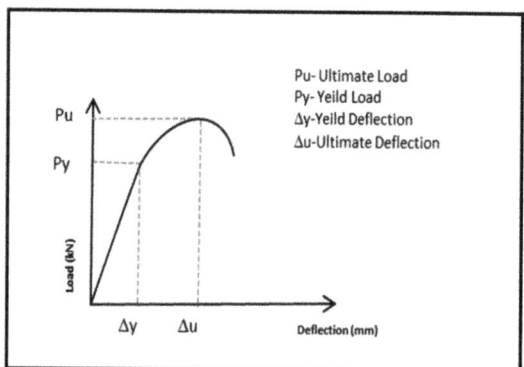

Figure 14. Displacement Ductility (Thamrin et al. [44]).

4. Theoretical Prediction

4.1. Equations Provided by ACI 44.1R-15

In this study, the theoretical flexural capacities (Mu) of steel-rod- and GFRP-rod-reinforced concrete slabs were computed based on the equations provided by ACI 440.1R-15 [25,46]. According to ACI 440.1R-15, the moment of resistance of a concrete slab reinforced with GFRP rods can be determined based on the internal force equilibrium, strain compatibility, and the controlling mode of failure. The predicted failure mode can be determined by comparing the reinforcement ratio in Equation (2) to the balanced reinforcement ratio in Equation (3).

$$\rho_f = \frac{A_f}{bd} \qquad (2)$$

where A_f is the area of GFRP rods, b is the width of the cross-section, and d is the distance from the extreme compression fiber to the centroid of the GFRP rods.

$$\rho_{fb} = 0.85\beta_1 \frac{f'_c}{f_{fu}} \frac{E_f \varepsilon_{cu}}{E_f \varepsilon_{cu} + f_{fu}} \qquad (3)$$

where f'_c is the specified compressive strength of concrete, f_{fu} is the ultimate tensile stress of GFRP bars, E_f is the elastic modulus of GFRP bars, and ε_{cu} is the ultimate strain in concrete (taken as 0.003).

The factor β_1 can be calculated using Equation (4).

$$\beta_1 = 0.85 - \frac{0.05(f'_c - 28)}{7} \geq 0.65 \qquad (4)$$

The predicted failure mode can be determined with the help of the following failure conditions. If ($\rho_f < \rho_{fb}$), then the condition of failure is due to tension. When ($\rho_{fb} < \rho_f < 1.4^*\rho_{fb}$), the balanced failure condition takes place with the crushing of concrete, followed by the rupture of the GFRP bar, and for ($\rho_f > 1.4^*\rho_{fb}$), the RC element fails due to compression failure. From the above conditions, all of the tested concrete slabs reinforced with GFRP rods failed due to compression failure. As $\rho_f > \rho_{fb}$, the slab is considered over-reinforced, the controlling limit state is the crushing of concrete, and the moment of resistance (M) can be calculated using Equation (5).

$$M = \rho_f f_f b d_2 \left(1 - 0.59 \frac{\rho_f f_f}{f'_c}\right) \quad (5)$$

where f_f is the tensile strength of the GFRP rods, and f_f can be calculated using Equation (6).

$$f_f = \sqrt{\frac{(E_f \varepsilon_{cu})^2}{4} + \frac{0.85\beta_1 f'_c}{\rho_f} E_f \varepsilon_{cu}} - 0.5 E_f \varepsilon_{cu} \leq f_{fu} \quad (6)$$

The moment resistance of the GRC and GRF slab specimens was obtained using ACI 440.1R-15, and the results are shown in Table 7. The theoretical moment of resistance obtained by slabs reinforced with GFRP rods is 4.48 kNm.

Table 7. Comparison of the moment of resistance for slabs reinforced with GFRP rods.

Specimen	Reinforcement Ratio (ρ_f)	Failure Mode	Moment of Resistance M,Exp (kN·m)	M,ACI (kN·m)	Percentage of Deviation (%)
GRC 1	1.44ρ_{fb}	Compression failure	3.87	4.48	15.9
GRC 2	1.44ρ_{fb}	Compression failure	3.75	4.48	19.6
GRF 1	1.44ρ_{fb}	Compression failure	4.24	4.48	5.84
GRF 2	1.44ρ_{fb}	Compression failure	4.04	4.48	11.0

4.2. Comparison of Experimental Results with Theoretical Predictions

The moment of resistance (M) of the slab reinforced with GFRP rods was calculated using ACI 440.1R-15, and experimental results are shown in Table 7. The moment of resistance calculated using ACI 440.1R-15 overestimated the ultimate moment carrying capacity of OPC/fly ash concrete slabs reinforced with GFRP bars by 17.81% and 8.46%, respectively. Figure 15 shows a graphical comparison of the moment of resistance predicted by ACI 440.1R 15 and experimental results.

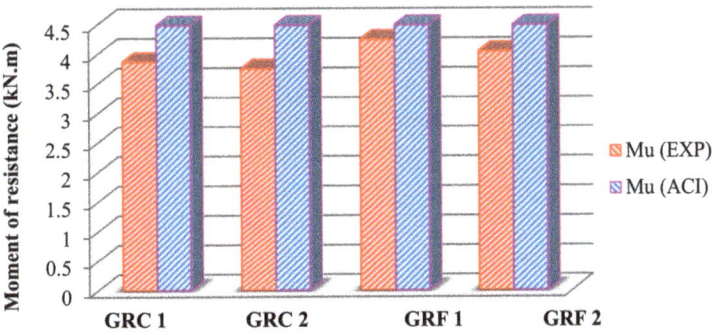

Figure 15. Moment of resistance of GFRP slabs.

5. Nonlinear Finite Element Analysis (NLFEA)

Nonlinear finite element analysis (NLFEA) was carried out using the ANSYS 2022-R1 [38] software, and the results were compared with the experimental results. From the experimental study, it was found that the fly ash concrete slab reinforced with GFRP rods has a higher load-carrying capacity than the other slab specimens. Therefore, the slab specimens SRF and GRF were considered for the finite element analysis.

5.1. Modeling

For modeling the concrete, the SOLID-65 element was used, which is an eight-noded solid element with three degrees of freedom at each node. For modeling the steel/GFRP rod, LINK-180, a two-noded element, was used. The support and the loading conditions of the model were created following the experimental setup, as shown in Figure 16.

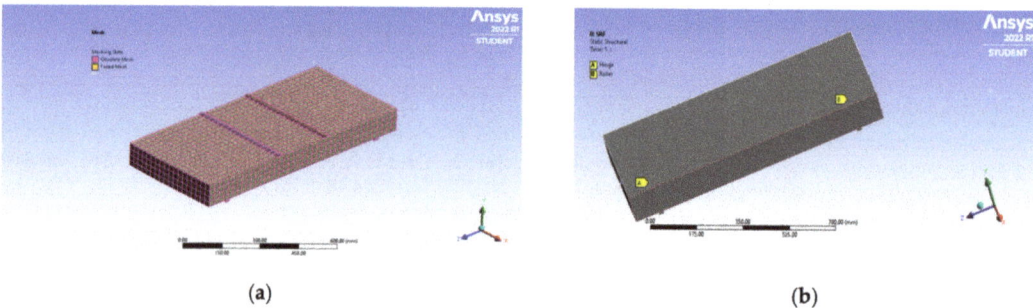

Figure 16. Slab model in ANSYS 2022-R1: (**a**) meshed model; (**b**) support conditions.

The ultimate deflections of the SRF and GRF slabs are shown in Figures 17 and 18, respectively. The ultimate loads of the slab specimens SRF and GRF were 27 kN and 30 kN, respectively, with ultimate deflections of 16.4 mm and 12.3 mm.

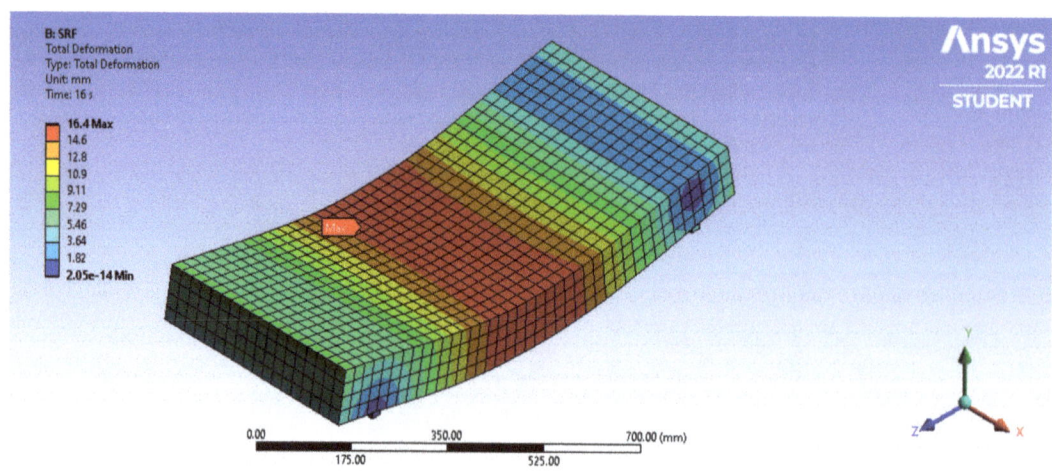

Figure 17. Ultimate deflection of SRF slab.

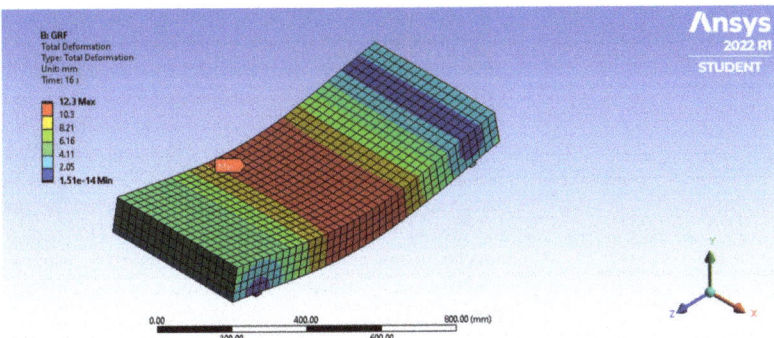

Figure 18. Ultimate deflection of GRF slab.

5.2. Comparison of Experimental Results with NLFEA Results

The comparison of experimental results with NLFEA results is shown in Table 8. The deviations of ultimate load and ultimate deflection from the experimental values are less than 10% for both the SRF and GRF slabs. Hence, ANSYS 2022-R1 [38] software can be used for the analysis of fly ash concrete slabs reinforced with GFRP bars. Figure 19 shows the comparison of the load–deflection behaviors of the SRF and GRF slabs obtained from the NLFEA to the experimental results. From the results, it is observed that both the experimental and numerical results are in good agreement.

Table 8. Comparison between experimental and numerical results.

Specimen Id.	Ultimate Load (kN)		Deflection at Mid-Span (mm)	
	Experimental	NLFEA (ANSYS)	Experimental	NLFEA (ANSYS)
SRF	28.5	27	17.9	16.4
GRF	31.8	30	13.6	12.3

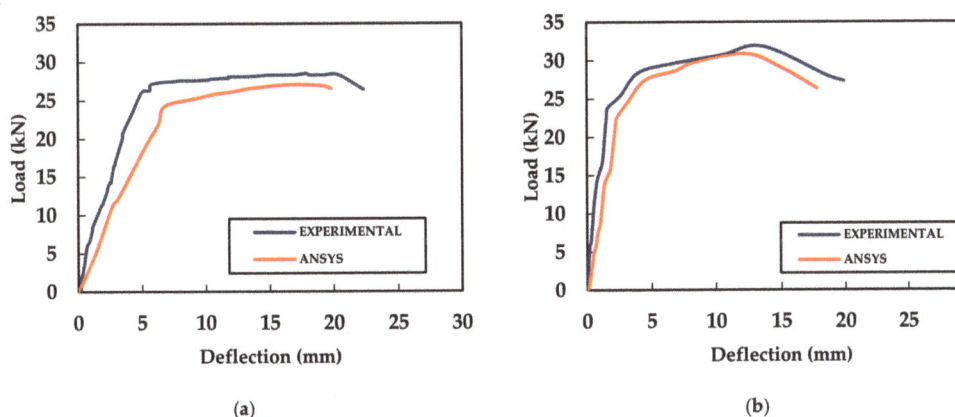

Figure 19. Comparison between experimental and numerical load–deflection behavior of (**a**) SRF and (**b**) GRF.

6. Conclusions

From this study, the following observations and conclusions can be made based on experiments conducted on OPC/fly ash concrete slabs reinforced with steel/GFRP bars.

1. The slab specimens SRF, GRC, and GRF showed 17%, 19%, and 30% increases in their ultimate load-carrying capacity when compared with the control specimen SRC.
2. The concrete surface strain was slightly higher for the fly ash concrete slab than for the OPC concrete slab, and the strain value of GFRP bars was higher than that of steel bars under the same loading. This shows that the flexural strength of the GFRP-reinforced concrete slab is higher than that of the steel-reinforced concrete slab.
3. The ultimate moment capacity of the GRF slab was 12% higher than that of the SRF slab, and the ultimate moment capacity of the GRC slab was 20% higher than that of the SRC slab. The slabs reinforced with GFRP rods are capable of exhibiting more curvature before failure when compared to the slabs reinforced with steel.
4. The average ductility of the steel rod was higher than that of the GFRP rod. Both the steel- and GFRP-reinforced slabs failed due to flexure with maximum strength and ductility.
5. The analytical equations given by ACI 440.1R-15 for calculating the moment of resistance overestimated the experimental results by 18%. Thus, ACI 440.1R-15 can be used for the design of concrete slabs reinforced with GFRP rods.
6. Only 10% deviation was observed between the experimental and the nonlinear finite element analysis (NLFEA) results. Hence, ANSYS 2022-R1 software can be used for the analysis of fly ash concrete slabs reinforced with GFRP bars.

Thus, it is observed that GFRP rods, when used as a replacement for steel rods in both conventional and fly ash concrete, provide improved strength compared to the one-way slab. In addition, the slabs reinforced with GFRP rods did not exhibit sudden failure, even though GFRP rods are brittle. The results of this investigation demonstrate the ability and the potential use of GFRP rods with fly ash concrete in slab elements.

Author Contributions: C.S.M. and S.M.: Conceptualized the model, established the empirical theorem, and conducted the experiments. P.S.J., B.G.A.G. and K.R.: Supervised the research as well as the analysis of results. P.S.J., C.S.M. and S.M.: Introduced the idea of static loading in this project, designed the slab, wrote, reviewed, and submitted the paper, and collaborated in and coordinated the research. P.S.J., B.G.A.G. and K.R.: Suggested and chose the journal for submission. P.S.J., B.G.A.G. and K.R.: Participated in the manuscript revision phase. All authors have read and agreed to the published version of the manuscript.

Funding: This research received no external funding.

Institutional Review Board Statement: Not applicable.

Informed Consent Statement: Not applicable.

Data Availability Statement: The data presented in this study are available on request from the corresponding author.

Conflicts of Interest: This manuscript has not been submitted to, nor is it under review by, another journal or other publishing venue. The authors have no affiliation with any organization with a direct or indirect financial interest in the subject matter discussed in the manuscript. The authors declare no conflict of interest.

References

1. Liew, K.; Sojobi, A.; Zhang, L. Green concrete: Prospects and challenges. *Constr. Build. Mater.* **2017**, *156*, 1063–1095. [CrossRef]
2. Barbuta, M.; Bucur, R.; Serbanoiu, A.; Scutarasu, S.; Burlacu, A. Combined Effect of Fly Ash and Fibers on Properties of Cement Concrete. *Procedia Eng.* **2017**, *181*, 280–284. [CrossRef]
3. Siddique, R. Effect of fine aggregate replacement with Class F fly ash on the mechanical properties of concrete. *Cem. Concr. Res.* **2003**, *33*, 539–547. [CrossRef]
4. Detwiler, R.J. *Substitution of Fly Ash for Cement or Aggregate in Concrete: Strength Development and Suppression of ASR*; RD127; Portland Cement Association: Skokie, IL, USA, 2002.
5. Siddique, R. Performance characteristics of high-volume Class F fly ash concrete. *Cem. Concr. Res.* **2004**, *34*, 487–493. [CrossRef]
6. Joanna, P.S.; Rooby, J.; Prabhavathy, A.; Preetha, R.; Pillai, C.S. Behaviour of reinforced concrete beams with 50 percentage fly ash international journal of civil engineering and technology. *Int. J. Civ. Eng. Technol.* **2013**, *4*, 36–48.
7. Malhotra, V.M. High-Performance, High Volume Fly Ash Concrete. *Concr. Int.* **2002**, *24*, 30–34.

8. Saha, A. Effect of class F fly ash on the durability properties of concrete. *Sustain. Environ. Res.* **2018**, *28*, 25–31. [CrossRef]
9. Thomas, M.D.A. *Optimizing the Use of Fly Ash in Concrete*; Portland Cement Association: Skokie, IL, USA, 2007.
10. Murali, M.; Mohammed, B.S.; Abdulkadir, I.; Liew, M.S.; Alaloul, W.S. Utilization of Crumb Rubber and High-Volume Fly Ash in Concrete for Environmental Sustainability: RSM-Based Modeling and Optimization. *Materials* **2021**, *14*, 3322. [CrossRef]
11. Hu, W.; Li, Y.; Yuan, H. Review of Experimental Studies on Application of FRP for Strengthening of Bridge Structures. *Adv. Mater. Sci. Eng.* **2020**, *2020*, 8682163. [CrossRef]
12. Mertol, H.C.; Rizkalla, S.; Scott, P.; Lees, J.M.; El-Hachal, R. Durability of concrete beams prestressed with CFRP. *Symp. Pap.* **2007**, *245*, 1–20.
13. Shave, J. The time has come for high strength, low maintenance fibre reinforced plastics. *New Civil Eng.* **2014**, 11.
14. Yooprasertchai, E.; Dithaem, R.; Arnamwong, T.; Sahamitmongkol, R.; Jadekittichoke, J.; Joyklad, P.; Hussain, Q. Remediation of Punching Shear Failure Using Glass Fiber Reinforced Polymer (GFRP) Rods. *Polymers* **2021**, *13*, 2369. [CrossRef] [PubMed]
15. Mussa, M.H.; Radzi, N.A.M.; Hamid, R.; Mutalib, A.A. Fire Resistance of High-Volume Fly Ash RC Slab Inclusion with Nano-Silica. *Materials* **2021**, *14*, 3311. [CrossRef] [PubMed]
16. Zaghloul, M.M.Y.; Mohamed, Y.S.; El-Gamal, H. Fatigue and tensile behaviors of fiber-reinforced thermosetting composites embedded with nanoparticles. *J. Compos. Mater.* **2018**, *53*, 709–718. [CrossRef]
17. Zaghloul, M.M.Y.; Steel, K.; Veidt, M.; Heitzmann, M.T. Wear behaviour of polymeric materials reinforced with man-made fibres: A comprehensive review about fibre volume fraction influence on wear performance. *J. Reinf. Plast. Compos.* **2022**, *41*, 215–241. [CrossRef]
18. Zaghloul, M.M.Y.; Zaghloul, M.M.Y. Influence of flame retardant magnesium hydroxide on the mechanical properties of high density polyethylene composites. *J. Reinf. Plast. Compos.* **2017**, *36*, 1802–1816. [CrossRef]
19. Zaghloul, M.Y.M.; Zaghloul, M.M.Y.; Zaghloul, M.M.Y. Developments in polyester composite materials—An in-depth review on natural fibres and nano fillers. *Compos. Struct.* **2021**, *278*, 114698. [CrossRef]
20. Mohamed, Y.S.; El-Gamal, H.; Zaghloul, M.M.Y. Micro-hardness behavior of fiber reinforced thermosetting composites embedded with cellulose nanocrystals. *Alex. Eng. J.* **2018**, *57*, 4113–4119. [CrossRef]
21. Zaghloul, M.M.Y.M. Mechanical properties of linear low-density polyethylene fire-retarded with melamine polyphosphate. *J. Appl. Polym. Sci.* **2018**, *135*, 46770. [CrossRef]
22. Zaghloul, M.M.Y.; Zaghloul, M.Y.M.; Zaghloul, M.M.Y. Experimental and modeling of mechanical-electrical behavior of polypropylene composites filled with graphite and MWCNT fillers. *Polym. Test.* **2017**, *63*, 467–474. [CrossRef]
23. Bedon, C.; Louter, C. Structural glass beams with embedded GFRP, CFRP or steel reinforcement rods: Comparative experimental, analytical and numerical investigations. *J. Build. Eng.* **2019**, *22*, 227–241. [CrossRef]
24. Jalal, A.; Hakim, L.; Shafiq, N. Mechanical and Post-Cracking Characteristics of Fiber Reinforced Concrete Containing Copper-Coated Steel and PVA Fibers in 100% Cement and Fly Ash Concrete. *Appl. Sci.* **2021**, *11*, 1048. [CrossRef]
25. ACI 440.1R-15; Guide for the Design and Construction of Concrete Reinforced with Fiber Reinforced Polymers (FRP) Bars; American Concrete Institute: Farmington Hills, MI, USA, 2015.
26. Balendran, R.; Rana, T.; Maqsood, T.; Tang, W. Application of FRP bars as reinforcement in civil engineering structures. *Struct. Surv.* **2002**, *20*, 62–72. [CrossRef]
27. Balafas, I.; Burgoyne, C. Economic design of beams with FRP rebars or prestress. *Mag. Concr. Res.* **2012**, *64*, 885–898. [CrossRef]
28. Hollaway, L. A review of the present and future utilisation of FRP composites in the civil infrastructure with reference to their important in-service properties. *Constr. Build. Mater.* **2010**, *24*, 2419–2445. [CrossRef]
29. Jabbar, S.; Farid, S. Replacement of steel rebars by GFRP rebars in the concrete structures. *Karbala Int. J. Mod. Sci.* **2018**, *4*, 216–227. [CrossRef]
30. Kemp, M.; Blowes, D. Concrete Reinforcement and Glass Fibre Reinforced Polymer. *Qld. Roads Ed.* **2011**, *11*, 40–48.
31. Borosnyói, A. Corrosion-resistant concrete structures with innovative Fibre Reinforced Polymer (FRP) materials. *Epa.-J. Silic. Based Compos. Mater.* **2013**, *65*, 26–31.
32. Sólyom, S.; Balázs, G.L.; Borosnyói, A. Material characteristics and bond tests for FRP rebars. *Concr. Struct.* **2015**, *16*, 38–45.
33. Patil, V.R. Experimental Study of Behavior of RCC Beam by Replacing Steel Bars with Glass Fiber Reinforced Polymer and Carbon Reinforced Fiber Polymer (GFRP). *Int. J. Innov. Res. Adv. Eng.* **2014**, *1*, 205–210.
34. Goonewardena, J.; Ghabraie, K.; Subhani, M. Flexural Performance of FRP-Reinforced Geopolymer Concrete Beam. *J. Compos. Sci.* **2020**, *4*, 187. [CrossRef]
35. Adam, M.A.; Erfan, A.M.; Habib, F.A.; El-Sayed, T.A. Structural Behavior of High-Strength Concrete Slabs Reinforced with GFRP Bars. *Polymers* **2021**, *13*, 2997. [CrossRef] [PubMed]
36. Jayajothi, P.; Kumutha, R.; Vijai, K. Finite element analysis of FRP strengthened RC beams using Ansys. *Asian J. Civ. Eng.* **2013**, *14*, 631–642.
37. Gherbi, A.; Dahmani, L.; Boudjemia, A. Study on two way reinforced concrete slab using Ansys with different boundary conditions and loading world academy of science. *Eng. Technol. Int. J. Civ. Environ. Eng.* **2018**, *12*, 1151–1156.
38. ANSYS 2022-R1; Manual Set. ANSYS Inc.: Canonsburg, PA, USA, 2022.
39. Gu, H.S.; Zhu, D.Y. Flexural Behaviours of Concrete Slab Reinforced with GFRP bars. *Adv. Mater. Res.* **2011**, *243–249*, 567–572. [CrossRef]

40. Moon, J.; Reda Taha, M.M.; Kim, J.J. Flexural Strengthening of RC Slabs Using a Hybrid FRP-UHPC System Including Shear Connector. *Adv. Mater. Sci. Eng.* **2017**, *2017*, 4387545. [CrossRef]
41. Dhipanaravind, S.; Sivagamasundari, R. Flexural Behaviour of Concrete One-way Slabs Reinforced with Hybrid FRP Bars. *Int. J. Appl. Eng.* **2018**, *13*, 4807–4815.
42. Venkatesan, G.; Raman, S.R.; Sekaran, M.C. Flexural behaviour of reinforced concrete beams using high volume fly ash concrete confinement in compression zone. *J. Civ. Eng. (IEB)* **2013**, *41*, 87–97.
43. El Zareef, M.; El Madawy, M. Effect of glass-fiber rods on the ductile behaviour of reinforced concrete beams. *Alex. Eng. J.* **2018**, *57*, 4071–4079. [CrossRef]
44. Thamrin, R.; Zaidir, Z.; Iwanda, D. Ductility Estimation for Flexural Concrete Beams Longitudinally Reinforced with Hybrid FRP–Steel Bars. *Polymers* **2022**, *14*, 1017. [CrossRef]
45. Gunes, O.; Lau, D.; Tuakta, C.; Büyüköztürk, O. Ductility of FRP–concrete systems: Investigations at different length scales. *Constr. Build. Mater.* **2013**, *49*, 915–925. [CrossRef]
46. Ahmed, H.; Jaf, D.; Yaseen, S. Comparison of the Flexural Performance and Behaviour of Fly-Ash-Based Geopolymer Concrete Beams Reinforced with CFRP and GFRP Bars. *Adv. Mater. Sci. Eng.* **2020**, *2020*, 3495276. [CrossRef]

Article

Finite Element Multi-Physics Analysis and Experimental Testing for Hollow Brick Solutions with Lightweight and Eco-Sustainable Cement Mix

Matteo Sambucci [1,2,*], Abbas Sibai [1], Luciano Fattore [3], Riccardo Martufi [3], Sabrina Lucibello [3,4] and Marco Valente [1,2]

1. Department of Chemical Engineering, Materials, Environment, Sapienza University of Rome, 00184 Rome, Italy; abbas.sibai@uniroma1.it (A.S.); marco.valente@uniroma1.it (M.V.)
2. INSTM Reference Laboratory for Engineering of Surface Treatments, Department of Chemical Engineering, Materials, Environment, Sapienza University of Rome, 00184 Rome, Italy
3. Center for Research and Services Saperi&Co, Sapienza University of Rome, 00185 Rome, Italy; luciano.fattore@uniroma1.it (L.F.); riccardo.martufi@uniroma1.it (R.M.); sabrina.lucibello@uniroma1.it (S.L.)
4. Department of Planning, Design, Technology for Architecture, Sapienza University of Rome, 00196 Rome, Italy
* Correspondence: matteo.sambucci@uniroma1.it; Tel.: +39-06-44585647

Abstract: Combining eco-sustainability and technological efficiency is one of the "hot" topics in the current construction and architectural sectors. In this work, recycled tire rubber aggregates and acoustically effective fractal cavities were combined in the design, modeling, and experimental characterization of lightweight concrete hollow bricks. After analyzing the structural and acoustic behavior of the brick models by finite element analysis as a function of the type of constituent concrete material (reference and rubberized cement mixes) and hollow inner geometry (circular- and fractal-shaped hollow designs), compressive tests and sound-absorption measurements were experimentally performed to evaluate the real performance of the developed prototypes. Compared to the traditional circular hollow pattern, fractal cavities improve the mechanical strength of the brick, its structural efficiency (strength-to-weight ratio), and the medium–high frequency noise damping. The use of ground waste tire rubber as a total concrete aggregate represents an eco-friendlier solution than the ordinary cementitious mix design, providing, at the same time, enhanced lightweight properties, mechanical ductility, and better sound attenuation. The near-compliance of rubber-concrete blocks with standard requirements and the value-added properties have demonstrated a good potential for incorporating waste rubber as aggregate for non-structural applications.

Keywords: ground waste tire rubber; hollow concrete brick; fractal; finite element analysis; compressive strength; acoustic absorption; eco-sustainability

1. Introduction

Our dependence on waste tires is clear; just think of the world of transport, in general. About 800 million tires worldwide reach their end of life each year, and much of the rubber from which they are made, which represents about 50% of their weight, is lost [1]. To effectively close the waste management cycle in a circular economy view, such material can be recycled as a granular product and intended for civil engineering applications. The key markets for material recovery are as follows [2]: (1) whole tires used to fabricate crash barriers, bumpers, or artificial reefs; (2) crumb rubber used to produce molded rubber products, flooring, or matting; and (3) powdered rubber used as modifiers to asphalt paving mixtures. During the last three decades, researchers investigated the possible use of ground waste tire rubber (GWTR) in concrete and mortars. Since the pioneering study on engineering properties of rubberized concrete conducted by Eldin and Senouci [3] in the

1990s, a large number of research activities have taken place, proving the possibility of creating alternative and eco-sustainable cementitious mixes, thanks to the presence of rubber particles recovered from discarded tires that replace the ordinary virgin aggregate. Besides the environmental benefits, the researchers suggested that cement-based composites incorporating GWTR can significantly enhance toughness and energy absorption [4], increase the mechanical strain capacity [5], achieve better noise attenuation performance [6,7], raise thermal insulation capacity [8], and improve freeze–thaw resistance with entrapped air voids [9] compared to plain concrete. However, it is also well documented that replacing the ordinary aggregates (sand and gravels) with GWTR adversely affects the concrete's mechanical properties, such as static stiffness and compressive strength, which can limit its use in structural applications [10,11].

The literature survey above shows that there is a potential for the use of rubber aggregates in building/architectural elements where the primary requirement is not mechanical strength but lightweight, thermal–acoustic efficiency, and durability, such as hollow bricks. Brick is the most basic precast unit for the construction of low-cost houses and multi-stored apartments. There are remarkable and noteworthy points going in favor of the use of these hollow components [12]:

- The dead load is much lower than for a solid block; due to this, one can structurally engineer them and reduce steel consumption in construction;
- The heat insulation of wall structures is achieved due to the inner cavities, which provide energy saving for all times. Similarly, hollowness results in improved sound attenuation;
- Low maintenance cost, minimal material requirements, and cost competitiveness with other materials make it a preferred material for today's building.

The conventional type of brick is made of fired clay [13]. Clay brick manufacturing is an energy-intensive process. It involves the consumption of a considerable amount of energy during the firing process, requiring temperatures between 1000 and 1200 °C, depending on the raw materials [14]. Apart from providing the above-listed features, rubber-concrete hollow bricks could eliminate this drawback, since no fuel and thermal treatment are necessary for their production. Except for the clinkering concerning the production of the cement binder, the curing and hardening of concrete bricks takes place at room temperature, without involving firing or other additional processing. At the same time, the presence of recycled rubber aggregates would bring valuable benefits from both an environmental and a technological point of view. Some attempts aimed at the design, development, and characterization of GWTR-added bricks were made in the past few years. Turgut and Yesilata [13] investigated the physical–mechanical and thermal performance of rubberized solid brick with varying sand-crumb rubber volumetric replacement (from 10% v/v to 70% v/v). The thermal insulation performances of these bricks are found to be better than their ordinary counterpart (percentage-wise insulating improvements up to 11%), and the mechanical properties of the samples satisfied the requirements reported in the international technical standards. Mohammed et al. [15] developed hollow concrete blocks by using 0% v/v, 10% v/v, 25% v/v, and 50% v/v GWTR as a replacement for fine mineral aggregate. The authors discovered that the samples can be produced as load-bearing hollow blocks, as well as lightweight hollow blocks, providing better thermal and acoustic performance in comparison with conventional hollow blocks. Fraile-Garcia et al. [16] examined the acoustic behavior of hollow bricks made of concrete doped with waste tire rubber (0%, 20%, and 30%). Elements with the maximum percentage of rubber in their composition provided a better response than the control samples (0% rubber) for low-frequency noise, both in the case of airborne insulation and impact sound insulation. In the first case, the improvement was up to 50%, whereas, in the second case, this percentage was 15%. Therefore, highly rubber-modified construction elements are convenient to isolate the low-frequency sounds, such as instruments or road traffic vehicles (heavy trucks, tractors, etc.) emissions. Al-Fakih et al. [17] analyzed the mechanical performance of interlocking masonry walls that were constructed using rubberized hollow bricks (10% v/v of crumb rubber). In contrast to conventional masonry walls under compressive loading, GWTR–cement interlocking walls showed increased ductility, experiencing measurable post-failure loads with significant displacement, due to the presence of crumb rubber, which allows for a large expansion of microcracks inside the specimens after failure.

Complementary to the material's composition, the internal geometry of the brick units assumes a crucial role in its structural, thermal, and acoustic performance [18]. Supported by digital design and finite-element modeling tools, several researchers were involved in studying optimized internal cavities to improve specific performance characteristics of the hollow bricks. A collection of some works on this research topic is reported in Table 1.

Table 1. Design and shape optimization of hollow concrete bricks: A brief overview.

Research Work	Aim of the Study	Major Remarks
Del Coz Díaz et al. [19]	Topological optimization of twelve hollow concrete block units, varying the number and shape of inner recesses, with the aim of reducing the brick's weight, keeping suitable structural properties.	A weight reduction close to 45% is obtained with respect to the classic concrete block, keeping comparable structural efficiency in terms of strength-to-weight ratio.
Al-Tamimi et al. [20]	Twenty-three brick designs with different hole arrangements and one solid model were studied for concrete material to reach the model with the optimum holes in terms of thermal-insulation efficiency.	Increasing the hollow ratio tends to decrease the heat transfer from outer to inner brick sides significantly. At the same hollow ratio, there was an effect of the shape of holes in reducing the thermal flow through the bricks: rectangular shapes were more thermo-effective than circular ones.
Sassine et al. [21]	Mechanical and thermal behaviors of ten concrete hollow-block configurations are simultaneously studied by varying the blocks' internal shape, aiming at determining the optimal hollow-block design and providing the optimal compromise between thermal insulation and mechanical strength.	Longitudinal bulkheads improve the thermal resistance of the blocks and, thus, reduce the heat flux passing through the element. The mechanical behavior varies slightly between the investigated models in vertical compression, reducing the influence of this parameter in the selection of the best design.
Valente et al. [22]	Mechanical performance of three types of hollow-brick designs, circular, square, and hexagonal holes, were numerically analyzed to select the best configuration for rubber–concrete mixes.	Circular and hexagonal hole designs offer the best result in terms of compressive strength. The "honeycomb" geometries have remarkable thermal and acoustic functionality; therefore, they have more attractive requirements for building.

There are a very limited number of works investigating functional brick configurations in terms of acoustic performance, which is one of the primary requirements in the design of building elements. Noise control has become an imperative engineering field in modern society, not only because of the recent recognition of noise as a serious health hazard, but also because the standard of living and the quality of life are becoming more important [23].

This study is focused on investigating the performance of GWTR–cement bricks topologically engineered with inner hollow layouts potentially designed for sound attenuating applications. Specifically, the influence of easy-to-design-and-manufacture fractal cavities was analyzed by finite element analysis (FEA), experimentally tested, and compared with hole shapes (circular) commonly used in brick technology. Fractal geometries were extensively studied in anti-noise applications, such as acoustic damping cavities for sound-absorbable systems based on Helmholtz's resonators [24–26]. Highly irregular cavity shapes increase the viscous resistance at the cavity boundary by adding geometrical features that impede the natural direction of airflow. Consequently, the sound wave, interacting with the cavity, experiences significant dissipation phenomena [26]. The key purpose of the present research was to model, develop, and characterize brick prototypes with double technological functionality, which means, deriving both from the rubberized concrete's characteristics (lightweight, toughness, thermo-acoustic peculiarities, eco-friendliness) and the component's design. The manuscript is structured in three parts:

- *Part 1*—Description and key properties of the cement mixes (reference and rubberized concretes) used for hollow bricks manufacturing;
- *Part 2*—Design, topological optimization, and finite element modeling (mechanical and acoustic analysis) of hollow brick prototypes based on circular and fractal inner cavities;
- *Part 3*—Production process of the designed hollow bricks, mechanical and acoustic experimental characterization, and FEA models validation.

2. Materials and Methods

2.1. Part 1: Raw Materials and Concrete Mixes Characterization

2.1.1. Raw Materials

Constituent materials for concrete mixes included a commercial Type II Portland limestone cement (strength class 42.5 R) supplied by Colacem (Gubbio, Italy), fine river sand, and GWTR aggregates. Two fractions of recycled tire rubber particles, 0–1 mm rubber powder (RP) and 1–3 mm rubber granules (RG), were provided by the European Tyre Recycling Association (ETRA, Brussels, Belgium) and manufactured by ambient mechanical shredding processing of waste tires. The polymeric aggregates were used to produce the rubber–cement mix as a volumetric replacement of the mineral aggregate. The river sand and GWTR particles involved in this research work are presented in Figure 1.

Figure 1. River sand and GWTR used in this study.

The density of the aggregates was measured by the pycnometer method. For sand, water was employed as a test fluid, in agreement with the EN 1097-7 [27] standard method. Concerning the testing on rubber, the standard protocol was slightly modified, employing denatured ethylic alcohol (Deterchimica, Viterbo, Italy) as a fluid of known density to minimize undesired floating phenomena. The water absorption was assessed as the ratio of the difference between the weight of the aggregates in saturated surface dry (SSD) condition and oven-dry condition (110 °C for 24 h) to the weight of oven-dry aggregates. The physical properties of the mineral and GWTR aggregates are shown in Table 2.

Table 2. Physical properties of river sand and GWTR.

Aggregate	Density (kg/m³)	Water Absorption (%)
River sand	2476	20.0
GWTR	1144	9.6

Figure 2 presents the size-grading analysis of rubber particles and sand, determined via the vibrating sieving method, following the DIN 51701 [28] standard.

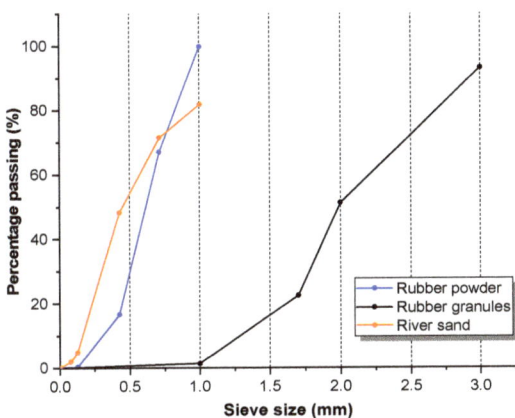

Figure 2. Particle size distribution of RP, RG, and river sand.

2.1.2. Mix Proportions and Samples Preparation

In this research, a single rubber–cement formulation (designed as RuC) that was obtained by totally replacing the sand with the two rubber fractions was investigated to produce the brick prototypes. Specifically, the mix design incorporated fine and coarse polymer aggregates in equal proportion (50% v/v RP–50% v/v RG). Such selected GWTR dosage was determined as "optimum" in previous research works conducted by the authors, where the influence of the tire aggregate size on the physical, mechanical, and microstructural performances of 3D printable cement [29,30] mixes and rubberized alkali-activated composites [31] was investigated in detail.

Reference concrete mix (REF), involving 0% v/v of GWTR, was also produced for comparison purposes. For REF mix, a water-to-cement ratio of 0.42 was chosen in agreement with common technical requirements to ensure proper cement mass hydration. In the rubberized formulation, the water dosage was adjusted to achieve proper fluidity and workability for mold-casting. The mix proportions of investigated formulations are listed in Table 3.

Table 3. Concrete mix proportions.

Sample ID	Cement (kg/L)	Water (kg/L)	Sand (kg/L)	RP (kg/L)	RG (kg/L)	w/c Ratio
REF	0.72	0.300	1.20	/	/	0.42
RuC	0.72	0.325	/	0.275	0.275	0.45

Dry components (cement, sand, and GWTR) were mixed inside a plastic tank to achieve a homogeneous blend. Tap water was gradually added to the mix until an adequate fluidity level for the pouring of the hydrated compound into the mold was reached. During the water addition, the batch was subjected to some vibration cycles, by a Giuliani IG/3 shaker machine (Giuliani Tecnologie, Turin, Italy), to improve the constituent mixing. Then the fresh mix was cast into rectangular plastic molds (110 mm × 185 mm × 50 mm). After casting, an additional vibration operation (2 min) was performed to expel any air bubbles embedded during the mixing and pouring. Firstly, the samples were cured in air for 24 h, and then the semi-hardened slabs were extracted from the molds and cured underwater for 28 days. After curing, for each concrete mix, a series of specimens were collected (Table 4) by wet sawing with an abrasive cutting disk, and they were intended for experimental testing. The materials characterization, described in detail below, provided the main physical, mechanical, and acoustic properties of the formulations under examination to be used as input data for the FEA analysis.

Table 4. Test samples for materials characterization.

Specimen Type	Number of Specimens per Test	Test
1 cm^3 cubes	3	Density
1 cm^3 cubes	3	Permeable porosity
1.5 cm × 1.5 cm × 10 cm beams	3	Three-point flexural test
1.5 cm × 1.5 cm × 3 cm prisms	3	Compressive test
5 cm × 5 cm × 2.5 cm blocks	1	Acoustic flow resistivity

2.1.3. Testing Program and Experimental Results

The density (ρ) of each specimen was measured by the hydrostatic weighing method described in ASTM D792 standard [32], using a ME54 analytical balance (Mettler Toledo, Columbus, OH, USA) equipped with a kit for gravimetric measurements.

The permeable porosity (Φ) was evaluated by vacuum saturation technique (ASTM C1202 standard [33]), using the experimental setup and test conditions presented in Figure 3a.

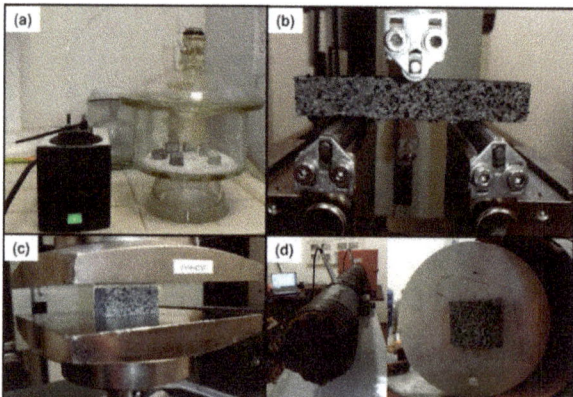

Figure 3. Graphical overview of the experimental program for materials characterization: (**a**) vacuum saturation method, (**b**) three-point flexural test, (**c**) compressive test, and (**d**) acoustic flow resistivity measurement.

The mechanical behavior of the samples was studied under flexural and uniaxial compression, using a Zwick-Roell Z10 (Zwick-Roell Group, Ulm, Germany) universal machine equipped with a 10 kN load cell. The bending test was performed in a three-point configuration (Figure 3b) with a support span of 60 mm, a cross head speed of 1 mm/min, and a pre-load of 5 N. Compressive test (Figure 3c) was run at 1 mm/min in displacement control. Flexural tensile strength (σ_t), compressive strength (σ_c), and compressive elastic modulus (E_c) were recorded and analyzed with TestXpert II software (Zwick-Roell Group, Ulm, Germany).

Measurements of the acoustic flow resistivity (R_f) were executed in an impedance tube (Figure 3d) in accordance with the test method proposed by Ingard and Dear [34]. More detailed information about the experimental setup can be found in Reference [35].

Table 5 provides the experimental results for ρ, Φ, σ_f, σ_c, E_c, and R_f of REF and RuC concrete mixes, including the average values with the standard deviation in brackets.

Table 5. Concrete mix proportions.

Sample ID	ρ (kg/m³)	Φ (%)	σ_t (MPa)	σ_c (MPa)	E_c (GPa)	R_f (N × s × m^{-4})
REF	2186 (18)	21.39 (0.25)	11.06 (1.48)	35.89 (7.71)	1.62 (0.47)	13,872
RuC	1281 (21)	22.82 (0.37)	1.65 (0.51)	4.93 (0.57)	0.24 (0.02)	19,862

The unit weight, mechanical strength, and stiffness of RuC mix were expected to be less than that of REF sample. The decrease is mainly attributed to the replacement of sand with lightweight polymer aggregates, the difference in deformability between rubber and cement paste that generates high-stress concentration at the interfacial transition zone (ITZ), leading to the formation of cracks in that region, the weak bonding between GWTR and cement matrix, and the entrapped air from the hydrophobic rubber particles, assisting the generation of internal porosity in the hardened material [36]. Although the addition of rubber would lead to an increase in the air void rate, experimental Φ-values demonstrated similar characteristics between REF and rubberized mixes. The proportion ratio between fine and coarse rubber fractions selected in this research provided beneficial effects in terms of microstructural properties. In this regard, some researchers have verified that incorporating varying sizes of GWTR improved the aggregate gradation so that a denser microstructure is produced to decrease the material's permeability [37,38]. R_f data showed a clear improvement in the acoustic properties of the cementitious mix by replacing the sand with the rubber fractions. GWTR aggregates would provide an additional sound attenuation mechanism related to their viscoelastic nature. When the rubber aggregate is vibrated, part of the energy is stored (elastic) and part is dissipated as heat (viscous) within the polymer. This viscous property results in loss of vibration energy as heat rather than being radiated as noise [39].

2.2. Part 2: Design, Modeling, and FEA of Hollow Bricks

COMSOL Multiphysics (COMSOL Inc., Stockholm, Sweden) computer aid engineering (CAI) software was employed for designing the hollow brick prototypes and modeling via FEA and their mechanical and acoustic behavior as a function of constituent concrete mix (REF and RuC mixes) and holes configuration.

2.2.1. Hollow Bricks Design

At first, COMSOL Multiphysics 3D geometry tool was used to design the brick models. The investigated models had the same dimension: rectangular base of 110 mm × 170 mm and height of 50 mm. Two different inner cavity designs were studied: circular hollow design (CHD) and fractal hollow design (FHD). Regarding the fractal design, a geometric configuration of easily modeling and manufacturing was investigated, so it could feasibly be scaled in brick prototype manufacturing. In this regard, the second-order Minkowski structure (Figure 4) was selected to produce the fractal pattern. This type of geometry was successfully implemented by the authors in previous research work for the development of acoustically active cavities in polymeric Helmholtz resonator prototypes intended for sound-absorbing interventions in automotive [40].

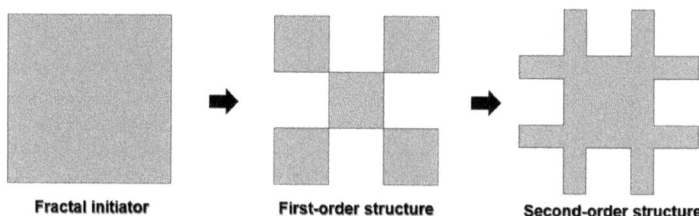

Figure 4. Iterative generation of the second-order Minkowski fractal geometry.

The size and number of the holes were defined in accordance with the technical indications reported in EN 771-1 standard [41] for half-solid bricks, considering a hole concentration ratio of 20–30%. The reference standard defines the specific design requirements as follows:

1. Hole area <1200 mm^2;
2. Minimum hole-external-perimeter distance >15 mm;
3. Minimum distance between adjacent holes >8 mm.

By adopting the above-reported design constraints, the circular and fractal cavities were arranged in the brick models, following a "honeycomb" pattern. Nagy and Orosz [42] demonstrated that this type of cavity arrangement extends the trajectory of the thermal flow inside the brick, enhancing the heat insulation performance. Table 6 summarizes the geometrical details of CHD and FHD models. The 2D layout and the 3D design of the bricks are presented in Figure 5a,b, respectively.

Table 6. Details of CHD and FHD brick models.

Brick Model	Number of Holes	Hole Area (mm^2)	Hole Concentration Ratio (%)
CHD	11	490.625	28
FHD	10	425	23

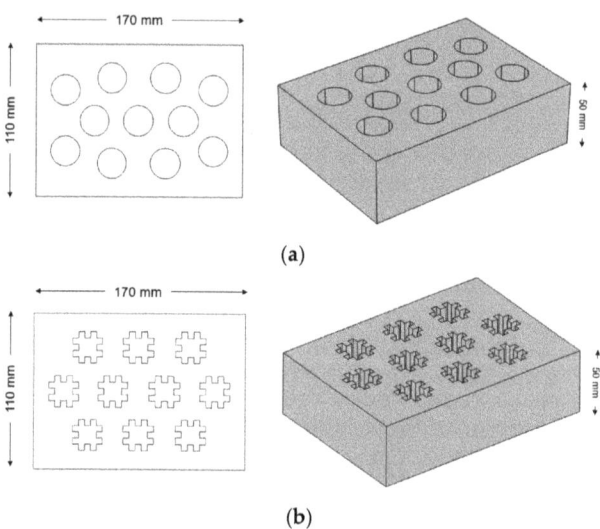

Figure 5. Hollow brick models: (a) CHD and (b) FHD.

2.2.2. FEA-Based Mechanical Analysis

Definition of the Problem

To preliminarily recognize the influence of the hollow design pattern on the mechanical performance of the brick, it was necessary to build a numerical FEA model, allowing a mechanical analysis of the component until the cracking state. The Structural Mechanics Module in COMSOL Multiphysics software was used for the calculation of the mechanical response of the hollow blocks under uniaxial compressive load regime. The numerical analysis consisted of simulating a static force-controlled compressive test, recording the stress–strain relationship and the mechanical parameters (compressive strength and modulus of elasticity) as a function of the brick's material and inner design. For this purpose, a compressive-load function (CLF) was defined as follows (Equation (1)):

$$CLF = 0.5 \times \xi \qquad (1)$$

where 0.5 (MPa) is the unit load recommended by the current European technical standard [43] for mechanical testing on brick masonry, and ξ is a multiplier parameter that defines the analysis resolution depending on the concrete mix implemented in the model. By considering the strength values obtained from the material characterization (Part 1), the following measurement ranges were selected: $0 < \xi < 30$ per 100 values and $0 < \xi < 5$ per 50 values for REF and RuC-based bricks, respectively). During the analysis, ξ gradually varied, simulating the application of an incremental compressive load perpendicular to the holes plain. The simulation reached convergence when the computed failure of the block occurred. For this purpose, a non-linear model was implemented in COMSOL. Compared to a linear elastic analysis, in which it is assumed that Hooke's law governs the material behavior and the stresses involved are relatively smaller than the brick strength [44], the non-linear modeling is able to trace the complete response of the component from the elastic range, through cracking and crushing, up to complete failure [45]. This method was successfully used for decades by structural engineers for analysis and strengthening of masonry units [44].

Mathematical Modeling

The micromechanics and the failure behavior of the brick were modeled by adopting the Willam–Warnke (WW) yield criterion. The WW failure surface can be written as a tri-parametric criterion, in accordance with Equation (2):

$$\frac{F(\sigma_{xp}, \sigma_{yp}, \sigma_{zp})}{\sigma_f} - S(\sigma_c, \sigma_t, \sigma_{bc}) \geq 0 \qquad (2)$$

where F is the function of the principal stress state (σ_{xp}, σ_{yp}, and σ_{zp}); S is the failure surface depending on σ_c, σ_t, and biaxial compressive strength (σ_{bc}); and σ_f is the uniaxial crushing strength (MPa). The graphical representation of failure surface in 3D principal stress space is illustrated by Figure 6, where parameters η, r_1, and r_2 refer to the relative magnitudes of principal stresses on the octahedral plane [46].

Cracking happens when the principal stress in any direction lies outside the failure surface. Brick failure occurs if all the principal stresses are compressive and lie outside the WW surface [44].

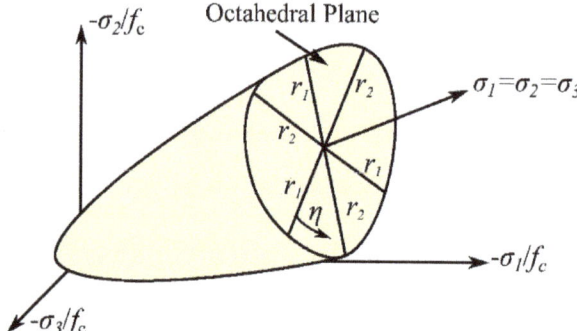

Figure 6. Three-dimensional WW failure surface in principal stress space (Reprinted with the permission from Ref. [46]. 2018, Elsevier Ltd.). In this representation, σ_1, σ_1, σ_3, and f_c are the three principal stresses and the uniaxial crushing strength, respectively.

Material Properties

The properties of the concrete mixes used in the FEA-based mechanical assay were taken from the laboratory tests described in *Part 1* of the manuscript and are listed below (Table 7):

Table 7. Material input properties for FEA-based mechanical investigation on brick models.

REF Concrete Mix			
Property	Value	Property Group	Evaluation
Density	2186 kg/m³	Basic	Experimental
Porosity	21.39%	Basic	Experimental
Elastic modulus	1.62 GPa	Basic	Experimental
Poisson's ratio (ν)	0.32	Basic	$\nu = 0.0895 + 0.0063 \times \sigma_c$ [47]
Compressive strength	35.89 MPa	WW model	Experimental
Tensile strength	11.06 MPa	WW model	Experimental
Biaxial compressive strength	43.07 MPa	WW model	$\sigma_{bc} = 1.2 \times \sigma_c$ [44]
RuC Concrete Mix			
Property	Value	Property Group	Evaluation
Density	1281 kg/m³	Basic	Experimental
Porosity	22.82%	Basic	Experimental
Elastic modulus	0.24 GPa	Basic	Experimental
Poisson's ratio (ν)	0.12	Basic	$\nu = 0.0895 + 0.0063 \times \sigma_c$ [45]
Compressive strength	4.93 MPa	WW model	Experimental
Tensile strength	1.65 MPa	WW model	Experimental
Biaxial compressive strength	5.92 MPa	WW model	$\sigma_{bc} = 1.2 \times \sigma_c$ [44]

Boundary Conditions

The boundary conditions selected in the analyses (Figure 7) involved a "boundary load" condition on the upper surface of the brick, applying a pressure defined by CLF (see Equation (1)). For the other boundaries of the model, the displacement in X, Y, and Z directions was constrained by setting "Prescribed displacement" equal to 0. With this condition, the translations and rotations of the selected boundary, along a specific spatial direction, are disabled.

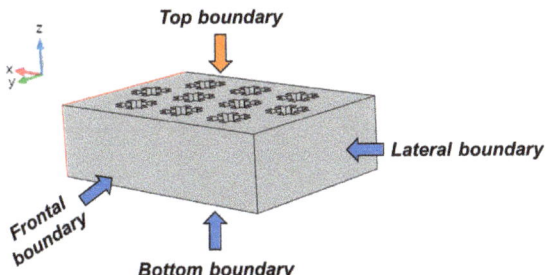

Boundary	Condition	Equation
Top	Boundary load	CLF
Lateral	Prescribed displacement in x direction	$u_{0x} = 0$
Frontal	Prescribed displacement in y direction	$u_{0y} = 0$
Bottom	Prescribed displacement in z direction	$u_{0z} = 0$

Figure 7. Schematic of the boundary conditions adopted in the FEA-based mechanical analysis.

A tetrahedral-shaped mesh was used for the numerical analysis. A "fine" element size (minimum mesh element size 3.5 mm and maximum mesh element size 28 mm) was chosen for meshing to achieve a good compromise among the results' accuracy and computational effort. Figure 8 illustrates the changes in WW damage index distribution on the brick surface for CHD (Figure 8a) and FHD (Figure 8b) models as ξ varies.

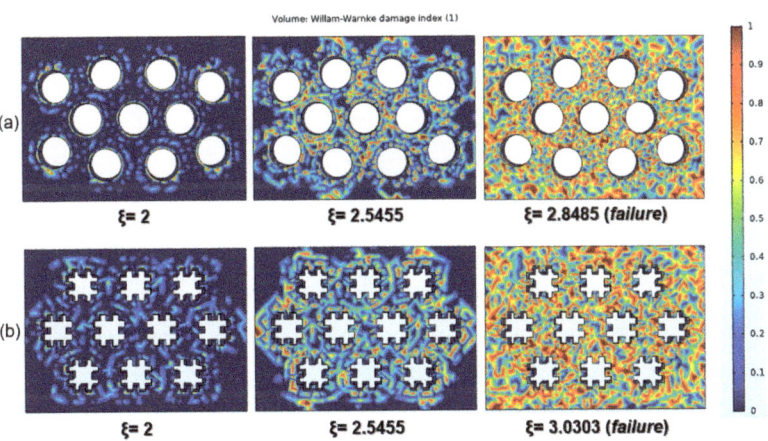

Figure 8. WW damage index distribution in (**a**) CHD model and (**b**) FHD model, considering RuC mix as constituent material.

2.2.3. FEA-Based Acoustic Analysis
Definition of the Problem

To explore the acoustic performance of the brick designs, a 2D FEA model was established through the Pressure Acoustic Module of COMSOL. The analysis aimed to characterize the absorption properties—more specifically, acoustic absorption coefficient (α)—of the hollow bricks in terms of sound frequency.

Mathematical Modeling

Two kinds of domains existed in the numerical model (Figure 9): (1) the background sound pressure field (Domain 1), which supplied a normal-incidence sound wave on the brick; and (2) the brick domain (Domain 2), which consisted of the air domain for the perforation holes and the concrete material matrix described by the Delany–Bazley (DB) poro-acoustic model. The DB model includes an empirical formulation for estimating acoustic impedance of porous materials. The acoustic impedance (Z_a) and wave-number (k) of the sound waves in a porous medium mainly depend on the frequency and on the static airflow resistivity (R_f) of the material. The expression of Z_a and k are as follows (Equations (3) and (4)):

$$Z_a = \rho_0 \times c_0 \left[1 + 0.057 \times \left(\frac{\rho_0 \times f}{R_f} \right)^{-0.754} - i0.087 \times \left(\frac{\rho_0 \times f}{R_f} \right)^{-0.732} \right] \quad (3)$$

$$k = \frac{2\pi f}{c_0} \times \left[1 + 0.0978 \times \left(\frac{\rho_0 \times f}{R_f} \right)^{-0.700} - i0.189 \times \left(\frac{\rho_0 \times f}{R_f} \right)^{-0.595} \right] \quad (4)$$

where ρ_0 is the air density (kg/m^3), c_0 is the sound velocity in air (m/s), f is the sound frequency (Hz), and i is the imaginary unit.

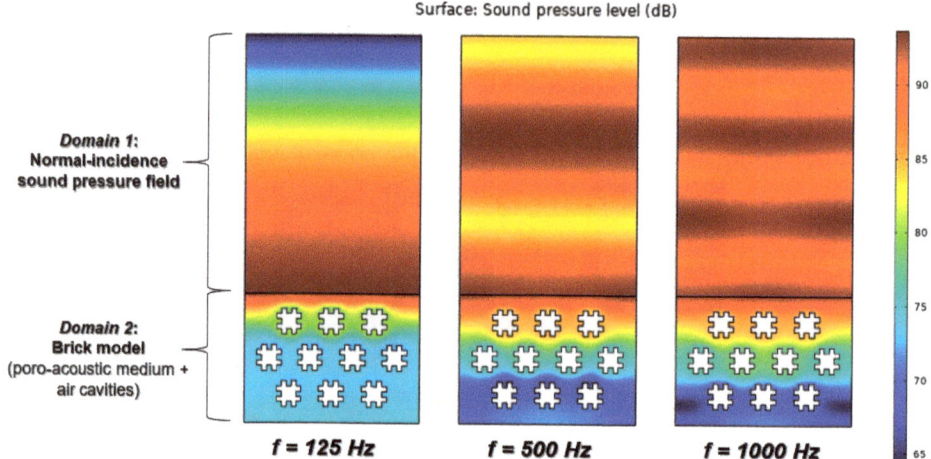

Figure 9. FEA-based acoustic model: sound-pressure-level variation as function of frequency in FHD brick, considering RuC mix as constituent material.

The incident background pressure field (P_i) is given as follows (Equation (5)):

$$P_i = e^{-i(k \cdot x)} \quad (5)$$

where k is the wave vector defining the propagation of the incident sound wave. The pressure, P, solved in the model is the total field, and the scattered field (P_{scat}) is given as $P_{scat} = P - P_i$. The α-coefficient, which represents the ratio of the absorbed and incident energy, is defined as follows (Equation (6)):

$$\alpha = 1 - |R|^2 \quad (6)$$

where R is the pressure reflection coefficient that gives the ratio of P_{scat} and P_{inc}. In the simulation, α-coefficient for CHD and FHD bricks was analyzed at different frequencies (up to 1000 Hz), in accordance with the permitted working range of the impedance tube used in the experimental validation.

Material Properties

The input parameters for concrete mixes required from the FEA model are presented in Table 8. The air-cavity domain was modeled by adopting the value available in COMSOL's material library.

Table 8. Material input properties for FEA-based acoustic investigation on brick models.

REF Concrete Mix			
Property	*Value*	*Property Group*	*Evaluation*
Density	2186 kg/m^3	Basic	Experimental
Porosity	21.39%	Basic	Experimental
Acoustic flow resistivity	13,872 N × s× m^{-4}	DB model	Experimental
RuC Concrete Mix			
Property	*Value*	*Property Group*	*Evaluation*
Density	1281 kg/m^3	Basic	Experimental
Porosity	22.82%	Basic	Experimental
Acoustic flow resistivity	19,862 N × s× m^{-4}	DB model	Experimental

Boundary Conditions

As to boundary conditions, the lateral boundaries are all set as rigid walls, owing to the periodicity of unit cells for normal sound incidence, and the bottom of the brick domain is also rigid according to reproduce the common test configuration in sound absorption measurements. Normal-sized tetrahedral mesh (minimum mesh element size of 0.375 mm and maximum mesh element size of 83.8 mm) was selected in the study.

2.3. Part 3: Hollow-Brick Production and Testing

2.3.1. Fabrication of the Brick Mold

A custom-made polypropylene (PP) vessel was employed as a master mold. Circular and fractal-shaped columns were implanted on the mold's base to realize the cavities in the brick prototypes. To accurately reproduce the geometry of the holes, especially in the case of fractal configuration, the columns were cut from a slab of high-density polystyrene (HDPS) by a computerized numerical control (CNC) milling machine (Falcon 1500, Valmec, Pescara, Italy). The cutting parameters are reported in Table 9. Figure 10 elucidates some phases of the cutting operations.

Table 9. Cutting parameters selected for CNC cutting of high-density polystyrene columns.

Cutter Type	Cutter Diameter	Cutting Depth	Spindle speed	Feed Rate
End mill	3 mm	2 mm	5000 rpm	30 mm/s

The correct alignment of the columns on the mold's base was made possible through the fabrication of masks that exactly reproduced the hollow surface of the designed bricks. Tailored cardboard masks (Figure 10a) were produced by high-precision laser-cutting technology (Figure 10b), using a Birio 1000 laser cutter (Birio, Naples, Italy). After positioning and gluing the columns, we removed the guide-masks were, and the mold was left to settle to allow the complete fixing of the columns on the vessel's base The completed and ready-to-use brick molds (FHD and CHD configurations) are illustrated in Figure 11c,d, respectively.

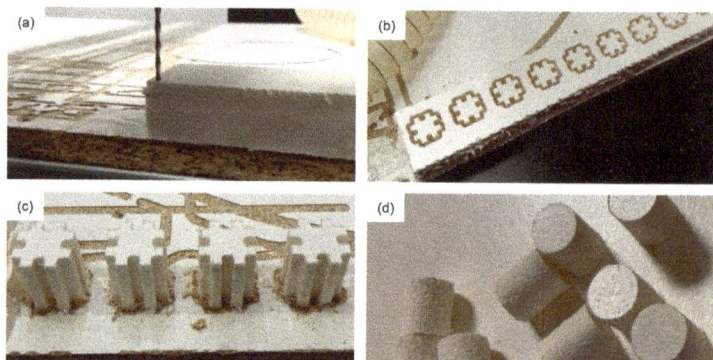

Figure 10. CNC cutting operation: (**a**) end mill, (**b**) array of fractal designs after cutting, and extraction of the (**c**) fractal and (**d**) circular columns.

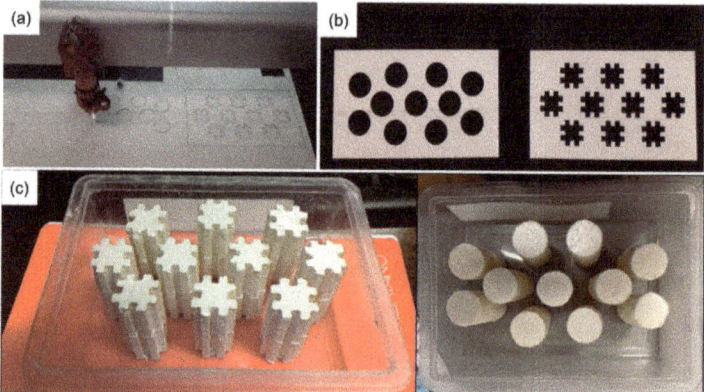

Figure 11. (**a**) Laser cutting of the guide-masks, (**b**) CHD and FHD guide masks, and (**c**) mold brick for FHD and CHD design.

2.3.2. Casting and Bricks Manufacturing

The concrete mixes preparation, casting procedure, and curing method followed the same protocol implemented in Section 2.1.2. After 28 days of curing, the bricks were demolded and the columns were removed both manually and by chemical etching with pure acetone. Then the surface of the brick specimens was polished by a diamond saw. For each brick design (CHD and FHD) and concrete mix (Figure 12), three replicates were produced.

2.3.3. Testing

The brick samples were mechanically characterized by compressive test (Figure 13a), using a Zwick-Roell Z150 (Zwick-Roell Group, Ulm, Germany) universal machine with a 150 kN load cell. The pre-load and loading rate were set to 20 N and 5 mm/min, respectively. Two brick specimens were tested for each combination (concrete material + inner design).

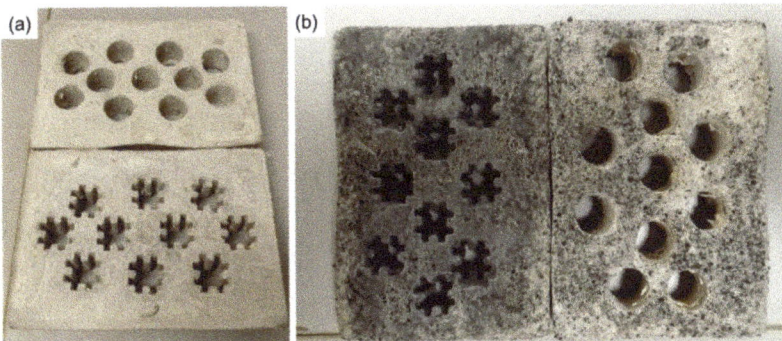

Figure 12. CHD and FHD hollow-brick prototypes: (**a**) REF mix and (**b**) RuC mix.

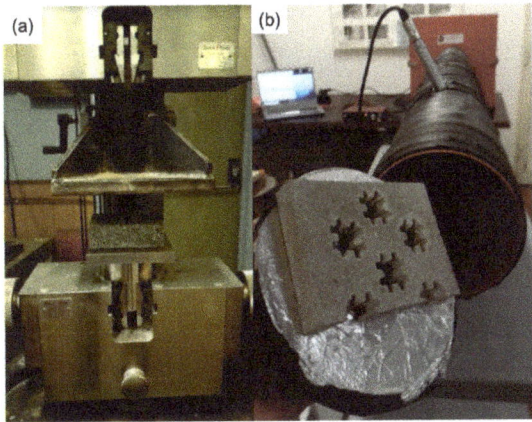

Figure 13. Experimental configurations for hollow brick characterization: (**a**) compressive test and (**b**) sound-absorption test.

The sound-absorption properties of the bricks were experimentally investigated by means of the impedance tube system by applying the standing wave ratio method (ASTM C384-95 standard [48]). The testing principle can be found in detail in Reference [35]. The measuring system included a 190 cm–long (L) Poly (vinyl chloride) (PVC) tube with an inner diameter (D) of 16 cm, a sound source (MPA30BT loudspeaker, Behringer, Willich, Germany), a $\frac{1}{4}$" condenser microphone (ECM800, Behringer, Willich, Germany), a Scarlett 2i4 audio interface (Focusrite, High Wycombe, UK), and Room EQ Wizard software (GIK Acoustic, Atlanta, GA, USA) test software. The sound source was mounted at one end of the tube, emitting a sine wave acoustic signal. Half-brick sample is placed at the other end fixed on a reflective termination (Figure 13b). The microphone moved along the length of the tube during the test, recording the sound pressure level at various frequencies. According to this system, the normal incidence α-coefficient was determined by Equation (7):

$$\alpha(f) = 1 - \left|\frac{SWR - 1}{SWR + 1}\right|^2 \tag{7}$$

where the standing wave ratio (SWR) index is defined as the ratio between the maximum and minimum sound pressure level values (in dB) measured along the tube for each acoustic frequency investigated. For the tube used in this study, the maximum operating frequency was 1270 Hz. Then α-measurements were performed in third-octave bands at 125, 250, 500, and 1000 Hz

3. Results

3.1. FEA-Based Mechanical Analysis

The results of the mechanical FEA study are presented in Figure 14a (load–strain curves) and Figure 14b (WW damage index–load curves).

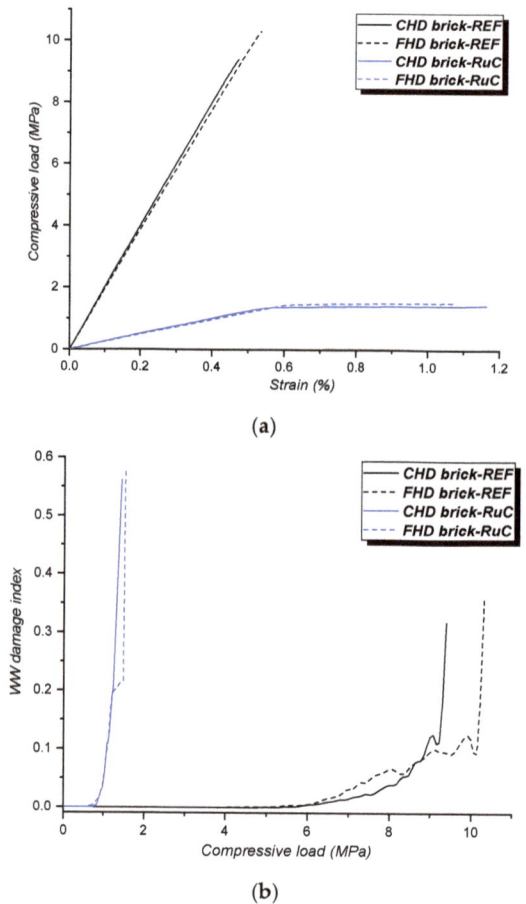

Figure 14. FEA-based mechanical results: (a) load–strain curves and (b) WW damage index–strain curves.

Regardless of the hollow brick design, a clear difference in mechanical behavior between the investigated concrete mixes can be observed. The FEA mechanical model implemented in this work faithfully reproduces the clear difference in the stress–strain relationship between ordinary and rubberized concrete. The former showed the typical brittle characteristic of ordinary cementitious materials. By considering RuC mix properties, the curve changes toward a more ductile behavior, showing lower mechanical strength and larger deformations compared to the plain concrete sample under the same loading conditions. From this evidence, it can be deduced that the features of the rubberized mix would be favorable in optimizing the strain capacity and, thus, the energy-absorption ability of the brick element. Therefore, an improvement to conventional concrete in terms of attained vibro-acoustic performance could be achieved [49], representing an attractive requirement for building applications. In addition to yielding at lower compressive loads, the brick prototypes in RuC mix experienced a higher damage index than those modeled with REF properties, reflecting the strength vulnerability of the rubberized mixes compared to plain concrete.

When analyzing the effect of the hollow configuration, we can see that slight differences were noted between CHD and FHD in terms of elasto-mechanical characteristics. Cavities' geometry had little influence on the mechanical stiffness of the brick model. For the REF mix, the computed elastic moduli (E_{FEA}) were 2.01 GPa for CHD and 1.95 GPa for FHD. For the RuC mix, E_{FEA}-values were 0.25 GPa and 0.24 GPa. The fractal design performed better in terms of mechanical strength. The simulated compressive strengths (σ_c^{FEA}) of the REF-modeled brick were 9.39 MPa and 10.30 MPa for CHD and FHD bricks, respectively. The σ_c^{FEA} values of the RuC-modeled brick were 1.42 MPa and 1.51 MPa for CHD and FHD bricks, respectively. Overall, an increase in mechanical-strength properties of 7% (RuC mix) and 10% (REF mix) was achieved when passing from CHD to FHD configurations. No supporting works in the literature are available on the mechanical influence of this type of fractal-like cavity to corroborate the results obtained. However, some research work verified that structural components topologically optimized with fractal geometries gained significantly in terms of load-bearing capacity [50,51]. The damage level of the brick model would seem to be less affected by the cavities' geometry. With the same material properties, the fractal design provided slightly higher WW indices than the circular configuration, while showing greater strength performance. The fractal cavity contains internal square hollow shapes, where their sharp edges would induce more stress concentration than the circular hole. This effect, although potentially deleterious to the mechanical performance of the brick, is well-balanced by the lower hole concentration ratio in FHD arrangement, which therefore exposes a greater load-bearing strength to the applied stress.

3.2. FEA-Based Acoustic Analysis

In addition to evaluating the acoustic response of the brick models as a function of the constituent concrete material and type of hollow pattern, an FEA acoustic simulation was also used to predict the noise attenuation characteristics in low-frequency regimes. Low-frequency sound absorption for noise mitigation remains challenging because the slow fluctuation of low-frequency acoustic waves leads to poor interaction between materials or structures and the viscous air medium, resulting in inefficient dissipation of sound energy [52]. Moreover, the fundamental frequencies of most engineering and civil structures are usually below 50 Hz; therefore, their attenuation must be addressed to avoid unwanted vibro-acoustic phenomena [53]. The low-frequency acoustic performances are difficult to analyze experimentally. Common measurement methods, including impedance tube or reverberation room, provide working ranges strictly related to the dimensions of the measuring apparatus. For instance, the minimum admitted frequency ($f_{min} \propto \frac{c_0}{L-D}$) of the impedance tube device used in this research was 150 Hz, thus not allowing accurate measurements in the low-frequency band. In this regard, the FEA model implemented in this work was also used to analyze the sound-absorbing properties of the modeled brick in this acoustic region of great engineering interest.

The low-frequency acoustic spectrum (Figure 15a) highlights the weakest sound-absorbing ability ($\alpha < 0.20$) and a negligible influence of the internal design of the brick on the acoustic performance. The constituent concrete material, on the other hand, provides a certain effect within 60 Hz. The physical–acoustic properties of RuC mix would seem to confer better attenuation performance than those of REF mix. The considerable increase in R_f following the replacement of sand with GWTR (see discussion in Section 2.1.3) could be the key factor in the brick prototype's higher low-frequency performance. FEA results find good agreement with previous studies [53,54] wherein the improved low-frequency noise-suppression capacity of rubber–concrete mixes has been demonstrated. At around

60 Hz, an inversion point was detected: α-curves grew continuously, converging to a value close to 0.9 (around 1000 Hz), indicating that high frequencies are more easily attenuated. In the high-frequency acoustic range (Figure 15b), the physical–acoustic properties of the REF mix ensured higher sound absorption rates than those of rubberized concrete. Above 500 Hz, passing from a circular hollow internal geometry to a fractal one, a very slight improvement in acoustic performance is appreciated by FEA modeling. It can be hypothesized that the different acoustic behavior of the two brick models is due to the cavity resonance sound-absorption mechanism. According to the thermo-viscous acoustic theory of irregular cavities, sound propagation in the air is related to the displacement and movement speed of small particles. When the shape of the sound-absorbing cavity is more complex, the surfaces are more reflective, and the smaller particles need to travel more distance. Therefore, the sound waves are reflected more frequently in the irregular space of the cavity wall, which consumes more sound energy and improves the sound absorption ability [55].

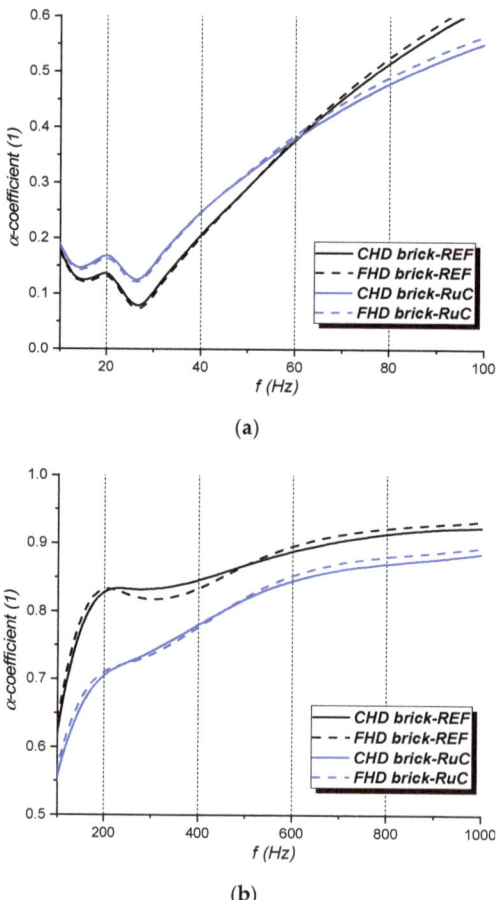

Figure 15. FEA-based acoustic results: (**a**) low-frequency α-coefficient curves and (**b**) middle-high-frequency α-coefficient curves.

3.3. Experimental Mechanical Testing

In accordance with the results predicted by the simulation, the mechanical tests reveal the marked difference in the load–strain relationship (Figure 16) between REF (brittle behavior) and RuC (ductile behavior) mixes, indicating, for the latter, a predictable deterioration in strength and an increase in post-peak strain capacity because of the influence of rubber particles. The curves, normalized with respect to the holes concentration ratio, also allow us to detect the difference between the hollow designs under study. Regardless of the type of concrete mix, the FHD sample provided higher stiffness and compressive strength than the CHD brick. The numerical analysis (Figure 14a) reports a similar trend in the compressive behavior of the bricks but underestimates the mechanical resistance values. Indeed, there is an average level of agreement between experimental and simulated compressive strength of 45% and 55% for REF and RuC-based bricks, respectively. This discrepancy can be attributed to several approximations considered in the building of the COMSOL model:

- The FEA model considers the material as homogeneous, neglecting the composite nature of the cementitious formulations under study and, therefore, the contribution of mineral (sand) and polymeric (GWTR) aggregates on the mechanical behavior of the model, including interface interactions, stress distribution, deformation mechanisms induced by the different nature of the aggregates, etc.;
- Some fundamental input properties for the WW failure criterion (such as ν and σ_{bc}) were obtained indirectly from constitutive models and may not reflect the real mechanical behavior of the material;
- Dimensional variations between the digital model and real brick prototype due to the hygrometric shrinkage of the material, the geometric accuracy of the cavities, and surface roughness can inevitably affect the mechanical response of the samples.

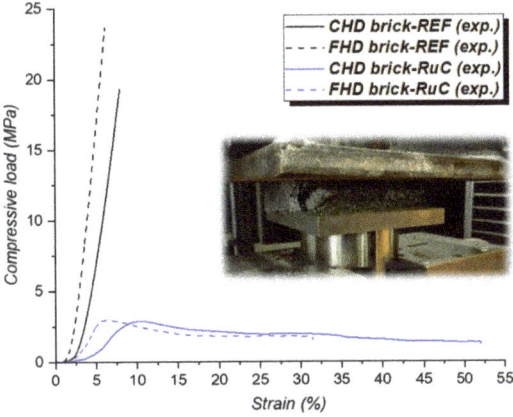

Figure 16. Experimental load–strain curves.

Figure 17 elucidates the average values of compressive strength obtained from mechanical testing. REF-based bricks satisfy the strength requirement for load-bearing masonry units. The minimum compressive strength value of 11.7 MPa is necessary for this case, as reported in the ASTM C90 standard [56]. The strengths dramatically decrease in the rubberized hollow brick, as expected. Both the bricks provide mechanical strengths very close to the minimum ASTM requirements [57] for non-load bearing hollow concrete masonry blocks (minimum strength requirement of 3.45 MPa). Furthermore, the histograms clearly show the increase in mechanical performance induced by the fractal-shaped hollow pattern. From CHD to FHD, increments of 22% and 18% are detected by considering REF mix and RuC mix, respectively, confirming the efficiency of fractal design on the mechanical behavior of the brick. To determine the best brick solutions from the structural point of view, a structural efficiency index (*SEI*) is defined. This value is computed as the ratio between the compressive strength (σ_c^{exp}), which is obtained from the experimental testing, and its weight (Equation (8)):

$$SEI = \frac{\sigma_c^{exp}}{Weight} \qquad (8)$$

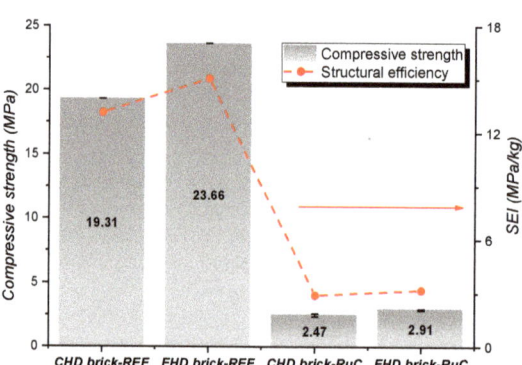

Figure 17. Compressive strength and *SEI* values for developed hollow bricks.

The greater *SEI* values found in FHD designs indicate the higher overall structural effectiveness of the block, which behaves better from a structural and handling point of view [19]. The results achieved represent a noteworthy starting point for researchers. With careful optimization and engineering of both GWTR–concrete mix and structural design, it is possible to reach a highly eco-sustainable brick (0% natural aggregates) with a mechanical performance suitable for applications in construction.

3.4. Experimental Acoustic Testing

The experimental sound–absorption curves (α vs. f) are presented in Figure 18.

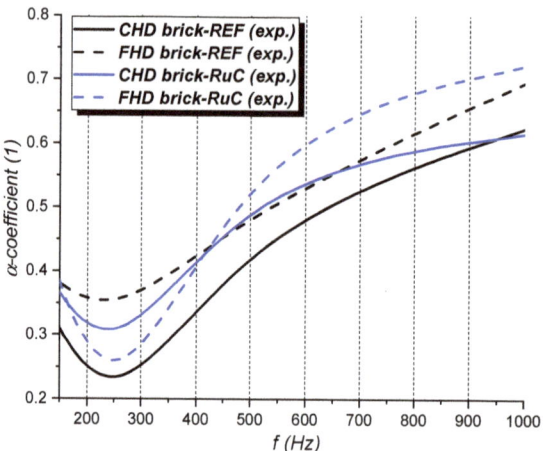

Figure 18. Experimental sound–absorption curves (α vs. f).

Considering the trend of α-curves, the experimental results follow with fairly good agreement the numerical analysis (Figure 15b). The acoustic absorption of the hollow brick prototypes tends to have the highest α-values as the frequency increases. In contrast to the simulation, however, lower attenuation levels are recorded. At 1000 Hz, α-coefficient varies between 0.62 (CHD brick-RuC) and 0.72 (FHD brick-RuC), compared to values close to

0.90 detected in the acoustic-based FEA modeling. The inconsistency between simulation and experimental analysis is primarily attributable to the poro-acoustic model adopted to describe the sound–material interaction and the acoustic impedance provided by the cementitious medium. DB's model was derived for fibrous sound-absorbing materials, such as glass wool and rock wool [58] and could induce uncertainties in the prediction of the acoustic response of concrete-based materials. In addition, the model required a limited number of input acoustic data referring to the material, affecting the accurate prediction of its real sound-absorbing performance. In this regard, it will be advisable, in future works, to implement more complete acoustic models suitable for concrete, including Neithalath's model [59] and Kim and Lee's model [60], which need a greater number of acoustic parameters describing the material's performance.

The experimental α-coefficients indicate a very good acoustic performance of the developed hollow bricks. Referring to the classification presented by Fediuk et al. [61], the samples exhibit sound-absorbing levels in the range of aerated and foamed concretes (α = 0.15–0.75), which represent the major solutions in building noise-attenuation interventions. Furthermore, the sound-absorbing ability of the proposed rubberized bricks is significantly higher than the experimental results obtained by Mohammed et al. [15], which found α-values less than 0.25 in concrete hollow blocks loaded with different amounts (10%, 25%, and 50%) of tire crumb rubber. In good agreement with FEA, the experimental α-curves highlight the better acoustic performance of FHD design above 400 Hz for both cement mixes. The sharpest incremental rate in sound absorption, moving from CHD to FHD, is recorded for RuC-based samples: +13% at 500 Hz and +18% at 1000 Hz. In the same frequency band, the GWTR–cement mix yields superior acoustic attenuation capacities compared to REF material, regardless of the internal geometry of the brick, indicating the acoustic functionality induced by the polymeric aggregate on the concrete's damping [62]. The improved acoustic efficiency of fractal-shaped cavities coupled with the sound abatement peculiarities deriving from the tire aggregate would provide attractive solutions for acoustic dampening at medium–high frequencies, which are of particular interest concerning the mitigation of traffic and urban-deriving noise [63].

4. Conclusions

In this study, ground waste tire rubber and fractal-shaped cavities were combined with the purpose of producing eco-friendly and lightweight hollow concrete bricks with improved acoustic efficiency and acceptable strength requirements for non-structural applications. The rubberized concrete mix was prepared by using two rubber fractions (0–1 mm rubber powder and 1–3 mm rubber granules) as a total aggregate. Firstly, its physical and mechanical properties were experimentally characterized and compared with an ordinary concrete mix (0% rubber) taken as a reference. A finite element analysis was implemented to build predictive models for the mechanical and acoustic absorption behavior of two kinds of brick models based on circular (traditional geometry) and Minkowski-like fractal cavities. Hollow-brick prototypes were manufactured and tested, evaluating the influence of constituent materials (reference and rubberized mixes) and inner hollow design on their mechanical and acoustic performance. The results revealed the following main findings:

- Regardless of the concrete mix, the fractal hollow pattern provides a significant improvement in the engineering performance of the brick in terms of mechanical strength, structural efficiency, and acoustic absorption over 500 Hz. Considering the rubber–cement mix as a constituting brick's material, an increase of 18% in compressive strength and 1000 Hz sound absorption coefficient can be achieved, moving from circular to fractal hole designs.
- Circular and fractal design bricks made up of ordinary concrete mix satisfy the minimum ASTM strength for load-bearing masonry units. A rubber–concrete mix involves a predictable loss in mechanical strength properties. However, the compressive strengths of rubberized blocks were very close to ASTM requirements for non-load-bearing applications.

- The samples investigated can be considered as "good" concrete sound absorbers. The rubber-functionalized cement mix yielded superior acoustic attenuation capacities compared to the reference material, regardless of the internal geometry of the brick, indicating the positive influence induced by the polymeric aggregates on the concrete's sound-absorbing efficiency.
- From the numerical analysis by finite element method, the physical–acoustic parameters of the investigated rubberized mixes would seem to predict better attenuation performances to the brick, even at low frequencies, which are noise events of great interest in engineering field.

In summary, the present study has proven that fractal geometries and waste aggregates can be successfully integrated into brick technology to obtain eco-friendlier solutions with enhanced structural and acoustic behavior. As a future trend, we plan to propose a careful optimization of the rubber–cement mix design to achieve mechanical performances that are fully within the strength requirements for lightweight masonry applications. In this regard, it will also be necessary to refine the finite element predictive models and to strengthen the connection between simulation and experiment for a more accurate optimization of the material and component's design.

Author Contributions: Conceptualization, M.V., M.S. and S.L.; methodology, M.V., M.S., A.S., L.F. and R.M.; software, M.S. and A.S.; validation M.S. and A.S.; investigation, M.S., A.S. and R.M.; resources, M.V., L.F. and S.L.; writing—original draft preparation, M.S. and A.S.; writing—review and editing, M.V., L.F., R.M. and S.L.; supervision, M.V., L.F. and S.L.; project administration, M.V. All authors have read and agreed to the published version of the manuscript.

Funding: This research received no external funding.

Data Availability Statement: Not applicable.

Acknowledgments: The authors would like to thank the technical staff of Saperi&Co center staff for supporting the design and production of the brick molds.

Conflicts of Interest: The authors declare no conflict of interest.

References

1. Valentini, F.; Dorigato, A.; Rigotti, D.; Pegoretti, A. Evaluation of the role of devulcanized rubber on the thermomechanical properties of expanded ethylene-propylene diene monomers composites. *Polym. Eng. Sci.* **2021**, *61*, 767–779. [CrossRef]
2. Shu, X.; Huang, B. Recycling of waste tire rubber in asphalt and portland cement concrete: An overview. *Constr. Build. Mater.* **2014**, *67*, 217–224. [CrossRef]
3. Eldin, N.N.; Senouci, A.B. Engineering properties of rubberized concrete. *Can. J. Civ. Eng.* **1992**, *19*, 912–923. [CrossRef]
4. Miller, N.M.; Tehrani, F.M. Mechanical properties of rubberized lightweight aggregate concrete. *Constr. Build. Mater.* **2017**, *147*, 264–271. [CrossRef]
5. Turatsinze, A.; Garros, M. On the modulus of elasticity and strain capacity of self-compacting concrete incorporating rubber aggregates. *Resour. Conserv. Recycl.* **2008**, *52*, 1209–1215. [CrossRef]
6. Ghizdăveț, Z.; Ștefan, B.M.; Nastac, D.; Vasile, O.; Bratu, M. Sound absorbing materials made by embedding crumb rubber waste in a concrete matrix. *Constr. Build. Mater.* **2016**, *124*, 755–763. [CrossRef]
7. Zhang, B.; Poon, C.S. Sound insulation properties of rubberized lightweight aggregate concrete. *J. Clean. Prod.* **2018**, *172*, 3176–3185. [CrossRef]
8. Guo, J.; Huang, M.; Huang, S.; Wang, S. An experimental study on mechanical and thermal insulation properties of rubberized concrete including its microstructure. *Appl. Sci.* **2019**, *9*, 2943. [CrossRef]
9. Richardson, A.; Coventry, K.; Dave, U.; Pienaar, J. Freeze/thaw performance of concrete using granulated rubber crumb. *J. Green Build.* **2011**, *6*, 83–92. [CrossRef]
10. Gerges, N.N.; Issa, C.A.; Fawaz, S.A. Rubber concrete: Mechanical and dynamical properties. *Case Stud. Constr. Mater.* **2018**, *9*, e00184. [CrossRef]
11. Jie, X.U.; Yao, Z.; Yang, G.; Han, Q. Research on crumb rubber concrete: From a multi-scale review. *Constr. Build. Mater.* **2020**, *232*, 117282. [CrossRef]
12. Thorat, P.K.; Papal, M.; Kacha, V.; Sarnobat, T.; Gaikwad, S. Hollow concrete blocks-A new trend. *Int. J. Eng. Res.* **2015**, *5*, 9–26.
13. Turgut, P.; Yesilata, B. Physico-mechanical and thermal performances of newly developed rubber-added bricks. *Energy Build.* **2008**, *40*, 679–688. [CrossRef]

14. Yüksek, İ.; Öztaş, S.K.; Tahtalı, G. The evaluation of fired clay brick production in terms of energy efficiency: A case study in Turkey. *Energy Effic.* **2020**, *13*, 1473–1483. [CrossRef]
15. Mohammed, B.S.; Hossain, K.M.A.; Swee, J.T.E.; Wong, G.; Abdullahi, M. Properties of crumb rubber hollow concrete block. *J. Clean. Prod.* **2012**, *23*, 57–67. [CrossRef]
16. Fraile-Garcia, E.; Ferreiro-Cabello, J.; Defez, B.; Peris-Fajanes, G. Acoustic Behavior of Hollow Blocks and Bricks Made of Concrete Doped with Waste-Tire Rubber. *Materials* **2016**, *9*, 962. [CrossRef]
17. Al-Fakih, A.; Wahab, M.A.; Mohammed, B.S.; Liew, M.S.; Zawawi, N.A.W.A.; As' ad, S. Experimental study on axial compressive behavior of rubberized interlocking masonry walls. *J. Build. Eng.* **2020**, *29*, 101107. [CrossRef]
18. Lourenço, P.B.; Vasconcelos, G.; Medeiros, P.; Gouveia, J. Vertically perforated clay brick masonry for loadbearing and non-loadbearing masonry walls. *Constr. Build. Mater.* **2010**, *24*, 2317–2330. [CrossRef]
19. Del Coz Díaz, J.J.; Nieto, P.G.; Rabanal, F.Á.; Martínez-Luengas, A.L. Design and shape optimization of a new type of hollow concrete masonry block using the finite element method. *Eng. Struct.* **2011**, *33*, 1–9. [CrossRef]
20. Al-Tamimi, A.S.; Al-Osta, M.A.; Al-Amoudi, O.S.B.; Ben-Mansour, R. Effect of geometry of holes on heat transfer of concrete masonry bricks using numerical analysis. *Arab. J. Sci. Eng.* **2017**, *42*, 3733–3749. [CrossRef]
21. Sassine, E.; Cherif, Y.; Dgheim, J.; Antczak, E. Investigation of the mechanical and thermal performances of concrete hollow blocks. *SN Appl. Sci.* **2020**, *2*, 1–17. [CrossRef]
22. Valente, M.; Sambucci, M.; Sibai, A.; Musacchi, E. Multi-physics analysis for rubber-cement applications in building and architectural fields: A preliminary analysis. *Sustainability* **2020**, *12*, 5993. [CrossRef]
23. Kim, H.; Hong, J.; Pyo, S. Acoustic characteristics of sound absorbable high performance concrete. *Appl. Acoust.* **2018**, *138*, 171–178. [CrossRef]
24. Sapoval, B.; Haeberlé, O.; Russ, S. Acoustical properties of irregular and fractal cavities. *J. Acoust. Soc. Am.* **1997**, *102*, 2014–2019. [CrossRef]
25. Hébert, B.; Sapoval, B.; Russ, S. Experimental study of a fractal acoustical cavity. *J. Acoust. Soc. Am.* **1999**, *105*, 1567–1574. [CrossRef]
26. Godbold, O. Investigating Broadband Acoustic Absorption Using Rapid Manufacturing. Ph.D. Thesis, Loughborough University, Loughborough, UK, 2008.
27. EN 1097-7; Tests for Mechanical and Physical Properties of Aggregates—Part 7: Determination of the Particle Density of Filler—Pycnometer Method. UNI Standards: Milan, Italy, 2008.
28. DIN 51701; Testing of Solid Fuels; Sampling and Sample Preparation, Sample Preparation. DIN Standards: Berlin, Germany, 2007.
29. Sambucci, M.; Valente, M.; Sibai, A.; Marini, D.; Quitadamo, A.; Musacchi, E. Rubber-Cement Composites for Additive Manufacturing: Physical, Mechanical and Thermo-Acoustic Characterization. In Proceedings of the Second RILEM International Conference on Concrete and Digital Fabrication, Eindhoven, The Netherlands, 6–9 July 2020; Springer: Cham, Switzerland, 2020; pp. 113–124. [CrossRef]
30. Sambucci, M.; Valente, M. Influence of Waste Tire Rubber Particles Size on the Microstructural, Mechanical, and Acoustic Insulation Properties of 3D-Printable Cement Mortars. *Civ. Eng. J.* **2021**, *7*, 937–952. [CrossRef]
31. Valente, M.; Sambucci, M.; Chougan, M.; Ghaffar, S.H. Reducing the emission of climate-altering substances in cementitious materials: A comparison between alkali-activated materials and Portland cement-based composites incorporating recycled tire rubber. *J. Clean. Prod.* **2022**, *333*, 130013. [CrossRef]
32. ASTM D792; Standard Test Methods for Density and Specific Gravity (Relative Density) of Plastics by Displacement. ASTM International: West Conshohocken, PA, USA, 1958.
33. ASTM C 1202; Standard Test. Method for Electrical Indication of Concrete's Ability to Resist Chloride Ion Penetration. ASTM International: West Conshohocken, PA, USA, 2019.
34. Ingard, K.U.; Dear, T.A. Measurement of acoustic flow resistance. *J. Sound Vib.* **1985**, *103*, 567–572. [CrossRef]
35. Sambucci, M.; Valente, M. Ground Waste Tire Rubber as a Total Replacement of Natural Aggregates in Concrete Mixes: Application for Lightweight Paving Blocks. *Materials* **2021**, *14*, 7493. [CrossRef]
36. Assaggaf, R.A.; Ali, M.R.; Al-Dulaijan, S.U.; Maslehuddin, M. Properties of concrete with untreated and treated crumb rubber–A review. *J. Mater. Res. Technol.* **2021**, *11*, 1753–1798. [CrossRef]
37. Si, R.; Guo, S.; Dai, Q. Durability performance of rubberized mortar and concrete with NaOH-Solution treated rubber particles. *Constr. Build. Mater.* **2017**, *153*, 496–505. [CrossRef]
38. Rezaifar, O.; Hasanzadeh, M.; Gholhaki, M. Concrete made with hybrid blends of crumb rubber and metakaolin: Optimization using Response Surface Method. *Constr. Build. Mater.* **2016**, *123*, 59–68. [CrossRef]
39. Geethamma, V.G.; Asaletha, R.; Kalarikkal, N.; Thomas, S. Vibration and sound damping in polymers. *Resonance* **2014**, *19*, 821–833. [CrossRef]
40. Sambucci, M.; Cecchini, F.; Nanni, F.; Pucacco, G.; Valente, M. Metamateriali fonoassorbenti sviluppati via 3D printing per interventi acustici nel settore automotive. *Riv. Ital. Di Acust.* **2020**, *44*, 1–23.
41. EN 771-1; Specification for Masonry Units—Part 1: Clay Masonry Units. UNI Standards: Milan, Italy, 2015.
42. Nagy, B.; Orosz, M. Optimized Thermal Performance Design of Filled Ceramic Masonry Blocks. *AMM* **2015**, *797*, 174–181. [CrossRef]

43. *EN 1996-1-1*; Eurocode 6—Design of Masonry Structures—Part 1-1: General Rules for Reinforced and Unreinforced Masonry Structures. UNI Standards: Milan, Italy, 2013.
44. Hejazi, M.; Soltani, Y. Parametric study of the effect of hollow spandrel (Konu) on structural behaviour of Persian brick masonry barrel vaults. *Eng. Fail. Anal.* **2020**, *118*, 104838. [CrossRef]
45. Lourenço, P.B. Computations on historic masonry structures. *Prog. Struct. Eng. Mater.* **2002**, *4*, 301–319. [CrossRef]
46. Kumar, P.; Srivastava, G. Effect of fire on in-plane and out-of-plane behavior of reinforced concrete frames with and without masonry infills. *Constr. Build. Mater.* **2018**, *167*, 82–95. [CrossRef]
47. Sideris, K.K.; Manita, P.; Sideris, K. Estimation of ultimate modulus of elasticity and Poisson ratio of normal concrete. *Cem. Concr. Compos.* **2004**, *26*, 623–631. [CrossRef]
48. *ASTM C384-95*; Standard Test Method for Impedance and Absorption of Acoustical Materials by the Impedance Tube Method. ASTM International: West Conshohocken, PA, USA, 1998.
49. El-Khoja, A. Mechanical, Thermal and Acoustic Properties of Rubberised Concrete Incorporating Nano Silica. Ph.D. Thesis, University of Bradford, Bradford, UK, 2019.
50. Nguyen-Van, V.; Wu, C.; Vogel, F.; Zhang, G.; Nguyen-Xuan, H.; Tran, P. Mechanical performance of fractal-like cementitious lightweight cellular structures: Numerical investigations. *Compos. Struct.* **2021**, *269*, 114050. [CrossRef]
51. Khoshhesab, M.M.; Li, Y. Mechanical behavior of 3D printed biomimetic Koch fractal contact and interlocking. *Extrem. Mech. Lett.* **2018**, *24*, 58–65. [CrossRef]
52. Cai, X.; Guo, Q.; Hu, G.; Yang, J. Ultrathin low-frequency sound absorbing panels based on coplanar spiral tubes or coplanar Helmholtz resonators. *Appl. Phys. Lett.* **2014**, *105*, 121901. [CrossRef]
53. Cheng, Z.; Shi, Z. Vibration attenuation properties of periodic rubber concrete panels. *Constr. Build. Mater.* **2014**, *50*, 257–265. [CrossRef]
54. Thakur, A.; Senthil, K.; Sharma, R.; Singh, A.P. Employment of crumb rubber tyre in concrete masonry bricks. *Mater. Today Proc.* **2020**, *32*, 553–559. [CrossRef]
55. Xie, S.; Yang, S.; Yan, H.; Li, Z. Sound absorption performance of a conch-imitating cavity structure. *Sci. Prog.* **2022**, *105*, 00368504221075167. [CrossRef] [PubMed]
56. *ASTM C 90*; Standard Specification for Load-Bearing Concrete Masonry Units. ASTM International: West Conshohocken, PA, USA, 2014.
57. *ASTM C129*; Standard Specification for Nonloadbearing Concrete Masonry Units. ASTM International: West Conshohocken, PA, USA, 2017.
58. Komatsu, T. Improvement of the Delany-Bazley and Miki models for fibrous sound-absorbing materials. *Acoust. Sci. Technol.* **2008**, *29*, 121–129. [CrossRef]
59. Neithalath, N.; Marolf, A.; Weiss, J.; Olek, J. Modeling the influence of pore structure on the acoustic absorption of enhanced porosity concrete. *J. Adv. Concr. Technol.* **2005**, *3*, 29–40. [CrossRef]
60. Kim, H.K.; Lee, H.K. Acoustic absorption modeling of porous concrete considering the gradation and shape of aggregates and void ratio. *J. Sound Vib.* **2010**, *329*, 866–879. [CrossRef]
61. Fediuk, R.; Amran, M.; Vatin, N.; Vasilev, Y.; Lesovik, V.; Ozbakkaloglu, T. Acoustic Properties of Innovative Concretes: A Review. *Materials* **2021**, *14*, 398. [CrossRef]
62. Bala, A.; Gupta, S. Thermal resistivity, sound absorption and vibration damping of concrete composite doped with waste tire Rubber: A review. *Constr. Build. Mater.* **2021**, *299*, 123939. [CrossRef]
63. Tie, T.S.; Mo, K.H.; Putra, A.; Loo, S.C.; Alengaram, U.J.; Ling, T.C. Sound absorption performance of modified concrete: A review. *J. Build. Eng.* **2020**, *30*, 101219. [CrossRef]

MDPI
St. Alban-Anlage 66
4052 Basel
Switzerland
www.mdpi.com

Journal of Composites Science Editorial Office
E-mail: jcs@mdpi.com
www.mdpi.com/journal/jcs

Disclaimer/Publisher's Note: The statements, opinions and data contained in all publications are solely those of the individual author(s) and contributor(s) and not of MDPI and/or the editor(s). MDPI and/or the editor(s) disclaim responsibility for any injury to people or property resulting from any ideas, methods, instructions or products referred to in the content.